KU-609-263

Contents

	Introduction	xi
	Family Tree of the Flies	xii

Part One

I	THE PATTERN OF FLIES	3
2	THE LIFE-HISTORY OF FLIES	12

Part Two

3	CRANE-FLIES	29
4	LAND-MIDGES	41
5	WATER-MIDGES	50
6	BLACK-FLIES	66
7	MOSQUITOES	71
8	HORSE-FLIES	96
9	SNIPE-FLIES	108
10	ROBBER-FLIES	119
11	BEE-FLIES	131
12	COFFIN-FLIES	147
13	HOVER-FLIES	155
14	COMPOST- AND DUNG-FLIES	170
15	VEGETABLE- AND FRUIT-FLIES	185
16	THE HOUSE-FLY AND ITS RELATIVES	201
17	BLOW-FLIES, BOTS AND WARBLES	213
18	PARASITES OF MAMMALS AND BIRDS	229

Part Three

19	FLIES AND MAN	241
20	SWARMS OF FLIES	262
21	THE PAST, PRESENT, AND FUTURE OF FLIES	274

	Further Reading about Flies	283
	Bibliography	285
	Index	301

The Natural History
of Flies

Harold Oldroyd

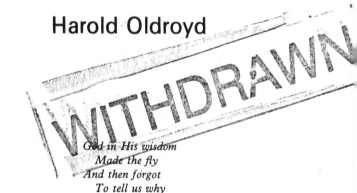

God in His wisdom
Made the fly
And then forgot
To tell us why

OGDEN NASH

The Norton Library

W · W · NORTON & COMPANY · INC ·

LIVERPOOL JMU LIBRARY

3 1111 01511 4620

THE WORLD NATURALIST / Editor: Richard Carrington

The Age of Reptiles by Edwin H. Colbert

The Natural History of Flies by Harold Oldroyd

Copyright © *1964 by Harold Oldroyd*

First published in the Norton Library 1966

ISBN 978-0-393-00375-8

Books That Live

The Norton imprint on a book means that in the publisher's
estimation it is a book not for a single season but for the years.
W. W. Norton & Company, Inc.

PRINTED IN THE UNITED STATES OF AMERICA

1 2 3 4 5 6 7 8 9 0

Plates

[*between pages 8 and 9*]

1   Heads of flies: (a) crane-fly; (b) female horse-fly; (c) house-fly feeding on a lump of sugar

2   The shape of flies: (a) a slender, loose-limbed crane-fly; (b) a sturdy, compact hover-fly; (c) dung-flies

3   The eyes of flies: (a) male cleg, *Haematopota pluvialis*; (b) female of the autumn fly, *Musca autumnalis*

4   Larvae highly adapted to their environment: (a) a Blepharocerid; (b) a Therevid; (c) a Pantophthalmid

[*between pages 24 and 25*]

5   Larvae of the onion fly, *Hylemyia antiqua*

6   An adult of the wheat bulb-fly, *Leptohylemyia coarctata*

7   A pupa and an empty larval skin of a culicine mosquito

8   An adult mosquito emerging from its pupal skin at the surface of the water

[*between pages 72 and 73*]

9   Two flower-loving flies; above, a hover-fly; below, a much smaller March-fly, family Bibionidae

10   Eggs of mosquitoes: (a) laid in a floating raft by a culicine mosquito; (b) laid separately by an anopheline mosquito, each egg equipped with floats

11   A male mosquito, *Culex molestus*, displaying very clearly the long hairs of the bushy antennae

12   Female yellow fever mosquitoes sucking human blood

[*between pages 104 and 105*]

13   A male adult of the New Zealand glow-worm, *Arachnocampa luminosa*

14   Cylinders of mud made by two African horse-flies

15   Robber-flies: (a) holding a blow-fly and sucking it dry; (b) newly emerged from the pupa seen on the left

16   A female bee-fly on a composite flower

[*between pages 152 and 153*]

17   Hover-flies in flight

18   A hover-fly newly emerged from its puparium, seen below it on the left

19   Head of the hover-fly *Volucella zonaria*, showing the nose-like prolongation of the face, an adaptation to feeding from flowers

20 A male of the Mediterranean fruit-fly, *Ceratitis capitata*, with characteristic knobbed bristles on the head

[between pages 216 and 217]

21 (a) A typical acalyptrate fly, *Stenodryomyza formosa*, from Japan. (b and c) typical calyptrate flies, the South African Tachinid *Billaea*, and the Japanese bluebottle *Calliphora lata*

22 Head of the yellow dung-fly, *Scatophaga stercoraria*, a fierce predator among small insects

23 (a) a tsetse-fly in the act of biting; (b) the blow-fly *Aldrichia grahami* on the entrails of a fish; (c) the greenbottle *Lucilia illustris* laying eggs on the head of a sardine

24 Eggs of carrion-feeding blow-flies: (a) laid among the hairs of a dead mole; (b) removed from the eye of a human corpse

[between pages 232 and 233]

25 Part of a human thigh, at the stage where larvae of *Piophila nigriceps* feed on it.

26 Flies living as larvae in the bodies of living mammals: (a) *Gasterophilus* in the stomach of a zebra; (b) warbles beneath the skin of a Tibetan gazelle, *Procapra picticaudata*; (c) *Gyrostigma* from the stomach of a black rhinoceros

27 Bat-parasites of the family Streblidae: (a) adult females of *Ascodipteron jonesi* embedded in the skin of a bat's wings; (b) close-up of the flies *in situ*; (c) an incrustation of puparia of *Trichobius phyllostoma*

28 *Musca sorbens* feeding on discharges from ulcerated skin: (a) on the legs of a child; (b) on a horse

[between pages 248 and 249]

29 (a) Blow-flies, *Chrysomyia pinguis*, sunning themselves on leaves; (b) the common fate of many muscoid flies is to be killed by an attack of the fungus *Empusa muscae*

30 (a) A piece of 'kungu-cake', made up entirely from the bodies of small midges, *Chaoborus edulis*; (b) a fly, *Muscina stabulans*, entirely covered with mites; (c) puparia of the Phorid fly *Paraspiniphora bergenstammi* in a milk bottle

31 Damage to an onion by larvae of the onion fly, *Hylemyia antiqua*, family Muscidae

32 A narcissus bulb damaged by larvae of the Syrphid fly *Merodon equestris*

Acknowledgements for photographs

J. Vanden Eeckhoudt: 1 (a) and (b) and (c); 2 (a); 16; 19; 29 (b). Harald Deering and Frank W. Lane: 2 (c); 9; 17. Walther Rohdich and Frank W. Lane: 3 (a). British Museum: 3 (b); 14; 30 (b). Nicholson: 4 (b). Lynwood M. Chase and Frank W. Lane: 7; 8. New Zealand Government: 13. Fred H. Wylie and Frank W. Lane: 15. R. A. Brabbins and Frank W. Lane: 18; 22. M. Mihara: 21 (a) and (c). Stuckenberg: 21 (b). S. A. Smith: 23 (a). R. Kato: 23 (b) and (c); 26 (a); 28 (a) and (b); 29 (a). Hugh Spencer and Frank W. Lane: 24 (a). W. E. Evans: 24 (b); 25. The remaining photographs are by the author.

Figures

1   Side view of the thorax of a horse-fly, *Tabanus*, to show the sizes of the
     three thoracic segments.   4
2   Head of the seaweed fly, *Coelopa frigida*.   6
3   Head of a male of the Dolichopodid fly *Megistostylus longicornis*.   9
4   Heads of two Streblid flies, parasites in the fur of bats.   10
5   The antenna of a tsetse-fly.   13
6   Egg of a Muscid fly.   15
7   The telescopic ovipositor of the acalyptrate fly *Palloptera muliebris*.   17
8   The ovipositor of a robber-fly which lays eggs in the heads of flowers.   18
9   The ovipositor of a robber-fly, compressed and strengthened until it will
     cut into the stems of plants and push an egg inside.   19
10   The 'rat-tailed maggot' of the Syrphid fly *Eristalis*.   20
11   The larva of the Ephydrid fly *Hydrellia nasturtii*.   21
12   Posterior spiracles of the larva of a Trypetid.   22
13   The 'phantom larva' of the midge *Chaoborus* (*Corethra*), an aquatic larva
     living permanently in deep water.   50
14   Head of a male midge, *Chaoborus plumicornis*.   59
15   The aquatic larva of *Simulium damnosum*.   68
16   The antenna of a male mosquito, showing the many-segmented flagellum
     with its whorls of long, sensory hairs.   79
17   A wingless Empid fly, *Apterodromia evansi*, from New Zealand.   95
18   *Anomalopteryx maritima*, an Ephydrid fly with reduced wings that is
     found on the remote sub-Antarctic island of Kerguelen.   95
19   Larva of the Rhagionid fly *Atherix*, another type of underwater crawler.   112
20   A hind leg of a robber-fly of the genus *Lagodias*, with long scaly fringes
     on the tarsal segments.   120
21   Head of the robber-fly *Proagonistes athletes*.   122
22   Wing of the fly *Nemestrinus*.   135
23   *Megalybus crassus*, a hump-backed Oncodid fly from Chile.   137
24   *Thambemyia pagdeni*, a Dolichopodid fly from Malaya, with hairy eyes
     and a grotesque proboscis.   142
25   Leg of the Dolichopodid fly *Hercostomus straeleni*, from Uganda.   144
26   Head of the Phorid fly *Megaselia meconicera*.   148
27   Proboscis of the hover-fly *Eristalis*, showing the sponge-like pseudo-
     tracheae at the tip.   156
28   The most wonderful fly in the world – *Sciadocera rufomaculata*.   167

29 Another sponge-like proboscis, that of the Sciomyzid *Elgiva albiseta*, used for mopping up liquid food.   170

30 A beetle-fly, Celyphidae, the enormous scutellum of the second thoracic segment completely covering the abdomen.   173

31 *Ozaenina diversa*, an Otitid fly of the sub-family Richardiinae.   178

32 Section through a leaf-mine.   194

33 Larva of the Muscid fly *Fannia canicularis*, with fringed processes suited to a life in a wet medium.   206

34 Larva of the Muscid fly *Phaonia keilini*, which preys on mosquito larvae.   208

35 An egg of *Gasterophilus* attached to the hair of a horse's leg, and with the larva about to emerge from it.   223

36 A bat-parasite of the family Streblidae.   234

37 A wingless bat-parasite of the family Nycteribiidae.   237

38 Three ways in which flies 'swarm' into houses.   269

39 The invasion of an empty house in London by larvae of the greenbottle *Lucilia*.   272

40 A map of the known distribution of Streblidae and the potential distribution.   278

Acknowledgements for figures

Figures 4, 5, and 40 are the work of Boris Jobling, and are reproduced by permission of the editors of Parasitology; figures 17 and 24 are the work of Harold Oldroyd, and are reproduced by permission of the Royal Entomological Society of London; all the other drawings were made specially for this book by Arthur Smith.

Introduction

Provided that we start with the study of living organisms in their natural environment, rather than with hypotheses about their relationships, we can proceed to every department of biological sciences.

C. F. A. PANTIN, 1962

FLIES ARE AMONG the most familiar of insects, and yet among the least understood. We talk about 'the fly', as we talk about 'the ant' and 'the frog', and then we have in mind one species, the common house-fly, *Musca domestica*. This highly adaptive insect happens to like the sort of litter that man produces wherever he lives, and while man has been domesticating the pig, the goat, the horse and the dog he has also unwillingly domesticated 'the fly'.

We notice other flies when they are a nuisance to us. Bluebottles and other blow-flies taint our meat; flies swarm round our heads and those of our grazing animals in summer; some bite, some buzz, some just disgust us with their habit of breeding in dung, sewage, carrion, even in living flesh. In fact, flies are a topic like drains, not to be discussed in polite society, to be left to those strange people who cultivate a professional interest in them.

It is a pity. There are some 80,000 different species of flies distributed throughout the world, and many more being discovered every day: a recent letter from Dr Elmo Hardy tells me that he is about to record 200 new species of fruit-flies (Drosophilidae) from the Hawaiian Islands. It is true that most of them breed in decaying organic matter of some kind, but we must remember that disgust is purely a human reaction. All animals must have protein food. The fact that many flies have learned to get it from the products of organic decay is a sign of their biological efficiency. Decay there must be. New organisms cannot be built unless old ones disintegrate. In exploiting sources of food that will always exist, flies have made their future secure.

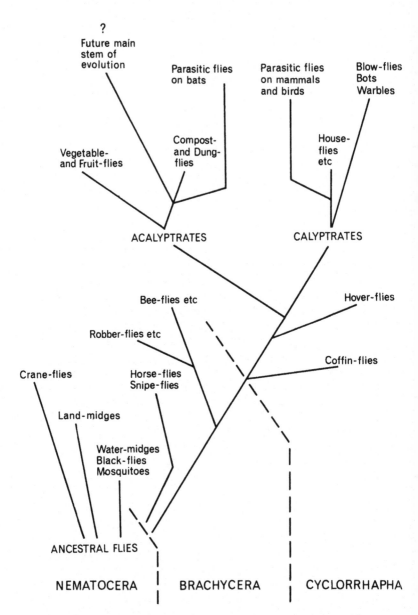

?

Future main
stem of
evolution

Parasitic flies
on bats

Parasitic flies
on mammals
and birds

Blow-flies
Bots
Warbles

Compost-
and Dung-
flies

House-
flies
etc

Vegetable-
and Fruit-flies

ACALYPTRATES

CALYPTRATES

Bee-flies etc

Hover-flies

Robber-flies etc

Horse-flies
Snipe-flies

Coffin-flies

Crane-flies

Land-midges

Water-midges
Black-flies
Mosquitoes

ANCESTRAL FLIES

NEMATOCERA

BRACHYCERA

CYCLORRHAPHA

A Family Tree of the Flies: showing how the various groups discussed in
successive chapters of this book may be related to each other, and how they fall
into the three major groups, or Sub-Orders.

Once this fact is accepted, the natural history of flies affords an unlimited field of study. So much is known already that the great problem is to compress it into a book. When I began, it was my intention to avoid the systematic approach, with its insistence on classification and arrangement. I wanted to talk in a general way about living flies, but unless the book is to be a series of anecdotes of the 'wonders' of fly-life it must have some theme that will give it coherence, and if it has a theme it must compare one group of flies with another.

One could take all the different kinds of habitat – cave-life, life in different kinds of water, predators, parasites, and so on – and enumerate the flies that follow each way of life. Monsieur Eugène Séguy has already done this in his invaluable book *La biologie des Diptères*, a mine of information on matters of detail. What I am attempting is something less complete, more of an impressionist picture of flies in evolution.

To discuss adequately a group so rich and varied it is necessary to particularize, mentioning the names of species, genera, families or larger groups. Unfortunately very few flies have vernacular names, and so scientific names cannot be avoided. I hope that the reader will be able to see where each of these fits into the general picture, but for any exact definition he must go to a systematic work such as Colyer and Hammond's *Flies of the British Isles*, Curran's *Families and Genera of North American Diptera*, Lindner's *Die Fliegen der Palaearktischen Region*, or Séguy's volumes in the *Faune de France* series.

Regrettably I must mention what are called 'the Meigen 1800 names'. These are a set of names that remained in obscurity from 1800 to 1908, and then were revived, thereby altering the scientific names of some of the most familiar families: thus the non-biting midges, family Chironomidae, became known as Tendipedidae. I have not used the 1800 names in this book, but I have included them in the index, since they are in current use by many people who write about flies.

The book is in three sections. Part One gives a general outline, first of adult flies and then of early stages, eggs, larvae, and pupae. Part Two is the main part of the book, and surveys the flies in sixteen groups, in what appears to be an ascending order of evolution, as seen from the viewpoint of natural history. I stress that this is not a work of classification, and that when I point out similarities of habit

I do not necessarily wish to maintain that the flies concerned are closely related. On the contrary, it cannot be overemphasized that flies have made the same evolutionary experiments again and again, and much of the interest in studying them is to see how different groups tackle similar problems.

Part Three is not as long as I would have liked it to be. Out of a number of interesting general topics concerning flies as a whole I have selected only three. One must obviously be *Flies and Man*, bringing together those scattered groups of flies which make a noticeable impact on human affairs. The second chapter of this section deals with what we call 'swarms' of flies, the different ways in which flies form aggregations, and how much truly communal behaviour is involved in them. Finally a brief speculation about what may be the future of flies.

It is a great pleasure to acknowledge the generous help of colleagues and other friends all over the world who have given me advice – knowingly or unknowingly – and lent me photographs to illustrate this book. In particular Dr S. Asahina, Dr Azevedo da Silva, Dr P. S. Corbet, Mr R. W. Crosskey, Mr C. E. Dyte, Dr W. E. Evans, Dr Paul Freeman, Dr A. J. Haddow, Mr B. Jobling, Dr R. Kato, Dr B. R. Laurence, Dr E. Lindner, Dr I. M. Mackerras, Dr A. J. Nicholson, Dr J. A. Reid, the Hon. Miriam Rothschild, Dr B. R. Stuckenberg, Professor O. Theodor, and Dr Z. Zumpt.

My colleagues Mr R. L. Coe and Mr K. G. V. Smith read the book in proof and offered many helpful suggestions.

For permission to quote I am indebted to Messrs Blackie & Son for lines from 'Spring', by Mary Howitt; to Messrs Collins for a sentence from *Fleas, Flukes and Cuckoos*, by the Hon. Miriam Rothschild and Dr Theresa Clay; to Messrs J. M. Dent & Sons for the poem 'The Fly', by Ogden Nash. The editors of *Parasitology* kindly agreed to the use of various drawings by Mr Boris Jobling, and Mr Arthur Smith has drawn a number specially for this book. The Trustees of the British Museum and The Council of The Royal Entomological Society of London kindly gave me permission to reproduce several figures from their publications.

Finally I owe a great deal to my wife for extensive help and encouragement at all stages of the work.

Part One

Chapter 1

The Pattern of Flies

They that crawl and they that fly
Shall end where they began.
THOMAS GRAY

MOST PEOPLE divide insects into those that crawl and those that fly, and refer to all the latter as flies. Dragonflies, mayflies, stoneflies, fireflies, ichneumon flies, all belong to different Orders of insects, at very different levels of evolution, and only distantly related to each other. None of these are true flies in the entomological sense.

The insects that come within the scope of this book are those belonging to the natural Order Diptera, and are recognized by the fact that they have only one pair of wings, those on the second, or middle segment of the thorax. The only other insects that might be confused with them in this respect are the males of certain scale-insects of the Order Homoptera. Many adult flies, however, have lost their wings in the course of evolution; how are we to recognize these?

Flies have adapted the second pair of wings, now disused for flight, into halteres, knobbed organs which act as gyroscopic stabilizers during flight, and which can be seen in plate 2a. When the true wings are missing the halteres are often present. If even these have been lost we can usually see that the thorax is adapted to take only one pair of wings, the middle segment having been greatly developed to accommodate the necessary flight muscles. All flies that are wingless today are descended from winged ancestors. This sounds fine in theory, but it must be admitted that in practice one recognizes wingless flies either by knowing by sight the family to which they belong, or at least knowing enough about flies to recognize the structure of the head and legs and the arrangement of the bristles. That is how one knows that the wingless insect of figure 36, from the fur of a bat, is a fly of the family Nycteribiidae and not a bug of the family Polyctenidae.

This Order Diptera includes insects of an extraordinary range of size and appearance. The largest may be almost 3 in long, with a similar wing-span: the smallest may be 50 times shorter, and 125,000 times smaller in bulk, hardly visible to the naked eye. Even this is only one-tenth of the range of size between a mouse and an elephant,

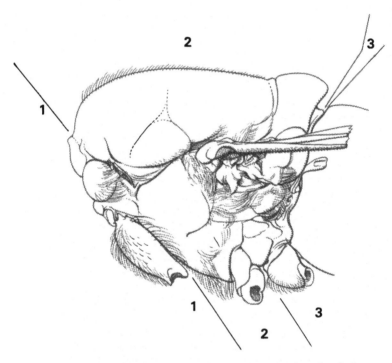

Figure 1. Side view of the thorax of a horse-fly, *Tabanus*, to show the relative sizes of the three thoracic segments. The enormous enlargement of the second segment on the upper, or dorsal side is to accommodate the muscles needed for flight.

yet it poses considerable problems for the insects at the two extremes. The very small flies are helpless against air-currents, and are carried high into the air by convection currents, and horizontally by winds. This is a hazard to them as individuals, but a stimulus to them as a group, since some or other of them are continually being carried into fresh environments. Prolific in numbers, as they must be to offset the great losses through accident, they have an excellent chance of producing a succession of mutations, from which natural selection in varying environments can evolve new genera and species.

On the other hand, the very big flies are generally a dying race. At all stages of their life-history their size makes them conspicuous to predators, and the low fecundity and slow development that often accompany large size make it difficult for them to maintain themselves, much less to spread. They may succeed locally, where conditions are ideal – the huge Pantophthalmidae, with their wood-boring larvae (plate 4c) are even a pest in some parts of tropical South America – but the fate of the dinosaurs awaits them: the terminus of an evolutionary line.

The daddy-long-legs and the house-fly represent opposite extremes of the Order Diptera. Crane-flies, with their gangling elongate bodies, narrow wings, absurdly slender legs, and slow flight, stand for the old-fashioned in flies, survivors from a primitive line. The house-fly and the bluebottle, compact, bristly, with broad wings buzzing at high speed, with a flight that is persistent and apparently purposeful, are aggressively advanced types. Yet even these have probably passed their peak, and the future evolution of flies probably lies with such small flies as *Drosophila*, the famous fruit-fly, because they have not yet committed themselves too far in any direction, and obviously retain a great evolutionary vitality.

In between the extremes lie some 80,000 species which have spread all over the world, from the Equator far towards the poles, in forest and in desert, high up towards the mountain-tops, on to remote islands, and even, though rarely, to the open sea. When I asked Dr Ian Mackerras, a distinguished dipterist of long standing, to suggest a theme about which this book could be woven he replied: 'The ubiquity of flies.' I considered this, but was defeated by the immensity of the theme. One can say: 'Flies are everywhere', or one can give an endless catalogue of types of environment and the flies that occur in them. So I decided instead to take as a theme the evolution of flies, using their natural history to show how each group represents a stage in the more efficient exploitation of natural resources.

The ancestral flies branched out into various lines of evolution, many of which became extinct. The living flies are members of branches that have managed to survive, some of them already in decline, some still in an expansive phase. Their different ways of life become more interesting to us if we can contrive to see them in an evolutionary pattern, to interpret the ways of adults and larvae as devices to get more and richer food, to breed more quickly and to leave more descendants.

The flies living today can with fair certainty be looked upon as two evolutionary lines. One, called the Sub-Order Nematocera, comprises the crane-flies and the various families of midges and gnats. The second line leads on to the house-fly and bluebottle, and has no convenient collective term. I shall call them the 'flies proper',

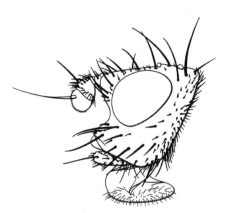

Figure 2. Head of the seaweed fly, *Coelopa frigida*. Compare the short antennae of this fly with those of the mosquito in figure 16, and the midge in figure 14. The bristles have no obvious function, as in figure 21.

because most of them are what the layman thinks of as a fly, rather than a midge or a gnat. This line is divided into two Sub-Orders, Brachycera and Cyclorrhapha, and we shall see later why they are given these names.

Midges and Gnats (Sub-Order Nematocera)

And the gnats are on the wing
Wheeling round in airy ring
MARY HOWITT

The expression 'midges and gnats' conjures up a picture of a swarm of tiny, fragile creatures, of the individually insignificant becoming effective because they assemble together and pursue a common purpose. Not all Nematocera gather in swarms, and we shall see in a later chapter that the appearance of common purpose is largely an illusion; but in the main the picture is of primitive creatures, moving about as a body, following who-knows-what, air-currents, or blind impulse? In contrast to these, the house-fly and the bluebottle appear as malignant creatures, absorbed in their own mysterious purpose.

There is much dispute about how the various families of Nematocera are related to each other, and the present book may seem to oversimplify this. As this is a work of natural history we are privileged to draw attention to similarities of habit and breeding-place, without necessarily implying that the flies concerned must be closely related. Nematocera will be dealt with in three groups. The first will comprise the crane-flies as a basic family, not necessarily ancestral to any other flies, but preserving many old-fashioned ways that other flies have grown out of; and in the same chapter I shall include several small families whose relationships are doubtful, and which are probably the remnants of evolutionary experiments that did not quite succeed.

Then comes the large group of land-midges, the little flies that flit round the compost-heap and rubbish dump, and in contrast to these the water-midges and gnats, including the mosquitoes. It is not quite accurate to call the first group terrestrial and the second aquatic, because many of them live in intermediate damp and boggy places, of varying degrees of wetness. Yet there seems to be a genuine difference in bias towards the land and towards the water which represents, I think, a divergence in the evolution of Diptera, and which also helps us to interpret the evolution of the other main stem of Diptera. So I shall call these 'earthy' and 'watery' groups respectively.

I should have liked to be able to call the first group simply midges, and the second simply gnats, but the common English names of families are quite inconsistent in this respect. Thus gall-midges and fungus gnats belong to the first group, and biting midges and gnats *tout court* to the second. Mosquitoes earn a chapter to themselves because of the information that exists about them, as well as of their medical importance.

Flies Proper (Sub-Orders Brachycera and Cyclorrhapha)

This rather clumsy term is used, for want of a better, to express the fact that members of the second great division of flies would never be spoken of, even by laymen, as midges or gnats. Though some of them are quite small, they lack the fragility of the midges. Even the slender ones have bulk, and in a curious way individuality. If we see one midge, or even one mosquito, we instinctively look for others; but one fly goes about its business in a seemingly purposeful way, and is self-sufficient.

The entire branch is correctly known as Brachycera, having short antennae, in contrast with the long, filamentous antenna characteristic of Nematocera. About two-thirds of all known flies, somewhere between 40,000 and 60,000 species throughout the world, belong to this branch, so obviously it is desirable to subdivide it again. Fortunately a natural division occurs between a smaller group of more primitive families, to which it is usual to restrict the name Brachycera, and a much larger group of families of more advanced evolution which have the name Cyclorrhapha, for reasons that we shall discuss in the next chapter.

Sub-Order Brachycera

> Little Fly
>   Thy summer's play
> My thoughtless hand
>   Has brushed away
>   WILLIAM BLAKE

These are 'old-fashioned' flies in the sense in which I have already used this term, with a structure that is obviously suited to their particular way of life: more modern flies have a uniform appearance that tells us little about their habits. Among Brachycera the adult flies lead very varied lives, and this is one of the few groups of flies of which the ways of the adults are more interesting than those of the larvae.

As among Nematocera, there is a 'watery' group of families and an 'earthy' one. The former consists of the horse-flies, those formidable biters from wet meadows, and one or two related families that are less well known. Collectively known as Tabanoidea, this group of families looks back towards the gnats and water-midges of the Nematocera. The 'earthy' group of Brachycera consists of predatory flies such as Asilidae (robber-flies), Empididae and Dolichopodidae, and of Bombyliidae (bee-flies) and some others, where the adults are flower-feeding.

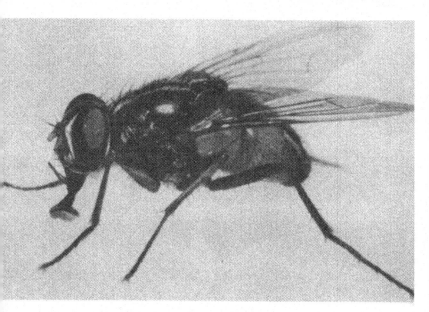

1. Heads of flies: (a) a crane-fly, with small eyes and primitive mouthparts; (b) a female horse-fly, with very large eyes, and with mouthparts adapted for both piercing and mopping up blood; (c) a house-fly feeding on a lump of sugar.

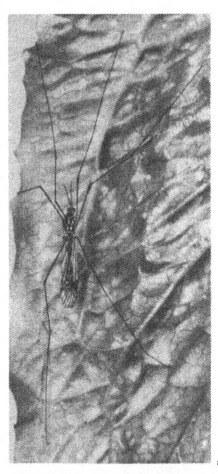

2. The shape of flies: (a) a slender, loose-limbed crane-fly; (b) a sturdy, compact hover-fly; (c) dung-flies, typical of bristly flies of more advanced evolution.

a

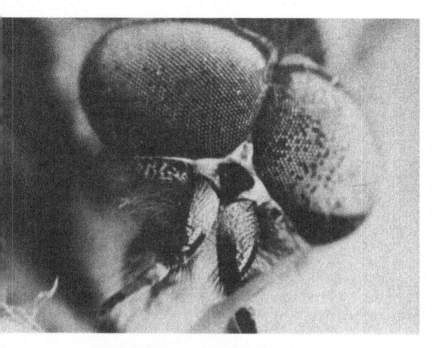

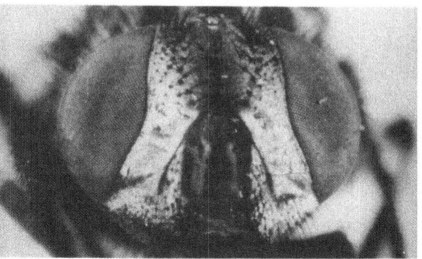

3. The eyes of flies: (a) in males the eyes may meet, with bigger facets facing upwards, as in this male cleg, *Haematopota pluvialis*; (b) in females the eyes are almost always separated, as in this female of the autumn fly, *Musca autumnalis*.

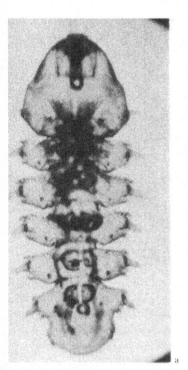
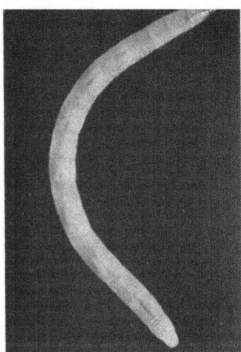
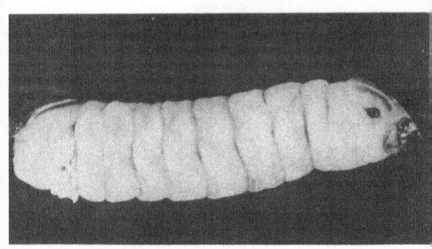

4. Larvae highly adapted to their environment: (a) a Blepharocerid,
which crawls over stones in rapidly moving water; (b) a Therevid,
a wormlike creature living in dry debris; (c) a Pantophthalmid,
a powerful wood-boring type.

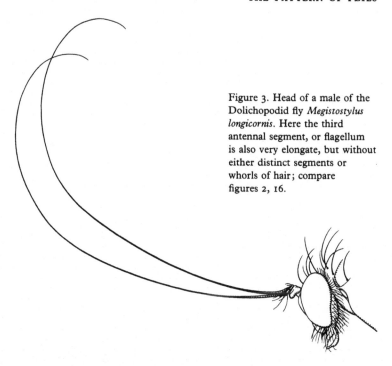

Figure 3. Head of a male of the Dolichopodid fly *Megistostylus longicornis*. Here the third antennal segment, or flagellum is also very elongate, but without either distinct segments or whorls of hair; compare figures 2, 16.

Between these two groups there is a transition more abrupt than is often realized. From this point on, all flies are essentially terrestrial, though many small groups have gone back more or less successfully towards the water, just as the seals and whales have evolved from the essentially terrestrial mammals.

Sub-Order Cyclorrhapha

Busy, curious, thirsty fly
Drink with me, and drink as I
WILLIAM OLDYS

The evolution of this Sub-Order was a big step forward in the history of flies, and several families remain that are relicts of tentative experiments in this direction. Phoridae, the coffin-flies (Chapter 12), and Syrphidae, the hover-flies (Chapter 13), are still outstandingly

9

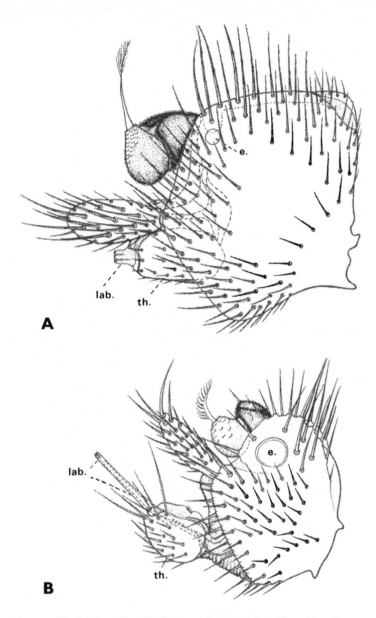

Figure 4. Heads of two Streblid flies, parasites in the fur of bats. Here the bristles may help to keep the head clear of the enveloping fur. The letters refer to anatomical structures with which we are not concerned here, except for the tiny eye (e), and the blood-sucking proboscis, with its base or theca (th) and its piercing labium (lab). From Jobling.

successful sidelines from the main flow of evolution in flies, and other, smaller examples are mentioned in those chapters.

The great mass of Cyclorrhapha, the typical 'higher' flies, fall into two groups, called the acalyptrates and the calyptrates, according as they lack or possess the 'calypters', or squamae, the wing-flaps that project over the halteres.

The acalyptrates are a confusing mass of flies, mostly small, rather bare, and comparatively featureless. Although a certain number are distinctive in structure or pattern, these have rarely any obvious relation to the way of life of the fly, as they have in the more primitive Brachycera. The acalyptrates have the appearance of a large, plastic mass of evolutionary material, which is very difficult to classify. Families in this group are much less clear than in the rest of the flies, and we cannot even enumerate them in a book such as this. They include fruit-flies, large (Trypetidae) and small (Drosophilidae), seaweed flies (Coelopidae), the cereal-feeding fruit-flies and gout flies (Chloropidae), and the newly aquatic shore-flies (Ephydridae). In two chapters I have tried to summarize the natural history of this complex of flies by dealing with them in four groups, according to the feeding habits of their larvae.

The calyptrates are bristly flies such as the house-fly and the bluebottle, and a great many of their relatives. They must have arisen from a precocious group of early acalyptrates, and have gone very much further already than any of their cousins. Some of the adult flies have developed structures for special purposes, notably the piercing proboscis of tsetse-flies and of the stable fly, *Stomoxys*, an entirely new experiment that is independent of the primitive biting mechanism of the mosquitoes and horse-flies. In the main, though, advances have been made by the larvae, which have shown a tremendous versatility in their choice of food material, as we shall see in the next chapter.

The flies of Chapter 18 are a group of three families of extremely advanced evolution, all of which have virtually eliminated the larval stage. They are known as Pupipara, because they do not release the larva until it is ready to form a pupa. The adult flies subsist entirely on the blood of birds and mammals: Hippoboscidae on both groups; Streblidae and Nycteribiidae only on bats.

Chapter 2

The Life-History of Flies

THE LIFE-HISTORY of an insect, like that of a vertebrate, is a continuous process of development, from the single cell of the fertilized, or occasionally unfertilized, egg to the fully mature adult. Such events as birth, hatching, moulting, and so on are merely incidents in this sequence, convenient landmarks by which to discuss it.

Beginning life as a single cell, an animal has to reach a certain level of organization before it can exist as an individual; until then it is an embryo. The state of immaturity at which the insect hatches from the egg varies greatly in different Orders of insects, and generally speaking the more highly evolved hatch at an earlier stage of development than their ancestors, and are crawling about at an age when more primitive insects are still an embryo in the egg.

Larvae of flies are extremely immature in appearance. They have no wings, of course, never have segmented walking legs, and have almost none of the external characters that we see in adult flies. When a larvae has obvious external structures such as those shown in plate 4, these are always suited to the requirements of larval life, for which they have been specially evolved. Most such structures have been evolved to meet the demands of aquatic life, the need to breathe, to swim, or just to cling against the pull of the current. As flies have advanced in evolution they have streamlined themselves more and more, and eventually evolved the maggot as a sort of universal model, a larva that could live in almost any medium.

It used to be said that the simplification in the course of evolution came about by gradual loss, and that flies once had larvae with legs, that could run about, like those of many beetles. The current view is expressed by Kim [166*]: 'Dipterous larvae are probably to be re-

* The numbers in square brackets throughout the book refer the reader to the bibliography at the back, in which each reference has its own number.

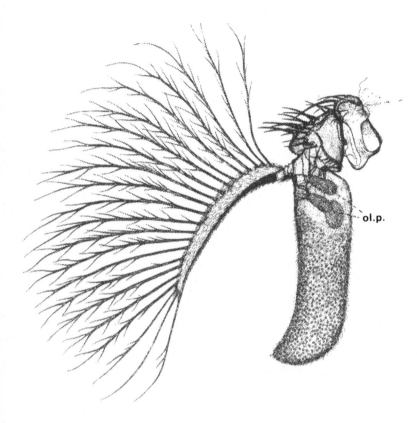

Figure 5. The antenna of a tsetse-fly. Note the elaborate articulation of the three segments; the olfactory pits (ol.p.) in the third segment; and the arista with branching hairs on the upper surface only. From Jobling.

garded as highly specialized examples of very early hatching, and not as derivatives of an oligopod type' (ie one with thoracic legs).

Why have they done this? What are the biological advantages of coming into the world as a soft, legless grub or maggot? By hatching from the egg while they are still plastic, and uncommitted to any adult structure they have been able to evolve along lines of their own, until a fly now really lives two lives in succession, one as a larva, and a second life as an adult. Larva and adult are more different from each other than many Orders of insects.

Biological advantages can be seen in this arrangement. One is that the insect has two periods in its life when it can enter into a competitive struggle, in two different worlds. In many flies the larval period seems to be the significant one, when the fly really gains an ascendancy over its rivals: this is true of flies as different as crane-flies and blow-flies. On the other hand in predaceous flies such as robber-flies the adult stage is the highly competitive one, and the larvae, though presumably successful in a quiet way, are scattered and furtive. Extremes are reached on the one hand in the warble-flies of cattle and deer, as well as in the non-biting midges, where the adult lives for a short time only for mating purposes; and on the other hand in the tsetse-flies and the parasitic flies (Chapter 18), where adult life is everything and the larva, nurtured within the body of the mother, has hardly any separate existence.

A second advantage is that the functions of feeding and of reproduction can be developed independently. Among flies the larva is the principal feeding-stage. The eggs have been laid in or near to a medium that provides plenty of food, often with no effort on the part of the larva itself, which lays in a stock sufficient for the rest of the life of the fly. The function of the adult fly is to mate, to mature the eggs, and to lay them in a suitable mass of food-material, all as quickly and efficiently as possible. Feeding by adult flies is supplementary to that of the larva. The adults need water to replace that lost in active flight, sugars to provide the energy for their flight, and extra protein if that provided by the larva is insufficient to provide for the eggs that the female is preparing to lay.

Flies that prey on other insects, or suck the blood of man and other vertebrate animals are making up this deficiency in their larval diet and it is one of the objectives of the present book to try to show how in different families of flies the balance has shifted to and fro between larva and adult in the abundance and richness of their food-material.

A third advantage of increasing divergence between larval and adult life is that the adults can venture out into a different world, making use of their highly developed power of flight, of their acute sense of smell and of their large eyes, which are constantly alert, even if their perception of detail is poor. Ranging actively about, exploring every type of environment, adult flies have colonized the earth, and discovered every conceivable medium to which their versatile larvae could adapt themselves.

The step from a larva to an adult that is so vastly different needs a rebuilding stage, called the pupa, to which we shall return later in this chapter. So flies have four stages in their life-history, egg, larva, pupa, and adult, and we may consider the first three of these briefly.

Every female fly produces eggs, though she may not always lay them. The number produced at one time ranges from one to about 250, with a maximum of perhaps 1,000 during the life of an individual. The greatest number are produced by some aquatic flies, like the water-midges, because the water is full of predators, and losses are very heavy; and by blow-flies which lay their eggs in freshly decaying materials, which are only in a suitable state for a short time. At the other extreme come the tsetse-fly and the so-called Pupipara,

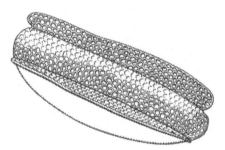

Figure 6. Egg of a Muscid fly, showing the ridges and the honeycomb sculpturing, which help in respiration when the egg is wetted by rain.

which mature only one egg at a time and nurture it internally until it is fully grown.

Eggs of flies are generally oval, white, or pale yellow, and featureless to the naked eye. Under the microscope they are often seen to have horns, discs, stalks, or just a raised network of ridges over the surface (figure 6), and Hinton [136] has recently demonstrated how these various devices protect the eggs against drowning after a shower of rain. The irregularities of the surface hold a thin film of air, which functions as a respiratory plastron or auxiliary lung, such as is found in many aquatic bugs and beetles. It not only provides a temporary supply of oxygen, but is able to replenish itself by diffusion of oxygen from the water. Ornamentation of the surface of an egg also increases surface tension and so may cause the egg to float, as we can see in the eggs of mosquitoes. Those of anopheline mosquitoes (plate 10b) float individually by means of air-cells; those of culicine mosquitoes adhere together in a raft, each egg having its own disc of fine filaments at the lower end (plate 10a).

Most female flies have a simple opening at the end of the egg-channel or oviduct, protected only by a pair of lobes. Such flies may either drop eggs at random to the ground, or stack them neatly into an orderly mass. When the eggs are pushed into a soft medium the whole tip of the abdomen may be inserted, or there may be a development of the last few segments of the abdomen into a tele-scopic ovipositor, with long, membraneous sections. The house-fly and many of its relatives have such an ovipositor, which can nor-mally be seen only after dissection. The relatively few flies that push their eggs into the interstices of flowers or leaves, into stems, or through the rind of a young fruit may have developed a rigid ovi-positor, the structure of which may give a clue to tell us where to look for the eggs. The Mediterranean fruit-fly shown in plate 20 has such an ovipositor.

The larvae of flies follow the same pattern of complexity that we noted in the adult flies: the larvae of Nematocera and Brachycera are distinctive, and can usually be named at least to the family; Cyclorrhapha have developed the maggot as a universal type. Those that live an aquatic life, such as Ephydridae, and individual members of other families, may have developed false legs, claws, suckers, extensible breathing-tube and so on, and come to resemble any other kind of aquatic larva in this respect. Those that live as internal parasites, like the bots and warbles, have evolved tough skins and strong spines. But most larvae of the acalyptrate and calyptrate groups of families are maggots with a pointed head end, almost with-out any sclerotized parts except for the mouth-hooks, and with a blunt posterior end which bears a pair of spiracles that can some-times be distinguished by the arrangements of their slits (figure 12, plate 5).

Identification of the larvae of flies is difficult. Hennig [119] has given a masterly summary of everything that is available for this purpose, with keys as far as they can be constructed, and references to all the published figures of dipterous larvae. One can usually hope to identify larvae of Nematocera and Brachycera to the family, often to the genus, sometimes to the species. Syrphid larvae have been fairly well studied. For the rest, including all the maggots, it is largely a matter of inspired guessing, guided as much by the circumstances in which the larva was found as by its inadequate structural features.

An illuminating study by Keilin [159] takes the problem of breathing as a key to the adaptation of the larvae of flies to their environment, and compares the development of the spiracles, the external breathing apertures of the tracheal system, in various

Figure 7. The telescopic ovipositor of the acalyptrate fly *Palloptera muliebris*, an instrument for delicately placing the egg in a small crevice.

groups. His description of the different places in which larvae of flies can live could hardly be bettered, and I venture to quote it in full:

There is hardly any type of medium capable of supporting life from which dipterous larvae have not been recorded. In fact they occur in every kind of watery medium, such as streams, rivers, waterfalls, ponds, lakes, brackish water, hot mineral springs, and the water-reservoir of terrestrial and epiphytic plants: they are found in a variety of dry and semi-fluid media such as sand, earth, mud, animal, and vegetable substances in different degrees of decomposition, excrements of animals, and in festered wounds of animals and plants. They are met with as scavengers in the nests of mammals, birds, wasps, bees, ants, and termites. As parasites they are found in different portions of plants, i.e. roots, stems, leaves, and flowers, boring galleries or forming galls. They are known also to live as true parasites of a great variety of animals such as Oligochaetes, Molluscs, Crustacea, Arachnids, Myriapods, practically every large Order of Insects, Amphibia, Reptiles, Birds and Mammals, including Man. Finally, in a certain number of species the larvae grow to different stages of development within the uterus of the mother, leaving it in some cases only just before pupation.

What can one add to that? Only, I think, to point out that very few larvae of flies can be truly called terrestrial in the sense of living exposed to light and air as, say, many caterpillars do. A few live openly on vegetation, but nearly all lie buried in some kind of medium that is always more or less humid.

The larval life of nearly all flies is interrupted by a number of moults, and it is not easy to answer simply the question: how many? We have seen that in effect flies leave the egg prematurely, and then telescope much of their later development, including all modifications towards adult structure, into the pupal period. Moults that take place during the exposed feeding period of the larva are most obvious, but those that occur either immediately after hatching, or immediately before pupation are the most difficult to observe.

In general Nematocera have the greatest number of obvious moults, with six or seven instars, or periods between moults, in black-flies and four in mosquitoes, whereas the maggots of the

Figure 8. The ovipositor of a robber-fly which lays eggs in the heads of flowers.

higher flies have only three apparent instars, the others being concealed inside the puparium. A few flies complete their larval life without moulting at all, examples being *Scenopinus* and the Muscid *Melanochelia*, according to Keilin [158].

We think of moults as the growing periods of animals that are confined within a rigid cuticle, like crabs and lobsters. This may be true of the most primitive larvae of flies, but already in *Chaoborus* certain parts of the body can be stretched, and the maggots of the Cyclorrhapha have a cuticle that is completely extensible. These last do most of their feeding in the course of the third instar, and grow considerably.

Perhaps nearest to the ancestral way of life are those Tipulid larvae that live in moss, and in the decaying wood of fallen trees. It may be that some groups of crane-flies progressed, via leaf-mould and the detritus at the foot of growing trees, to the soil of the open grasslands; while others moved via the vegetation of swamps, and the margins of ponds and streams, to a fully aquatic life in still water and in torrents, in fresh water, brackish and even salt. All these habitats are colonized by different groups of crane-flies, and some have joined the associations of insect larvae that live in the liquids held in the cavities of certain plants, Bromeliads, *Nepenthes*, etc.

No other single family of flies quite equals the versatility of Tipulidae in its choice of larval habitats. The other Nematocera seem to have chosen between two diverging ways. The families that I have called the land-midges have moved towards the land, and a larval life concealed in the soil, in vegetable detritus and leaf-mould, which is what I refer to throughout this book as 'compost', in fungus, or in plant-galls. The families that I have called water-midges and gnats have inclined more towards an aquatic life, though, as I shall point out from time to time, most of these larvae really live in a marginal

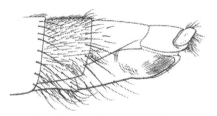

Figure 9. The ovipositor of a robber-fly, compressed and strengthened until it will cut into the stems of plants and push an egg inside.

zone, at the surface of water, or in a thin film flowing over a surface. Black-flies are among the few that have entirely severed their connection with the atmosphere during larval life.

Aquatic larvae as a rule are carnivorous, or feed on algae. The terrestrial larvae of this group, those of certain Chironomidae and Ceratopogonidae, live in dung, decaying animal matter or the sap of wounded trees – all highly concentrated media, as might be expected if they have come from aquatic ancestors of carnivorous habit.

Among the larvae of Brachycera we find a rather similar divergence between a 'watery' and an 'earthy' group of larvae. Indeed the families related to the horse-flies (which are the subjects of Chapters 8 and 9) have many affinities with the water-midges and gnats, whereas the robber-flies and 'earthy' Brachycera are the long transition towards the higher flies. As long as we remain in the Sub-Order Brachycera the larvae are varied and in general suited to their way of life, though in the families Empididae and Dolichopodidae the larvae are already becoming streamlined and rather featureless. Syrphidae and families round about also have adaptive larvae, and in versatility the family Syrphidae is nearly as inventive as the Tipulidae. But after this, in the acalyptrates and calyptrates, maggot-like larvae are the rule.

The maggot has no hardened head-capsule, and nearly always the black mouth-hooks can be seen through the transparent cuticle. At most there are spiracles only on the thorax and at the tip of the abdomen. The small modifications that exist in certain families may have biological significance – Keilin has suggested how the peculiar spiracles of some of the parasitic larvae and those that are reared internally, may suit the clogging, almost airless fluids in which the

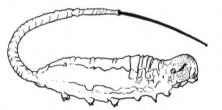

Figure 10. The 'rat-tailed maggot' of the Syrphid fly *Eristalis*, a larva which creeps about under water and reaches to the air with its extensible posterior siphon.

larvae live – but often they must be evolutionary gestures without obvious purpose.

Between them, the acalyptrate and calyptrate families have exploited almost all the possible larval habitats. It seems likely that this line of evolution, like others, began among vegetable debris, leaf-mould and other natural equivalents of the compost-heap. This is the universal breeding-medium, which provides so much: which conceals, protects against strong light and enemies, mitigates the effect of heat and frost, of sudden deluge and – even more menacing to insect-larvae – of the insidious, drying wind. One mass of material may offer a range of temperatures and a variety of food-materials. This sounds like an estate-agent's description of a desirable property, and it seems surprising if flies started out in such a medium that they should ever have moved away from it.

The reason may have been the search for protein in a more concentrated form. The midges that breed in vegetable debris are numerous, but individually small. Bigger adults need bigger larvae to build up their food-supply for them, especially if the adults cannot help themselves by taking a meal of blood. This line of flies had early lost the primitive piercing mouthparts, and so far only a very few of them have developed a substitute mechanism. So the pressure in this group is towards a more and more nutritive larval food.

Dung is an obvious improvement, in which the vegetable cells are crushed and broken, the proteins predigested, and much of the water extracted. We ourselves put down dung in our gardens as a concentrated food for plants: the larvae of many calyptrate and acalyptrate flies found it for themselves. Urine further increases the nitrogen content, and so the droppings of birds and bats are attractive to flies.

From the excreta of vertebrates it is a short step to attacking the vertebrates themselves. First nasal and other fluids, discharges from

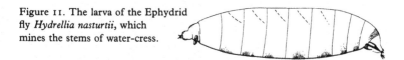

Figure 11. The larva of the Ephydrid fly *Hydrellia nasturtii*, which mines the stems of water-cress.

wounds or sores, dead cells, pus; then a gradual transition to lacerating living tissues. This progress is shown by the sheep blow-fly, *Lucilia*, which attacks sheep via the soiled wool round the anus, and by the screw-worm *Callitroga*, which lays eggs in a discharging nose or a running sore, and ends by devouring much of the living tissue. There is perhaps a distinction to be made between the act of drawing blood from a capillary or from a pool in lacerated tissues, and that of feeding on the tissues themselves. The bloodsucker is a true parasite, living at the expense of the host in what could be, in theory at least, a permanent relationship. The tissue-feeder is a carnivore that just happens to be much smaller than its victim.

A long time ago, Keilin [157] suggested that the maggot of the Cyclorrhapha is a type of larva that seems most suited to a parasitic life, although in practice such larvae live in a great range of organic materials. He concluded that the maggot might originally have been developed for a parasitic life, and hence that all Cyclorrhapha may be descended from a parasitic ancestor. '*Il sera donc possible que toutes les larves des Diptères Cyclorrhaphes qui sont libres actuellement ne le soient que secondairement; elles se seraient rédaptées à cette vie libre, par les moyens qu'elles ont acquis après leur passage par la vie parasitaire.*'

After a survey of the natural history of the cyclorrhaphous flies such as we make in later chapters of this book, it is difficult to accept this view. A progress from compost-feeding, via dung and carrion, to final parasitism seems more plausible. Mr C. E. Dyte has suggested to me that the maggot may be an example of pre-adaptation, as discussed by Carter in his book on *Animal Evolution*, a form of evolution in which a structure originally evolved to suit one set of requirements proves itself efficient in quite a different one. Thus the maggot, with its mouth-hooks and its posterior spiracles, would be excellent for feeding in dung, with its lack of oxygen, but this would also give

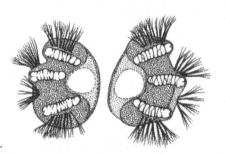

Figure 12. Posterior spiracles of the larva of a Trypetid. No one knows what purpose is served by the elaborate branching structures.

such larvae, as it were, a predisposition to exploit living tissues, where the physiological requirements may be similar.

The calyptrate flies are fairly easy to rationalize, with the housefly group retaining a variety of diets, from attacking growing vegetables to bloodsucking, and with the blow-fly line of evolution concentrating more grimly upon a diet of animal protein, dead or alive. It is extremely difficult to speak generally about the large assortment of flies belonging to the acalyptrate families. In this book, to make it possible to say something about them in a reasonably few words, I have treated them in four groups. Chapter 14 deals with a basic group of compost-feeders, living in decaying vegetable matter, centring round the Dryomyzidae, and a group based upon the Sepsidae that have moved into sewage, dung, dead animal matter, and – with the Sciomyzidae attacking snails – towards a carnivorous diet. In Chapter 15, Trypetidae, Chloropidae and others make a group of plant-breeding flies, the females of which often have a rigid ovipositor with which they push their eggs into growing plants; and *Drosophila* is one of a group of sweet-toothed flies which seek the rich decay of fermenting, yeast-producing fruits, vegetables, and fungi.

When the larva has built up its food reserves sufficiently the time has come to make the transformation into an adult fly, and this is achieved in one big step by means of a massive reconstruction in the *pupa*. This is usually a quiescent stage, though the pupae of many mosquitoes swim actively, and Hinton [134] showed that the pupa of the black-fly *Simulium* both feeds and spins a cocoon. But these are exceptional activities.

The pupae of Nematocera and nearly all Brachycera are curious mummy-like objects, with external bulges in which are developed the head, wings, legs, antennae and so on of the adult fly inside. Often the adult features are shown clearly enough for the family to be recognized. The pupae also have well-developed adaptive structures to suit the needs of the pupal environment. Aquatic pupae, and those that live in media deficient in oxygen, have 'horns' or 'ears', or other extensions from the prothoracic spiracles, and these may even be sharp and stiff enough to pierce the tissues of submerged plants and draw oxygen from them. Whereas the most important spiracles in larvae are those at the tip of the abdomen, most pupae breathe through the prothoracic spiracles, which are also of major importance to the adult fly. Many pupae lie in the soil, and some Brachycera, notably robber-flies and bee-flies, have powerful backwardly projecting spines with which they work their way to the surface when the adult is about to emerge (plate 15). Very strong spines on the head may be used to break out of the surface layer, even sometimes from burrows in wood.

Those families of more advanced flies that have adopted a maggot-like larval stage have also simplified the process of pupation. They have ceased to break out of the last larval skin, which hardens and darkens, and forms a *puparium*. Thus the true pupa of these flies, most of the Cyclorrhapha, remains hidden from view. There is now no question of any movement during the pupal period, and when the adult fly emerges it has not only to break out of the puparium – which it does by lifting off a circular cap at one end, from which these flies get the name Cyclorrhapha – but also has to make its own way to the surface of the soil or any other material in which it is buried. The soft-crumpled, immature fly achieves both these objects by inflating a large sac in the head called the ptilinum.

A puparium is quite different from a *cocoon*, which is a protective shell made from silk, mud, masticated wood, sand, or any other extraneous material. It is this that in butterflies and moths is called a chrysalis. A cocoon is constructed by a larva just before pupation, and is a special adaptation that has arisen many times, quite independently, in members of different families of flies. A very few Cyclorrhapha make a cocoon, but they still form the usual puparium inside it.

Among Brachycera the Stratiomyidae are exceptional in not discarding the larval skin when they pupate, but this seems to be a protective device specially evolved in that family, and not related to the puparium of Cyclorrhapha.

Whether the pupal stage is passed in water or in a dry medium, the adult flies almost always emerge into the air and avoid being wetted. Some black-flies of the family Simuliidae, the most completely aquatic of flies, actually emerge as adults under water, swim to the surface, and fly off immediately. This is an outstanding performance, because most other flies are not only fatally trapped if they are caught in a water-film, but when they emerge from the pupa they require anything from minutes to hours to become sufficiently hardened to be able to fly. Seaweed flies, of the acalyptrate family Coelopidae are greasy and water-repellent, and have the ability at any time of their adult life to take off from the surface of the water, but they are exceptional.

More typical is the gall-midge that Barnes watched as it emerged from the pupa, and which took thirty minutes to free itself. Even then safety is not assured: a most pathetic sight is to see a bluebottle that has emerged into a space too small to allow it to expand, and which has died still crumpled into the folds in which it was neatly packed into its puparium.

A recent paper by R. E. Snodgrass [268] takes a different view of the relationship between larva and adult. He rightly insists that the adult is the 'real' insect, and the larva only a temporary transformation, but he then argues that a larval stage was developed to make up for the shortcomings of the adult: 'The moths and butterflies, for example, have developed a proboscis for feeding on nectar, a highly inadequate diet. It becomes the duty of the larva, therefore, to provide the essential food elements. . . .'

It is difficult to go with this eminent morphologist when he writes: 'The wingless young mosquito could not follow the bloodsucking

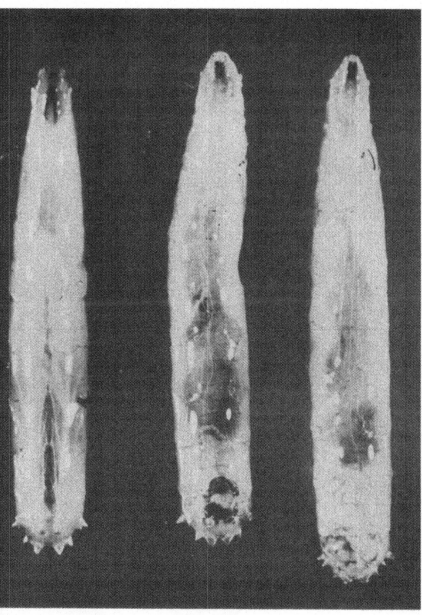

5. Larvae of the onion fly, *Hylemyia antiqua*. The larvae of the more advanced flies are maggots, with few external structures except for a pair of posterior spiracles and a few papillae surrounding them.

6. An adult of the wheat bulb-fly, *Leptohylemyia coarctata*.
Comparison of this with plate 5 illustrates what is meant by saying
that flies have a complete metamorphosis.

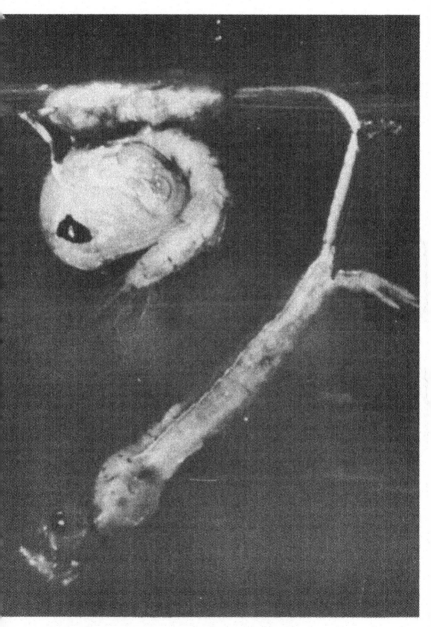

7. A pupa and an empty larval skin of a culicine mosquito. Each reaches to the air/water surface with its spiracles, the larva with the posterior pair and the pupa with those of the thorax.

8. An adult mosquito emerging from its pupal skin at the
surface of the water.

habit of its mother or the nectar-feeding of its father. *Hence* [my italics] it has taken to the water where it has acquired a body structure suitable to aquatic life, and has developed a very special feeding mechanism of its own.' This is strictly a *non sequitur*. It seems to me more helpful to look upon the larva as an ancillary stage, relieving the adult of most of the laborious and protracted feeding needed during the life of an individual. Those adults that have continued in the primitive blood-sucking habit, like the mosquitoes and black-flies, and those that have reverted to predaceous or bloodsucking habits, like the robber-flies and the tsetse-flies, are thereby supplementing an inadequate larval diet.

In other words the adults, where necessary, make up for the deficiencies of the larva, not the opposite.

Part Two

Chapter 3

Crane-Flies

THE DADDY-LONG-LEGS might reasonably be called the father of flies. This familiar, harmless insect is looked upon with affection, particularly in temperate countries, where it is part of the brief summer scene. It is particularly the big crane-flies that are called by this name, and many people assume that the little crane-flies that they disturb from the grass of damp meadows are merely young ones. Being winged insects, they are of course fully grown adults of other species.

The bigger crane-flies belong, in the main, to one sub-family Tipulinae, the long-palped crane-flies, and the smaller ones to the Limoniinae, or short-palped crane-flies. Some recent authorities regard these as two distinct families, but that is of little moment for our present purpose.

Crane-flies are among the drolls of the insect world. Their abdomen and wings are too narrow, their legs absurdly too long. They have what I have earlier called an 'old-fashioned' look, each part of the body retaining much of what one thinks must have been the basic structure of the first flies. Within this one family there are hints of many trends, particularly in feeding-habits and larval habitat, which appear again in flies of much later evolution.

Although the adult stage of crane-flies is the one that is most familiar to us, it is biologically less important than the larva. Indeed, not very much is known about the details of the adult life of most species. They sit about on exposed surfaces, but they are not such ardent sun-lovers as are many flies. Rather one associates crane-flies with cool, damp places, and a great many escape notice because they are resting under leaves and between the stems of grasses.

Reduction or loss of wings occurs sporadically in many different families of flies, and crane-flies provide examples of this. The most conspicuous example is the snow-fly, *Chionea*, which can be seen above the snow-line in Europe and North America, at temperatures as low as −10°C. During the summer months it shelters from the heat in hollows in the soil, and particularly in the nests of wasps and of small mammals, where its larva lives. Marchand [195] noted that even the warmth of his hand was sufficient to repel these cold-loving flies.

The main occupations of adult crane-flies, as of the majority of adult flies, are mating and egg-laying. Feeding is less important, and probably water is the most pressing need, helped out with a little sugar from the nectar of flowers. Many crane-flies, including the familiar daddy-long-legs of the sub-family Tipulinae have the face drawn out into a rather asinine snout, with the mouthparts at its tip (plate 1a). There are no mandibles, and the maxillae are modified and reduced, so that neither chewing nor piercing is possible. The maxillary palpi are usually prominent, and it is from these that the division into long-palped and short-palped crane-flies is taken. They obviously serve to detect food by smell and touch. That the snout is adapted to flower-feeding is borne out by the fact that genera that are otherwise only distantly related each have the proboscis drawn out into a long, slender tube, which they use to extract nectar from correspondingly deep, tubular flowers.

In *Geranomyia*, of the tribe Limoniini, the snout is itself very long, and then the sucking mouthparts are drawn out still further. In *Elephantomyia* and other genera the entire snout has become a fine, tubular proboscis, many times the length of the head, and the actual mouthparts at its tip are greatly reduced in size.

It is easy to point to an adaptation like this and say that it is an example of how flowering plants and flies evolved together. It is less easy to see why they did so. Knab [168] who described the habits of these particular crane-flies, says, 'the flowers of this group of plants are well understood to be so organized that they are profitable only to insects which are specialized to reach the nectar at the bottom of the deep and narrow tube, this latter protecting the honey from undesirable marauders and rain'. He goes on to speak of the flowers concerned, mostly Compositae, as those 'which offer a superior quality of honey to those insects organized to obtain it'. I do not know whether either of these assumptions about the flowers is any more than mere conjecture.

The elongate proboscis of butterflies and of bees is capable of being extended when necessary, but it does not prevent these insects from feeding from shallow flowers, or from open sources. The various genera of flies that have developed an elongate, tubular proboscis – we shall see a number of these, scattered through the Nematocera and Brachycera – are less well served, because their relatively inflexible proboscis, while it opens up one source of food to them, actually reduces their ability to feed at close range. In Diptera this development has the appearance of 'one-track evolution', that is of a trend which cannot be reversed after it has passed a certain level of specialization.

The evolution of a flower-feeding proboscis by flies has not been entirely without a future. In some families, those in which mandibles and maxillae have been retained as part of the sucking-tube, it has been possible to adapt the proboscis for piercing the skin of vertebrates and sucking their blood. This is a step that adult Tipulidae have never taken. They presumably lost their mandibles and maxillae at an early stage. Neither have they shown the evolutionary ingenuity of, say, the tsetse-flies (described in Chapter 16).

Mating habits of crane-flies also set a pattern for Diptera in general. Much depends on the timing of emergence from the pupa. If there is a general emergence of both sexes, mating may take place at once, sometimes when the females have scarcely freed themselves of the pupal skin, and certainly before they have hardened. If emergences are more scattered, and particularly if the males come out before the females, then the males may be less easy to find. Sometimes they may sit beside the pupae until females come out. The males of the bigger daddy-long-legs of the genus *Tipula* usually search about for a female, flying a little and walking a lot, over tree-trunks, and under leaves. The smaller Limoniinae, the short-palped crane-flies, are conspicuous for their mating swarms: another practice that is very common among the smaller flies of many families.

In another chapter of this book something is said about swarms and other gregarious activities of flies. It is sufficient at this point to note that the swarms of crane-flies consist chiefly of males, and when a female appears she is pounced upon by a male, who rides with her to a nearby support, a branch or a leaf. This pattern is not invariable and there are some that search as well as swarm, thus illustrating the gradual emergence of the habit; but among the dancing swarms that we have already come to think of as rather typical of nematocerous flies, we must include those of many of the smaller crane-flies.

Another mating device common among flies, and exhibited by crane-flies, is that of rhythmically waving some part of the body that is brightly coloured, sometimes wings, often a specially developed part of the leg or foot. The crane-flies, with their exceptionally long legs, make free use of this activity. They often hang by the forelegs, or the fore and middle legs, with the hind legs dangling and waving. Collectors in the tropics have particularly noted this habit, and the associated trick of choosing spiders' webs as places from which to hang. Miss Evelyn Cheesman, in her book on *Hunting Insects in the South Seas*, speaks with admiration of the crane-fly *Doaneomyia tahitiensis* which has conspicuous white bands on the legs, and which dances in dark gullies on the island of Tahiti. This fly also hangs upside down in spiders' webs, and is neither trapped by the web, nor attacked by the spider. 'It is a curious habit for any insect to hang suspended in this fashion, and why they should do so in a place so fraught with danger as a spider's web is incomprehensible. On some islands there are spiders with white bands to their legs, which at first sight could be mistaken for these [crane-flies], but whether the spider mimics the [crane-fly], or the other way is not plain, for it is difficult to see what advantage such mimicry would be to either.'

It seems possible that the association may give protection to the fly. Crane-flies are slow-flying, and vulnerable to many predators, especially if they make themselves conspicuous by waving white legs in the air. This is probably a device that attracts the attention of the other sex, and is most often an adornment of the male, as in some other families of flies, notably Dolichopodidae. Flying predators such as Asilidae and Empididae are attracted by movement, but they are likely to be foiled by a spiders' web, which acts as a small-scale balloon-barrage.

On the other hand, many crane-flies go through the rhythmical motion of 'bobbing', raising and lowering the body by bending the legs, again for no apparent reason. Some observers have tried to associate the motion of bobbing with egg-laying, and with feeding, but it seems to occur when neither is possible. A correspondence some years ago in the *Entomologist's Monthly Magazine* established that both bobbing and hanging from spiders' webs are widespread habits all over the world – in Samoa, Buxton described such an association between *Trentopohlia* and Pholcid spiders – but no one has explained them satisfactorily.

Apart from these few general habits there is little to be said about the ways of adult crane-flies. The life-histories of the various species are determined, in the main, by the requirements of their larvae and so, like human parents, crane-flies live where it is best for the children. They have not made use of the divergence between larva and adult to colonize two different habitats, as have many of the more advanced flies.

The larvae of true crane-flies, belonging that is to the family Tipulidae and not to one of the other families that we are also considering in this chapter, can nearly always be recognized by having a single pair of spiracles posteriorly, surrounded by a disc of fleshy lobes, like a flower modelled in plasticine. The head is fairly complete, but partly excavated posteriorly, and it can be retracted into the thorax. The larvae of the one or two related families will be mentioned later.

As we have seen earlier, consideration of the habits of the larvae, and of the course of their evolution in adaptation to different food-materials, is a very good guide to the past history of a group of flies. Among their very varied larval habits, crane-flies still retain examples of the various trends that we have been able to follow in Diptera as a whole.

Keilin [159] in the paper we have already discussed, points out that the law of irreversibility of evolution applied to the spiracles of dipterous larvae will show the correct relationships of otherwise anomalous groups. All larvae of Tipulidae have posterior spiracles only, or none at all, and in Keilin's view this is the prime adaptation to life in water, or in a medium where respiration is difficult. Crane-flies therefore belong to the 'watery' half of the Nematocera – with the gnats rather than with the land-midges.

We believe that flies as a group have arisen from ancestors whose larvae, though terrestrial, lived in wet moss and so had, as it were, a foot in both worlds. In considering the larval habits of the crane-flies we may begin with a group, the Cylindrotominae, which divides similarly into two. Larvae of the genera *Triogma* and *Phalacrocera* live among moss submerged in moving water. They have long, fingerlike processes along the body which, helped by the extreme sluggishness of the larva, become covered with a mass of algae and other organisms, like a pier with barnacles. The larvae of *Liogma* and *Cylindrotoma*, in contrast, are rather like legless caterpillars, and eat leaves in a caterpillar-like way. In this they are unique among larvae of flies, and seem like a curious echo of their distant relatives,

33

the Lepidoptera. These larvae, though technically terrestrial, feed on plants of a boggy habitat, and take on the green colour of their food material. The aquatic forms have an active pupa, which clings to submerged vegetation by means of hooks at the tip of the abdomen, and reaches the surface air through long breathing horns arising from the thorax. The pupae of the terrestrial forms cling to vegetation in the open.

It has at times been suggested that Cylindrotominae might be the stem from which the whole of the Nematocera and Brachycera arose. This is not now thought to be likely, but they diverge very early along the line of evolution, and they do in some ways throw back to the ancestral flies. Alexander thinks that they flourished in the early and middle Tertiary period, and are now in considerable decline.

Of the two big groups of crane-flies we may look first at the Tipulinae, the daddy-long-legs proper. They include the biggest known crane-flies, some tropical species of *Ctenacroscelis* reaching nearly 100 mm in wing span, and descend to a smallish *Microtipula* only 7 mm long. Their larvae are rather plump and fleshy, the posterior spiracles surrounded by a crown of six lobes. The well-known 'leather-jackets' of our lawns and playing-fields are among the most terrestrial of Tipulid larvae. They feed principally just below the surface of the soil, eating the roots of plants of all kinds. They are abundant in old grassland, and can do a great deal of damage to any young crop, including lawn-grass, where they eat enough of the roots to cause a bare, brown patch.

Although terrestrial, and provided with a tough skin, leather-jackets are moisture-loving, and their appearance above ground is mostly confined to times when the air close to the surface is moist. Warm, humid evenings in summer bring them out, as well as evenings when a heavy dew has fallen. At these times leather-jackets will eat young plants at surface level, like cutworms.

Evolution in a different direction from the boggy areas of their origin has led many Tipulinae into moss on dry rocks, and into the decaying wood of fallen logs and the stumps of felled trees. *Dolichopeza* and *Oropeza* are examples of moss-dwellers, their larvae and pupae green in life. They are among the leg-swinging frequenters of spiders' webs that we have just mentioned. Species from a number of genera breed in rotting wood of varying degrees of softness. Most of them, species of *Tipula*, *Dictenidia*, and others, including the beautiful orange and black *Ctenophora*, have the usual simple ovipositor, and can only deposit their eggs on the surface, or push them into soft pulp; but *Tanyptera*, another beautifully coloured red and black fly, has a slim, pencil-like ovipositor, and looks very much like one of the ichneumons that lay eggs in similar places. As its appearance suggests, it lays eggs in wood that is less decayed, and is the extreme in the direction of a wood-boring fly. This is a larval habitat that has not been greatly exploited by Diptera, though a few isolated flies have colonized it. Alexander records that larvae of *Tanyptera frontalis* were found in old maple-wood which, while not entirely sound, was still so hard that it had to be cut with a hatchet.

The larvae of the genus *Tipula* itself have a variety of habitats, besides eating the roots of our lawn-grass and seedling plants. In fact one species or another can be found in all the habitats from decaying wood, through moss, earth, and mud, to a more or less completely aquatic life. Even the aquatic larvae breathe air of course, and can remain submerged only for a limited period; though the genus *Antocha*, as we shall see later, has overcome even this disability. The six lobes surrounding the posterior spiracles often have a fringe of hairs or fine filaments, which help to trap a film of air when submerged and so create a respiratory plastron (see above). These larvae may also have blood-gills, finger-like lobes which are not respiratory in function, but help to maintain osmotic pressure and the balance of salts between the blood and the surrounding water.

The second, and larger section of the family, the Limoniinae or short-palped crane-flies, as I mentioned earlier, are treated as a separate family by some authors. The technical systematic arguments do not concern us here, but we may note that in their biology and habits the short-palped crane-flies cover a range as wide as, indeed wider than, that of the daddy-long-legs proper. We can see again the same situations, and similar larval adaptations to meet them. The two groups have certainly evolved independently and often on parallel or convergent lines.

In general we can say that the short-palped crane-flies have all the versatility of the long-palped genera, and more. In place of the genus *Tipula*, sampling all the habitats with one species or another, we have *Dicranomyia* going even further by having the only leaf-mining crane-fly. *Limonia* (*Dicranomyia*) *kauaiensis* Grimshaw makes mines in the leaves of *Cyrtandra paludosa* and other species of the genus, in the Hawaiian Islands.

35

Leaf-mining, and the formation of galls on growing plants, are extremes of the type of feeding that begins with terrestrial larvae attacking the roots, and have arisen independently in a number of widely separated families of flies as we shall see later. The fact that the only leaf-miner among the crane-flies should be confined to the Hawaiian Islands, and should be abundant there – D. E. Hardy says of this fly that 'this is one of the most common species taken in vegetation in the mountains of some of the islands' – is consistent with what we know of the past history of other extremely local flies (see Chapter 21).

The short-palped crane-flies have gone further, too, in the other direction, in adapting themselves to life in the water. We have seen that the large, open spiracles at the tip of the abdomen, themselves a modification making it easier to breathe in watery places, compel the larva to reach the air periodically, even if only at long intervals. If this is prevented, then diffusion of nitrogen from the open spiracles will gradually flood the tracheal system. This limitation has been overcome in a small group of crane-flies, the genus *Antocha* and its relatives, by closing the tracheal system completely. Exchange of oxygen takes place by diffusion through the thin cuticle of tracheal gills, lobes at the tip of the abdomen where tracheae are abundant. The pupae have long-branched filaments of a similar function arising from the thoracic spiracles.

These larvae live in well-aerated water, often rapidly moving, much the same as we shall see later in the black-flies of the family Simuliidae. Indeed, the pupae of the two families are very similar in appearance. The larvae of *Antocha*, however, live in silken tubes or cases open at both ends, and often covered with fine debris, small stones, and so on. This at once suggests comparison with the case-bearing caddis-flies of the Order Trichoptera, one of the Orders belonging to the Panorpoid Complex from which the Diptera too are supposed to have arisen.

A number of other larvae of crane-flies live in silken cases, or construct them when they are about to pupate, but these other larvae all have the usual pair of open spiracles posteriorly.

Not to be outdone by any other family, crane-flies have also produced a marine member, or at least a littoral one. We have already mentioned the genus *Geranomyia*, with its long proboscis and its habit of feeding from long, tubular flowers. The larvae of *Geranomyia* live under water, feeding on algal growth, diatoms, etc; sometimes they make silken cases, but emerge from them more readily than do the larvae of *Antocha*, coming to the surface for air, and even feeding on exposed wet rocks. It is as a development of this habit that the larvae of *Geranomyia unicolor* manage to feed on algae on marine structures that are washed by the tide. Instead of moving themselves in and out of the water as do their freshwater relatives, these larvae of the intertidal zone stay put, and are submerged and exposed alternately, twice every day. The disc at the tip of the abdomen which bears the spiracles is surrounded by a whiskery fringe of hairs, which help in trapping a plastron of air over the spiracles during the periods of immersion. Larvae living in salt water have no blood-gills: their body fluids are more nearly of the same osmotic pressure as the surrounding water, and exchange of salts and dissolved waste products is easier.

Another direction in which the short-palped crane-flies have gone further than their bigger relatives is in moving towards a carnivorous diet. The larvae of Tipulinae are essentially vegetarian, eating moss, algae, diatoms, rotting and macerated wood, and living cells from the roots of growing plants, and even trees. It has been reported that these larvae will on occasion eat small animals, even earthworms, if they encounter them in the course of feeding, but this is probably not a significant part of their diet. As we shall see, some other larvae, notably those of the robber-flies of the family Asilidae, have the same habit.

The sub-family Limoniinae, however, includes many species that are wholeheartedly carnivorous, especially in the tribes Hexatomini and Pediciini. Their mandibles are curved and sharp, and are used to attack other small animals: dragonfly nymphs, larvae of other flies, and small worms. They can take prey almost as long as themselves and sometimes are cannibalistic on their own species. They tear and swallow their prey, and may have a roughened area in the gullet that helps them to keep it down, like the backward-facing teeth of sharks and snakes. It is said that some of these larvae can pierce the human skin and give a painful bite.

Carnivorous larvae, of course, are immensely more active than vegetarian ones. Some of the moss-feeders curl up like caterpillars, and hardly move at all, while the tube-living *Antocha* can move actively if stimulated, but hardly ever leave their silken tubes. The carnivorous larvae, on the other hand, are very active, and are slender and wormlike in shape, in strong contrast to the fleshy, obese, rather repulsive leather-jackets. This is the reverse of human experience where vegetarians tend to be thin, and hearty steak-eaters to put on weight. Perhaps if we still had to hunt our meat before we ate it our shape would be different.

The account by Crisp and Lloyd [53] of a community of Tipulid larvae in a patch of woodland mud gives an illuminating cross-section of habits in this family.

There are a few crane-fly-like insects which are sometimes considered to be closely related, though current opinion is rather against this view. They clearly represent more or less abortive evolutionary experiments, and are interesting for the light that they throw, either on the origins of crane-flies, or on the transition towards other groups of flies.

The larvae of *Ptychoptera* and *Bittacomorpha*, members of the family Ptychopteridae, are remarkable for their elongate, telescopic breathing-tube, which carries the hind spiracles up to the surface to reach the air. They also have well-developed creeping welts, swellings or projections from some of the body-segments, especially on the underside, which act as false legs, and enable the larva to crawl about on the sub-stratum while holding its tube towards the surface. This structure is very suitable for a larva that lives in a shallow liquid or semi-liquid medium that is deficient in oxygen: stagnant pools, foul with organic refuse; wet, decaying vegetation; manurial refuse such as is found round cowsheds and farmyards. These habitats have been colonized by *Ptychoptera* in Nematocera, and in Cyclorrhapha by the 'rat-tailed larvae' of certain hover-flies (Syrphidae), and by the Ephydrid *Ptychomyza*. The larvae of these flies are remarkably alike, though the adult flies are very dissimilar.

Somewhat like the Ptychopteridae are the Tanyderidae, a small group of twenty-five species of which the larva is known in only one. *Protoplasa fitchii* is one of the rarest of crane-flies, found only in North America. It is featured in textbooks because the adult fly has five branches to the radial vein of the wing: a hypothetically primitive arrangement almost never found in any real fly. The larva was for long unknown, and then misidentified, but was finally described by Alexander in 1930. It lives in sand in shallow water and has two pairs of spiracles, one on the thorax and one on the tip of the abdomen. It thus far agrees with theory, because the so-called amphipneustic condition is an intermediate stage towards the single pair of spiracles of the Tipulids, and is also the general condition in the higher flies, especially Cyclorrhapha. In other ways *Protoplasa* is a sort of off-beat crane-fly. The Tanyderidae as a family are a relict group, lingering on in the tip of South America, in Australasia, Malaya, Japan, Turkestan, and North America. One species was recently discovered in the extreme south of Africa, where other relicts of this kind are increasingly being found.

The 'winter-gnats' of the family Trichoceridae (Petauristidae) are familiar to people who live in temperate countries, because they carry to an extreme the liking of the crane-fly for cool, damp conditions. Although they occur in summer in such places, they are overlooked, or confused with true crane-flies; in the winter they come out and dance in the open on every sunny day. They are also common in caves, gorges, and grotto-like places, and in mines. They have been said to occur 600 ft below the surface in a mine. Their swarms are male mating swarms, such as we have described, and the larvae breed in decaying vegetable matter of a compost-like nature, occasionally damaging stored root-crops and tubers. As we shall see, this habit is characteristic of the group of Nematocera centring round the Mycetophilidae, the 'land-midges' of this book.

Into the family Anisopodidae (Rhyphidae) are placed a few, rather ill-assorted flies that vaguely link the crane-flies on the one hand, with the midges on another, and with the lower families of Brachycera on a third. There are about fifty species of *Anisopus* scattered about the world, but the genus is chiefly known from the two European species, *A. cinctus* and *A. fenestralis*. The latter is one of several quite different flies that we commonly call 'window-flies', because they occur indoors, and are first noticed crawling up a window, trying to get out. *A. fenestralis* is not unlike a small crane-fly, but also much resembles some members of the family Rhagionidae at the beginning of the Brachycera. The larvae of *Anisopus* feed in decaying matter of a rather higher organic content than the true crane-flies, preferring it enriched with dung, sewage, and other products of animal origin, or with the yeasts of fermentation. In the house this fly will breed in debris round the kitchen sink and in neglected dishcloths. Out of doors the adult flies sometimes occur in dancing swarms.

Mycetobia is a midge-like fly, the larva of which lives in decaying wood, and in the fermenting sap that flows from wounds in living trees. This genus has been the entomological equivalent of a stateless person; having at one time been placed with the midges of the family Mycetophilidae, and ejected from there, it is now grouped with *Anisopus*, but the arrangement is not entirely a happy one. The fact that Edwards, in his review of this family for the *Genera Insectorum*, mentions only two living species (and two fossil ones), and dismisses five other names as wrongly placed in the family, shows how much *Mycetobia* has been misunderstood: and when he concludes 'nothing of interest is known concerning their habits' we too lose interest.

The crane-flies as a group are world-wide, and over the years an immense number of species has been described, perhaps as many as 10,000. Only an insignificant fraction of these species has been bred, or their habits studied. The Tipulinae and Cylindrotomini, Tanyderidae and perhaps such genera as *Mycetobia*, are groups that have had their heyday and are now in decline. They seem to survive best in cool, temperate regions, and species of the genus *Tipula* penetrate far to the north, and high on snow-capped mountains. So do some of the Limoniini, though perhaps in smaller numbers. Short-palped crane-flies are probably more abundant in the tropics, and many remain to be discovered there.

With their slow flight they are very vulnerable, and are heavily attacked by predators of all kinds, from ants to birds. Predatory flies such as Asilidae and Empididae kill very many crane-flies. Larvae are on the whole less vulnerable than the adult flies, though when leather-jackets expose themselves on the surface they are devoured in great numbers by birds, and the mud-loving forms by water-fowl.

Alexander remarks that Tipulinae and Cylindrotominae are not found in small, remote islands, where only Limoniini have penetrated. He seems to think that the larger, relatively more powerful flies have, as it were, 'escaped', but it seems more likely that it is the smaller crane-flies that are distributed passively by wind. The large daddy-long-legs find it difficult to maintain control in strong wind, and, if they were carried bodily away, would probably suffer fatal damage on the way.

The vulnerability of crane-flies in windy conditions is reflected in the large number of species with reduced wings, or no wings at all, which are found on remote islands, on high mountains, and towards the poles where high winds prevail.

Chapter 4

Land-Midges

AFTER THE DIVERSITY of the crane-flies, the rest of the nematocerous flies can be considered in two big groups: the terrestrial ones or land-midges, and the aquatic ones, the water-midges or gnats; though as I have explained, the common English names of the families are inconsistent in the use of 'midge' and 'gnat'. The mosquitoes belong to the second group, but so much is known about them, and their practical importance is so great, that they deserve a chapter to themselves.

First then, the land-midges of the super-family Mycetophiloidea. These are flies of the compost-heap, flitting about generally in shady places, always to be found where there is gently decaying vegetation, dung, rotten wood, or a growth of mould or fungus. That part of the garden that is out of sight of the house, that odd bit of ground just off the path, where unwanted rubbish is dumped, that neglected piece of woodland with fallen and fungus-covered trees – these are the haunts of this group of flies.

The adult flies are all harmless creatures, and have no equipment for biting. They have long antennae like a string of minute beads, and the wing-veins are few, with hardly any cross-veins. In spite of the fact that they look so featureless and all alike, about 2,000 species of Mycetophilidae and over 300 of Sciaridae are known to exist. These two families used to be united, and are still so treated in some books. Most of these have been discovered in the cool, temperate regions of both hemispheres. It is likely that they are less common in hot countries, but it is certain that many such small and fragile flies must exist in shady, steamy areas of the tropical rain-forest, as well as in the temperate forests that occur on the higher parts of tropical mountain ranges. Collectors seldom bother much about insects that are so difficult to preserve well, and so uninteresting to look at when you have them. Only a few of the biggest are collectors' specimens.

I commented earlier upon the unrivalled opportunities for evolution that are offered to a larva by a compost-like medium. One of the many advantages of an almost uniform temperature, often above that of the surroundings, and an unlimited supply of vegetable food, is a short larval life, and two to three weeks is generally enough, at any rate in the warm months of the year. This means that such flies can pass through perhaps six, eight, or even ten generations between spring and autumn; whereas many aquatic larvae may be delayed by low water-temperature, and many soil-living larvae by the difficulty of finding enough suitable food. The advantage in speed of mutation, selection, and evolution is obvious.

Larvae of Mycetophilidae and Sciaridae are essentially terrestrial, and usually have eight pairs of open spiracles, one pair on the thorax and seven on the abdomen. They are slender, worm-like creatures, only a few millimetres long, with a small but distinct dark head, and often showing the spiracles as a row of tiny spots like portholes along the body. They are typically vegetable-feeders, but some of them prefer to live in growing fungi, which they tunnel until the firm mass of the fungus begins to liquefy and rot sets in. In English, Mycetophilidae are known as fungus-gnats, though such essentially harmless flies might better be called fungus-midges. The larvae sometimes damage commercial crops of mushrooms, and the adults may be a nuisance if they breed in large numbers in rotting potatoes or other vegetable refuse.

From ancestors living in such compost-like materials, evolution has commonly progressed in three directions. Some descendants have moved into wetter localities, and become at least semi-aquatic; other representatives have moved into drier localities such as under bark, in dead wood, and in soil; yet others have become carnivorous. Some Mycetophilidae and Sciaridae do each of these things. Both families stray into shallow water among reeds and other water-plants; more of them go the other way, their larvae then feeding under bark, in dead wood and in nests of birds. In all these places they probably feed on vegetable matter or dung or on the moulds and fungi that flourish there.

One bizarre locality has come to light in London in recent years, when large numbers of adult *Sciara* made a nuisance of themselves in offices that were furnished in a modern, utilitarian style, and would seem to offer none of the quiet, damp corners that such flies like. My colleague Dr Paul Freeman finally found that the larvae were living in the peaty material used as a heat-insulator and fire-resisting lining to the office safe! Only minute holes gave access to this material, which must have become decayed with time and condensed moisture, and bred the moulds upon which the *Sciara* larvae can breed.

This is a rather extreme example of the indoor habits of *Sciara*, which readily breeds in any dark, damp corner, cellar or outhouse. The larvae of these two families will tackle almost any kind of debris, with the Sciaridae perhaps inclining towards a higher concentration of animal proteins, Mycetophilidae perhaps more attracted by moulds and fungal spores. Thus they invade the burrows of small insects, and the nests of birds, wasps, ants, and termites. But lest this should make them seem entirely furtive, both families have members that live a fully exposed, terrestrial life.

I have already briefly mentioned the peripatetic 'army worm', the larva of *Sciara militaris*. The same popular name is also given to the caterpillars of certain Noctuid moths, which have a similar habit of moving about in bands. The *Sciara* larvae march harmlessly across the leaf-mould of a forest floor as a band of worm-like insects an inch or so broad, and 10–15 ft long. No one knows why they should do so.

Terrestrial Mycetophilidae can be spectacular too, in their own way. The well-equipped head of the larva has labial glands which produce silk, and this is put to ingenious uses. Spread web-like over the surface of a fungus, or in the form of silken tunnels in the soil, it helps the larva to move about quickly in a difficult medium. With the addition of particles of soil, small stones, and so on, the larva often builds a cocoon in which to pupate. Some carnivorous larvae, especially those of the sub-family Ceroplatinae, follow the example of the spiders, and use their web to catch other small animals as prey.

They may even go further, and convert their web into a light-trap. The famous 'Glow-Worm Cave' at Waitomo, in New Zealand, is visited by boatloads of tourists who come to see the colony of luminous larvae of *Arachnocampa luminosa* on the ceiling of the cave. This midge is widely distributed through both islands of New Zealand, living in caves, tunnels, and steep, damp ravines. Waitomo is particularly suitable for tourists who wish to see the display, because they can drift along silently by boat. The larva can switch off its light if it is disturbed, and it is said that even talking loudly may bring this about.

Similar 'glow-worms' – of course quite unrelated to the true glow-worms, which are beetles – have been reported from caves in Tasmania and New South Wales. It is thought that the glow attracts insects into contact with a mass of hanging threads forming part of the web of the larva: these threads are sticky, and have droplets that perhaps are poisonous, containing oxalic acid, but which are not themselves luminous, though they appear so in the flashlight photographs. The prey is mostly Chironomid midges breeding in pools in the caves and ravines concerned.

There are other luminous larvae of Mycetophilidae, and that of the Japanese *Ceroplatus nipponensis* shows perhaps how the New Zealand glow-worm may have evolved. The Japanese larvae live in a web over fungus-infested wood, for example underneath a wooden bridge that is wet from a rushing stream (Kato [156]). The larva is only faintly luminous, with a pale blue light that has to be looked for with care: moreover it is not carnivorous, but feeds on the spores of the fungus. It is interesting to wonder whether the New Zealand glow-worm first became carnivorous, and then evolved a stronger light because this brought more food; or whether its light became stronger through changes in its internal chemistry and accidentally attracted prey, which tempted the midge to change its diet.

Arachnocampa, the New Zealand midge, produces its light in the swollen ends of the Malpighian tubules, the excretory organs that are the insect's equivalent of the kidney, and which in this instance are provided with a reflector ingeniously adapted from the lining of a trachea. The pupa remains slung in the web, and also has its light: even the adult female retains a feeble light for part of its life, though the male has none.

Ceroplatus, in contrast, has no special luminous organ, but has luminous granules distributed throughout its fat-body. It is tempting to see the arrangement in *Arachnocampa* as an improved version of this, making use of the accidental production of 'cold light', concentrating the luminous patches until the light has become bright enough to be useful to the insect.

An equally remarkable feature of these larvae of Ceroplatinae is that they have no functional spiracles, a condition normally found in larvae that have no ready access to air, and have to rely on dissolved oxygen. It is an extreme adaptation of aquatic larvae, and seems out of place in those which are fully terrestrial. It may be a further pointer to the evolution of this group through the spray-living forms, because aquatic larvae of some Psychodids live under such conditions, as we shall see later.

The family Bibionidae, though related to the two previous families, has become more thoroughly terrestrial, both as adults and as larvae. The adult flies are bigger and heavier, and some have a distinctly armoured appearance, complete with formidable spurs and spines on the legs and head (plate 9). The adults are the first flies we have yet met that are predominantly flower-lovers, and hoverers in the sun, both charming habits that are shared by a great many flies, and which offset to some extent the less pleasant habits of the majority.

They are called 'March-flies' in North America, and *Bibio marci* has been known in Europe as 'St Mark's fly', both names associating the Bibionids with spring, and a return to sunshine and flowers. They overdo it, of course. *Dilophus febrilis* and other small Bibionidae swarm over early flowering shrubs, and can cause much annoyance to persons who are also enjoying the pleasures of their garden after the winter, but in orchards these same flies are very welcome, because they supplement the work of the bees in pollinating the fruit-blossom.

Male Bibionids have in an exaggerated form the enlargement of the upper facets of the eyes that I mentioned in Chapter 1 as possibly being associated with the habit of dancing in swarms. Bibionids certainly do this.

The larvae of Bibionids are among the most terrestrial, and probably the most primitive, of any flies. They have strong chewing mouthparts, and eat most kinds of vegetable food. In the soil they are often gregarious, and may be found as a mass of grubs in a pocket in the soil. In such concentrations their feeding may damage the roots of growing crops, and it is said that they are sometimes introduced to cultivated land in manure or leaf-mould spread as humus. They also breed in dung, and in caves.

As is appropriate to their terrestrial habits, Bibionid larvae have a very complete set of open spiracles, two thoracic pairs and eight on the abdomen. The larvae of the related family Scatopsidae have one fewer, lacking the second thoracic pair. Superficially rather like Bibionids, Scatopsidae are structurally and biologically linked on the one hand with the Sciaridae that we have already mentioned, and on the other with the gall-midges, which also belong to this group. Like all flies whose scientific name begins with 'Scat . . .', they are predominantly dung-feeders, and this leads them to live in the nests of other insects and larger animals, in the sort of association that is known as 'commensal', or feeding at the same table. Colyer and

Hammond [47] record that they have frequently seen the minute adults of *Scatopse transversalis* running actively about among ants on an old, rotten elm stump, an example of the clear distinction ants can draw between benevolent and malevolent intruders.

Scatopse notata is a less welcome intruder into human dwellings, partly because of its larval habits, and partly because it is one of a number of flies that are gregarious, and apt to appear 'not in single spies, but in battalions'.

The fifth family of this terrestrial group, though allied to the others in a number of details significant to the systematist, has gone a long way on a line of evolution of its own, ending up as the unique family Cecidomyiidae, the gall-midges. In almost every aspect of their biology the gall-midges manage to do things differently from any other flies.

Adult gall-midges are tiny, fragile flies, with very few of the wing-veins that are the great stand-by of the systematist, and with antennae that are a miracle of elaboration on the most minute scale, but possibly more of a confusion than a help in the classification of the family. The biggest handicap to understanding this family, however, is that most of the larvae live in the tissues of growing plants, which they provoke into making a swelling known as a gall.

It is futile to argue whether adult or larval structure is the more significant in classifying insects, and as far as possible we ought to study both. Cecidomyiidae and a few other families, notably the acalyptrate families Agromyzidae and Trypetidae, introduce a third confusing factor. Cecidomyiidae and some Trypetidae make galls, and most larvae of Agromyzidae tunnel in leaves and make complicated mines. For a long time people have made collections of plants deformed by these operations, kept them in botanical presses, drawn and described them. This is obviously the right approach for the horticulturalist, who wants to find out what is disfiguring his shrubs, and how to stop it. The late Dr H. F. Barnes published a series of volumes on the gall-midges of cultivated plants, each volume covering a practical group such as trees, or root-crops. These volumes contain an immense amount of information about the biology of each individual species, even though the practical treatment seems often to be the same: 'remove the affected parts and burn them'.

The trouble about studying a family from its galls and its cocoons, is that these are formations of plant-tissue, and obviously must owe at least as much to the idiosyncrasies of the plant as they do to the insect. The entomological study of Cecidomyiidae has lagged far behind that of the plant-galls, and behind that of other families of flies. The larvae are among the most featureless of all maggots. A coherent understanding of this family is therefore one of the big gaps in our knowledge of the flies of the world. What we do know of them suggests that they would well repay study.

Not all larvae of Cecidomyiidae live in galls, however. A few still follow the ancestral habit of feeding in rotting vegetable matter, and in dung. Some are feeders in fungus, especially fungoid growths covering other plants or woodwork, in moulds, mosses and liverwort. Featureless larvae of a pink colour should be suspected of being Cecidomyiidae. They are said to suck the vegetable juices, and also to take any microscopic animals they may encounter – a habit that we have met before.

Since Cecidomyiid larvae have such tiny, insignificant mouth-parts, it has been a puzzle how they could feed from plant tissues. It has to be assumed that they suck the juices, since they can scarcely be thought to bite and chew. Recent investigators [240] turned all the formidable equipment of the age upon the tiny larva of the Hessian fly – a pest of wheat – making a cinemicrographical study of the head of the feeding larva, and at the same time recording the sound of its feeding with a high-fidelity amplifier and an oscilloscope. They discovered that the larva made its head into a cup-shaped hollow and applied it intermittently to the wheat-stem, while making a sucking noise!

Some Cecidomyid larvae are truly carnivorous: eg the predatory *Aphidoletes* pierces the body of aphids and sucks them dry, in the same way as the larvae of some hover-flies of the family Syrphidae, as we shall see later. The adult female of *Xylodiplosis* has an ovipositor with which it pushes its eggs into crevices in freshly cut stumps of oak-trees, and presumably also into the raw ends of naturally broken branches. When the larva is fully fed it comes to the exposed surface and jumps to the ground by bending its body and flinging itself into the air: a method of avoiding a tedious crawl that is often practised by dipterous larvae.

The transition to the predominant gall-making Cecidomyiidae is perhaps through certain genera that attack the roots of plants, and which may be present in large numbers, enough to kill the plant. This is an obvious link with the similar habit of the larvae of Bibionidae.

Since the larvae of gall-midges have almost no visible structure except the sternal spatula, a clove-shaped plate sometimes called a 'breast-bone', it is not surprising that there has been much speculation about the function of this. It has been suggested that it might be used during the feeding phase of the larva, as an aid to feeding, or to locomotion or to both. Possibly it might be a special tool for scraping at woody tissue. Pitcher [234] made a detailed study of this question, and pointed out that the spatula does not appear generally until the larva has almost or quite finished feeding, and that it would seem to serve some purpose connected either with cutting a way out of a closed gall, or with afterwards digging into the soil before pupating. He studied the amount of abrasion of the spatula after the larva had tunnelled into soils of varying composition, after it had made a cocoon, but had not yet pupated inside it. His conclusion was that the abrasion was caused by using the spatula to dig through the soil.

One further distinction of Cecidomyiidae is that species in a few different genera are able to reproduce by *paedogenesis*. This word comes from the Greek for a child, and means that an animal becomes sexually mature, and produces offspring, without itself ever reaching adult form. The reproductive cells are of course set apart from the somatic, or body-building, cells at an early embryonic stage, and they usually remain dormant until the body has become adult. These exceptional larvae have a few large eggs, in number up to about thirty-five, which can be seen through the thin cuticle. These soon become larvae, and literally devour their parent from inside. They in their turn suffer the same fate, so that for several generations the adult, with its different ways and its different problems, is entirely eliminated. For the time being the midge becomes like the aphid, rapidly reproducing through its asexual generations in the summer, and never leaving its birthplace. In this way, from five very large eggs that were originally laid under loose bark by an ovipositing female may arise a great mass of larvae. Such masses of larvae of *Miastor* may commonly be found under the bark of cut logs.

After a time, the daughter larvae abruptly cease to produce eggs of their own, but go on to pupate in the usual way. All the larvae that arise from one original egg are necessarily of the same sex, but both male and female larvae are paedogenetic, and so normal sexual reproduction can be resumed by the mass. An interesting point is that the paedogenetic larvae do not have the sternal spatula, but this reappears in larva when they are about to pupate, which confirms Pitcher's view of the time when this structure is needed.

Möhn [204] has recently described a gall-midge, *Tekomyia populi* from Germany, that is paedogenetic as a pupa, producing its daughter larvae without completing its metamorphosis into an adult fly. He claims that this is the first case of true pupal paedogenesis in flies. Certain Chironomid midges in which this is said to occur are only apparently paedogenetic: the adult female lays her eggs while she is still enclosed within her pupal skin, and is therefore a 'pharate adult'.

The reproductive stage of *Tekomyia* was called a 'hemipupa' by Möhn, and another hemipupa, that of *Henria psalliotae*, is described and figured by Wyatt [312]. This latest example is a Cecidomyiid that infests mushroom compost in England, and is potentially a pest, particularly if this method of reproduction allows it to multiply more rapidly.

Several members of this group of families form a pupa within the last larval skin, in a manner more characteristic of the higher flies. Colless [321] mentions *Mayetiola* and *Chortomyia* in the Cecidomyidae, *Penthetria holosericea* in Bibionidae, some Scatopsidae and members of Colless's new family of fungus-breeding midges, Perissommatidae.

Chapter 5

Water-Midges

THIS LINE of evolution is a very important one in the natural history of flies. As we have seen, movement into the water generally has certain well-marked consequences. Firstly, the practical problems of life under water – of breathing, of swimming, or of clinging to a fixed support against the force of the current – lead to the development of showy external structures, which make such larvae attractive to study and to photograph, and which provide biologists with many a pretext for airing their knowledge of physics and chemistry. Secondly, most aquatic insects turn carnivorous eventually. Some browse on submerged plants or feed on mosses and diatoms or mine

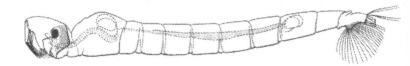

Figure 13. The 'phantom larva' of the midge *Chaoborus* (*Corethra*), an aquatic larva living permanently in deep water. Note the pairs of hydrostatic air-sacs fore and aft, the carnivorous mouthparts, and the posterior blood-gills, which are used not for respiration but to manipulate salt-balance.

in water plants, and these form the starting-points of food-chains, like those on land, in which primary feeders are eaten by small secondary feeders, and these in turn are preyed upon by a succession of bigger carnivores.

In the water carnivores predominate, and by far the most probable fate of any aquatic larva is to be eaten, though perhaps only because other hazards are less than on land. Terrestrial insects as well as aquatic ones lay very large numbers of eggs, and we shall have more to say later about the popular speculation as to what would happen if all the progeny of a single pair of house-flies were to survive for a whole summer. In practice, mercifully, nearly all are lost through starvation, drought, or alternatively drowning, cold, or heat, or asphyxiation in an airless organic medium. It is more difficult to move about in a terrestrial medium, soil, compost, and so on, and so predatory larvae have to search harder to find food, often taking a long time to feed themselves up to the point of pupation. That is probably why many terrestrial larvae can be found with their intestines mainly filled with the vegetable matter that surrounds them, and only occasional animal remains to show that they take a tit-bit when they can: much as primitive tribes live on roots, berries, and so on, but will kill and eat any small animals that they can catch.

In water everything circulates more freely, and conditions are generally more stable. In permanent waters, at any rate, abrupt changes in the physical conditions do not occur, or they fluctuate in a rhythmical way by tides or seasons. There is no all-embracing mass of solid food-material, but the water is full of minute organisms, plant and animal, which can be strained out with little effort by a sedentary larva. Thus filter-feeding is common, either just by sitting in a moving current of water, or by sweeping it through the mouth with mouth-brushes, or both. Such a diet is highly nutritious – think of the huge whales that feed entirely on planktonic animals, each of which is infinitesimal compared with the size of the whale. But sedentary or slow-moving larvae in clear water are very vulnerable, and it is not surprising that other predators make away with many of them.

The first family along this line, and not so heavily committed to the water as many of the others, is the family Psychodidae, which has an aquatic half and a half that is apparently terrestrial. The aquatic half, the sub-family Psychodinae, are the moth-flies or owl-midges, commonly to be seen on the inside of the windows of houses, and abundant about farms, cow-sheds, caves, and any place where organic matter is present. The adult flies have pointed, leaf-shaped wings, covered with hairs, and so look rather like small moths, especially as they often hold the wings together like the roof of a house, as many moths do.

These moth-flies have aquatic larvae, which live mostly in puddly, shallow, semi-fluid conditions, with a high organic content, and often deficient in oxygen. Crisp and Lloyd [53] found larvae of

Psychoda, Pericoma, and *Telmatoscopus* to be an element of the fauna of the patch of woodland mud that they investigated, and they point out how difficult it is to see the larvae because the abdominal plates that are characteristic of many larvae of this family may be black and very close-set. There are two pairs of spiracles, those on the thorax raised up on short pegs, and the posterior pair at the tip of a rigid siphon, which ends in four lobes with water-repellent hairs.

Such larvae mostly crawl about on or near the surface of the mud, and some of them have spines, or feathered processes along the body, which help in this movement. Others live rather on the edges of mud, in the bottoms of ditches, and in crevices near by. It is characteristic of Psychodid larvae, a primitive family, that they are really only semi-aquatic, and only a few of them venture into deep water. Those which do so, notably *Ulomyia, Maruina,* and *Pericoma californica,* have gone the whole way, and live in torrents and rapids, with special sucker-like organs with which they cling to the rocks, like the family Blepharoceridae, which we shall meet later. Even these Psychodid larvae are often not below the water-line, but hang on where they are constantly soaked in spray.

This clinging type of larva was called the 'rheophile' or current-loving type by Feuerborn, in contrast to the more usual 'rheophobe', or current-hating type. This latter flourishes in the bacterial beds of sewage-works, where it is one of the essential elements in the biological cycle that breaks down the waste into water-soluble compounds. Lloyd and his research team at Leeds, about twenty years ago made a detailed study of the exact operation of the purifying process [183]. Two species of *Psychoda, severini,* and *alternata,* are particularly important there, and live in the beds of gravel over which a trickle of water is maintained by moving distributors. The larvae feed upon the rich growth of algae and fungi, with bacteria and Protozoa, that covers the immense surface area of the pebbles. Without their action, and that of the Chironomid larva and worms that also feed on it, this growth would quickly clog the filters. The rapid growth of population after each time the filter has been cleaned, which makes these larvae particularly effective, is helped along by parthenogenesis.

Two points of special interest to us arise from this work at Leeds. One is that under certain conditions the larvae may become carnivorous, and eat any of their companions in the filter. This is of practical importance in sewage disposal, because if food became temporarily scarce the culture might consume itself, and subsequent algal growth would be uncontrolled, and clog the filter; it is of general interest because it is a habit that often occurs among fly-larvae of many groups. The other point of interest is that the larvae (and the pupae too) are 'rheophobe', and a moderate increase in the force of the water will wash them away from their precarious hold.

Larvae of moth-flies go further in the landwards directions than towards the water. Some breed in dung, and in rotting vegetation, both fully wet media, but the larvae of *Trichomyia* burrow into the rotting wood of fallen trees. They are elongate and relatively smooth, with strongly chitinized mandibles, and without posterior siphon, or the spines and other processes proper to the aquatic larvae of this family. Keilin, who studied these larvae, insists that they actively tunnel into the wood, and do not merely occupy burrows made by beetles and other insects.

As we have said, the adult moth-flies are non-biting and entirely harmless, though they can be a pest if they occur in great numbers; like the adults of other midges, such as the Chironomidae, they breed in rivers like the Nile, and can cause asthmatical outbreaks in hot, dry countries.

Only one genus of this sub-family sucks the blood of vertebrate animals. *Sycorax silacea* has been reported to attack the edible frog in France, where this frog is specially subject to observation, and the fly may transmit a parasitic worm. But in the second half of the family Psychodidae the picture changes abruptly, and we come up against some of the world's most troublesome disease-carriers.

Psychodid flies belonging to the sub-family Phlebotominae are the true 'sand-flies', though this name is often used, especially in conversation, for the biting midges of the family Ceratopogonidae.

Phlebotomine sand-flies are found in hot countries, that is throughout the tropics of the world, and in sub-tropical and warm-temperate regions, as far north as Manchuria and Paris, and southwards to the Cape of Good Hope, into the northern part of Australia and over all South America except the cold southern tip. They have not as yet penetrated to the Hawaiian islands. About 350 species are known, of which only the adult females suck blood. We may note these as the first 'biting flies' that we have encountered, and try to see them in relation to other Nematocera.

Sand-flies are tiny, fragile flies with their wings more elongate but less hairy than those of the moth-flies and with a longer proboscis.

53

Adult *Phlebotomus* are about at dusk and at night, and are little noticed by anyone who is not looking for them, though their efforts as carriers of disease soon attract attention. They fly only weakly, for short distances, and are very easily disturbed when they are feeding. Their larvae are hidden away in crevices in the soil, sometimes 20–30 cm down, in cracks of masonry, between stones, and so on. It is a paradox of this group that the adult sand-flies, as this name suggests, are associated in the mind of their victim principally with hot and arid places, but the larvae in fact are dependent on moisture to a degree that is unusual in an apparently terrestrial animal. However humid the air, sand-fly larvae and pupae die very quickly unless liquid water is present, and on contact with their skin.

Sand-flies are now known to be commoner in tropical forest than had previously been realized. Many hide in the crevices of the great buttress roots of the high forest trees.

We have seen that the other Psychodidae are all aquatic as larvae. Apparently the larvae and pupae of *Phlebotomus* are also really aquatic, but they have colonized a microclimate that exists in crevices, where condensation of atmospheric moisture provides the droplets of liquid water that they require. Even in the hottest climate this water persists. The food of the larvae is organic: decaying vegetable matter, dead insects, and particularly the dung of small animals. Such breeding-places are common round human habitations, and so the sand-flies are provided simultaneously with a breeding-place and a source of blood for the adult females.

This is similar to the association between fleas and their hosts. The adult fleas bite vertebrate animals, and the larvae live in crevices round their habitations, feeding on organic debris. The larvae of fleas are very similar to those of primitive Nematocera, and for this reason, mainly, fleas are considered to be wingless descendants of a primitive fly. D. E. Hardy says in his *Insects of Hawaii*: 'Some of the Psychodids are among the most primitive of Diptera, and genera such as *Nemopalpus* and *Bruchomyia* are obviously close to the basic stock from which the Order Diptera arose.'

It looks as if sand-flies and fleas might be the results of two parallel experiments in living at the expense of a domesticated vertebrate animal: ie one that has a den or lair upon which its life is centred. The perhaps limited food in the microclimate of the larval crevice is then supplemented by the adult's sucking the blood of the involuntary host.

The legless larvae of sand-flies are distinguished by having long, isolated hairs, with a few especially long ones posteriorly. The skin of the last larval stage is to be seen attached to the tip of the abdomen of the pupa, and still bears these terminal long hairs. Both features are also characteristic of the larvae and pupae of fleas.

Not far removed, biologically at least, from the Psychodidae are the family Ceratopogonidae, the most notorious members of which are the 'biting midges'. These are the insects, sometimes mistakenly called sand-flies, which can make summer evenings by the water or in the garden such a misery. They are very tiny indeed: those whose wing-length reaches 2·5 mm, or $\frac{1}{10}$ in are big ones, and some scarcely attain a length of 1 mm. They are therefore seldom noticed until they begin to pierce one's skin, but they quickly make up for their small size with a most formidable bite.

Only the females have a piercing proboscis, with a pair of mandibles like the blades of scissors, quite similar to those of *Phlebotomus*, the sand-flies of the previous family, and to Simuliidae, the black-flies of the next. At any rate when they are piercing the skin of a vertebrate animal, these blades are not used in a cutting fashion, but rather to stab the victim, like a heroine in a melodrama defending her honour.

Only three genera are known to suck the blood of warm-blooded animals: *Culicoides*, *Lasiohelea*, and *Leptoconops*. Most genera suck the juices of flowers, or feed upon other small insects near their own size, including non-biting midges and small mayflies. Downes has investigated the relative importance of nectar, pollen, and blood in the diet of these and other flies, and his views are of considerable general interest. Of more immediate concern from the aspect of natural history are the epicurean tastes that some flies of this family have developed.

Thus some of the flies of the large genus *Forcipomyia* do not crudely and greedily suck dry their insect victim, but cling to the wing, and delicately pierce one of the wing-veins. They may be found attached to adult Neuroptera – alder-flies, lace-wing flies, and so on – and to Lepidoptera, and flies of the genus *Pterobosca*, which cling to the wings of bigger crane-flies and dragon-flies and may have their legs specially adapted for hanging on like an aerial hobo. Forsius says of biting midges that attack the lace-wing *Chrysopa*: 'The midges are often very firmly attached to the hosts, and do not readily release their hold, even if the hosts are taken by sweeping, or caught with the fingers.' In one case a midge was found the day after capture sucking a dead *Chrysopa* in a test-tube. If the midges are shut up in a vial, together with a *Chrysopa*, they are usually torn to pieces and eaten by the latter. So perhaps the trick of feeding from the wings rather than the body of one's victim shows prudence as well as refinement of taste.

Yet other *Forcipomyia*, still satisfied with cold, invertebrate blood, but wanting it in greater quantities, attack the caterpillars of moths and of sawflies, and sometimes even bite frogs. *Atrichopogon* has developed a liking for the blood of blister-beetles and oil-beetles of the family Meloidae, surely a perverted taste, since these beetles are notorious for the cantharidin that they carry in their blood. Such a tiny fly is able to pierce so tough a victim only by attacking the thin membrane between the abdominal segments.

It is as biters of man and his animals that this family of flies are best known. The diseases that they carry will be discussed later, but their mere nuisance value is enormous. They can seriously impede activity out of doors, especially at dusk, in the early morning, and on humid, sultry days. Nor are they limited to one climatic zone. They are notorious in northern countries like Scotland; they also effectively ruined my breakfast on a verandah facing a beautiful valley of the African rain-forest, only a few miles from the Equator.

A red disc soon appears round the site of a bite of *Culicoides*, and may be followed by a swelling, and in some subjects by a large, clear blister. In spite of the irritation it is best not to scratch this blister, because it is easily broken, and can be very troublesome to heal. Jobling long ago recommended using a crystal of washing soda as a relief to the bite.

On the credit side, for what it is worth, these little flies sometimes mass in large numbers on flowers. These swarms consist entirely of pollen-loving females, and they must play an important part in the cross-fertilization of spring flowers. Perhaps we should reckon on the credit side, too, that *Culicoides anophelis* chases mosquitoes that have gorged themselves on blood, human or animal, and robs them of a little of it by piercing the abdomen: a literal example of the biter bit.

The larvae of Ceratopogonidae are apneustic: that is, they have no open spiracles, and can get their oxygen only by diffusion from surrounding water, or a watery medium. They thus have the appearance of a family with aquatic ancestors, whose terrestrial descendants are colonizing media that are really alien to them.

This family, like the Psychodidae, falls fairly readily into two sections, though some genera are intermediate, and emphasize that this is only a convenient device for comparative purposes. The truly aquatic larvae are long and worm-like, and move along by wriggling like tiny eels. They live mostly round the margins of ponds and streams, and are responsible for the troublesome *Culicoides* that we have mentioned. They do not confine themselves to fresh water, but feed also in brackish, or even salt, more especially that of pools and lakes inland, where the salinity has risen through evaporation.

The adult flies of the aquatic group have the wings either bare, or covered with microscopic hairs that fall into a pattern. Thus *Leptoconops*, which occurs in desert and semi-desert areas all over the world, and has a bad reputation as the 'Bodega gnat' and the 'valley black gnat', is really a beautiful insect, the females shining black with milky wings, and the larvae orange. Its larvae live in wet soil, and on the margins of salt lakes as well as of tidal waters.

The 'terrestrial' group of biting midges, as we have seen, is really an offshoot of the aquatic ones, and its larvae live in soggy places, in decaying leaves, manure – ie a mixture of dung and vegetable matter – under loose bark, in rotting fungi, in fact in all the standard compost-like materials. That this is a return to an ancestral habitat is shown by the fact that these larvae have no spiracles. Although they have modified their shape, becoming more flattened, and varied in form to suit their individual habitat, they have not managed to recover the open spiracles that their ancestors lost when they took to the water.

Some larvae of this group have it both ways, by living in the small accumulations of stagnant water that accumulate in the cavities of plants: in the axils of the epiphytic Bromeliads of tropical America, the urns of the Oriental and Australian pitcher plants, *Nepenthes* and *Sarracenia* and in temperate countries in the common teasel, *Dipsacus silvestris*. These cavities become very foul with organic decay, and provide food for a variety of saprophagous larvae, which in turn are preyed upon by carnivores.

Ceratopogonidae are a family that it is difficult to generalize about, principally because they are so versatile, and make the most of this half-world between the land and the water. Thus the single genus *Forcipomyia* contains species that live in compost-like materials, in ants' nests, in rotten wood, in damp moss and algae, and in the Bromeliads that we have just mentioned. *Dasyhelea confinis* has been reported from *Nepenthes mirabilis* in Sumatra, while other *Dasyhelea* larvae live in the material running down the trunk from ulcerated wounds in trees. Although without spiracles, and adapted to living in water, these larvae cannot swim, and will drown if completely submerged: an unexpected fate that overtakes many so-called aquatic larvae of flies if you carry them home in a tube full of water!

Saunders suggests that the uniform factor in the larval habitats of *Forcipomyia* is 'some dark enclosed cavity where the atmosphere approaches saturation, and where moulds and fungi are abundant'.

The distinction between biting midges and non-biting midges is one that is easily appreciated in practice! Entomologically, the biting midges, family Ceratopogonidae, include many that devour other insects, as well as those that attack warm-blooded animals. The non-biting midges, in contrast, have reduced mouthparts, or effectively none at all, and are generally said not to feed as adults. In these flies the larva has to accumulate enough reserves to provide for the adult as well.

These two groups of midges used to be classified as the contrasting halves of one family Chironomidae, but nowadays this name is restricted to the non-biting midges. Indeed some students think that Ceratopogonidae are more closely related to the Simuliidae, which we shall meet presently, than they are to the Chironomidae.

Be that as it may, Chironomidae in the restricted sense are a distinct biological unit. The adult flies are very familiar to us all. They are the midges that rise and fall in lazy swarms on summer evenings, and in England, where the state of the weather is the foremost topic of conversation, they are believed to promise a fine day tomorrow. Like many other pieces of weather lore, this has a basis in truth, because the swarms of midges are very much affected by wind-strength and air-currents. They are one of the best examples of a mating swarm of male flies, serving as a conspicuous rallying point for the species locally, and keeping itself in position by reference to some fixed point, such as a tree or post. I hope to give more attention to these and other swarms in a later section of this book.

Adult Chironomidae are interesting mainly for their group activities. Individually they are rather dull creatures, like soft-bodied and almost colourless crane-flies. The males have fine, bushy antennae (cf. figure 14), and large eyes, and look rather like male mosquitoes, from which they can easily be distinguished by not having the long proboscis that even the non-biting male mosquitoes carry before them. The utter insignificance of the individual midge is underlined by D. J. Lewis's statement about *Tanytarsus lewisi*: '18,000 dried bodies weigh about one gramme, and more than half a million weigh an ounce.'

This is one family of flies in which the variety and interest lie

Figure 14. Head of a male midge, *Chaoborus plumicornis*. Note the antennae with their long, sensitive hairs; the eyes wrapped round the antennae; and the long, segmented palpi, characteristic of the Sub-Order Nematocera.

principally in the young stages and their biology. The adults have little to offer beyond their swarming habit. The life of the adult is believed to be very short. Emergence from the pupa usually takes place early in the day, and the flies sit about on vegetation, logs, or tree-trunks near the water, while the body hardens. Towards evening, those that have not been snapped up by predators take part in a mating flight, the swarm of males rising and falling, while individual

59

females fly in, find a mate, and go off with him to the surrounding vegetation. Probably the males die within the next day or so, and the females live only long enough to mature and lay their eggs.

If the mating flight is delayed by bad weather, rain or high winds, possibly many of the flies can survive for a few days by keeping still under leaves and stones, and so conserving their limited resources of food and energy.

Swarming flights are characteristic of late afternoon and evening, and continue until it is almost completely dark. This is especially so in high latitudes, where summer twilight is very long, and under these conditions the midges will come to lighted windows. Those which fly in the colder months of the year dance more towards the middle of the day, when the air is warmest: not only for the direct stimulation of its warmth, but also with the help of the convection currents that it produces. Like gliders, the midges must have some turbulence to give them lift, but they have little strength of flight, and in anything more than a light breeze they either move into the lee of some obstacle, riding the eddies behind it, or settle altogether.

Among Chironomids, as among crane-flies, some species look for their mates on the vegetation, and not in flight; and a few are flightless. These last belong to the sub-family Clunioninae, and seem to be an offshoot of the Orthocladiinae that have spread to the edge of the sea, and beyond. There is a transition from fully winged species, which fly from rock to rock, down to those which have their wings reduced in area, and functionless. Most of these live near low-water mark and run about on the exposed rocks at low tide, the females being fertilized by winged males flying in at a low level. Some observers have suggested that these midges have been able to colonize this habitat by cutting their adult life down to the interval between one tide and the next, but things do not seem to be quite so desperate. The flies may be able to shelter in crevices which trap small pockets of air, and *Clunio adriaticus* is said even to penetrate below the water surface, and live among colonies of mussels.

The extreme of this line of evolution is reached in the Pacific with the genus *Pontomyia*, a truly pelagic insect, the male of which can run about on the surface of the sea, supported by the surface tension at the tips of its middle and hind feet, and using the tips of its wings as paddles. It can also swim below the surface, using the long first and third legs and possibly the narrow wings, in penguin-fashion. The larvae were found by Buxton in the marine plant *Halophila ovalis*, living in a mud tube, and feeding on diatoms. The female is a worm-like creature, without wings, and with vestigial legs, like the females of the Streblid fly *Ascodipteron* (plate 27). The female *Pontomyia* stays put, and waits for the male to visit her, whereas the female *Ascodipteron* becomes immobile only after pairing.

The problem of finding a mate during a brief and insecure adult life has led certain midges of this family to the expedient of parthenogenesis: ie to laying eggs that have not been fertilized. All the offspring produced parthenogenetically are females, and in a few such species males are not known to exist at all. *Chironomus grimmi* and *Tanytarsus boiemicus*, and a few others, have pushed this process to its logical conclusion, and the female may lay her eggs before she has broken out from her pupal skin. Earlier, when discussing Cecidomyiidae, I mentioned that this phenomenon in Chironomidae has been referred to as pupal paedogenesis, but that it is nothing more than precocious egg-laying by the pharate adult female. Nevertheless the parthenogenesis of the Chironomidae has the same value as the paedogenesis of the Cecidomyiidae, in allowing a population of flies to build up rapidly in a new locality, without having to spend time in going through the full cycle of normal development.

Aquatic Chironomidae lay their eggs into the water, often in masses, or in 'strings'. The latter are like long capillary tubes packed with eggs, lying more or less lengthwise. Several tubes may be twisted together, joined up into a continuous band, or coiled round inside a larger gelatinous tube, like the element of a very small electric fire. Such masses of eggs can be found floating in pond water, and vary considerably between different species.

It used to be thought that all Chironomidae had aquatic larvae, but this is not so. There are many terrestrial larvae, living in moss, humus, and compost-like materials, dung, rotting wood, and in soil at the roots of grasses and other plants. Larvae of *Spaniotoma furcata* are sometimes a pest in greenhouses, where they attack seedlings.

The non-biting midges with terrestrial larvae do not form a natural group within the family Chironomidae, but are isolated genera, and even species, that have taken to the land independently. Like the aquatic larvae, they have no open spiracles, and sometimes two closely allied species may be one aquatic and the other terrestrial. Edwards suggested that those with terrestrial larvae tend to be winter insects, but did not indicate a connection between these two attributes, nor suggest which came first in evolution.

The great interest in this family of flies is in the habits of the water-living larvae. In shape these are like small worms, rather plump, with a well-developed head that cannot be withdrawn into the body. The rest of the body is generally smooth and cylindrical, except for a two-fold proleg (false leg, or pseudopod) on the prothorax, and a cluster of appendages at the rear of the body: paired prolegs, and often two sets of gills. There are tracheal gills, which increase the diffusion of oxygen into the closed tracheal system, and blood-gills, which are believed mainly to adjust the balance of salts in the blood in relation to the surrounding water.

As a final refinement of adaptation, some Chironomid larvae have a red pigment in the blood containing haemoglobin, the same substance that is used in the blood of vertebrates, including man, to take up and release oxygen. These red larvae are the 'bloodworms' that occasionally cause alarm when they appear in the domestic water-supply. Miss Walshe [298] kept some of these larvae in the laboratory, and she has given an illuminating account of their way of life, which may be taken as characteristic of the many water-living Chironomid midges. She describes tube-making in *Chironomus plumosus* as follows:

In the laboratory, larvae put into an aquarium containing several centimetres of sieved mud immediately burrowed into it, and established their tubes within a few hours. Such tubes vary somewhat in proportions: those of final instar larvae usually extend to a depth of 3–6 cm in the mud, and have an internal diameter of 2–3 mm. Their walls are consolidated with salivary secretion, the larva applying this by means of its mouthparts and its hooked anterior proleg. The larval tubes once established within an aquarium seem to be permanent, and the larva pupates within its tube. Under normal conditions, larvae never completely leave their tubes, although they occasionally project from them to gather food from the surface of the mud. They always hold on to the mouth of the tube with the posterior prolegs, and withdraw hastily when disturbed.

The larva spends a large part of its time in a resting position in the tube, but periodically 'a series of rhythmic dorso-ventral undulations passes backwards along the body, driving water posteriorly', and thus maintaining a supply of dissolved oxygen to the larva.

The principal method of feeding is the filtering of planktonic organisms and particles from a stream of water passing through the tube. The larva spins a cone of salivary secretion across the lumen of the tube, the apex of the cone being sometimes connected to the mouth [of the larva] by a thread. After filtering for some time, the larva eats the net, with its enclosed particles, and starts again. The entire process takes less than two minutes, of which half is spent in irrigating the tube, and half in eating and spinning the net. It may be repeated many times, and feeding is sometimes continuous for many hours. The whole forms a very stereotyped behaviour pattern, which is easily recognized.

In addition, the larva on rare occasions will come out of its tube as far as the posterior prolegs, and scrape the surface mud, laying down a net of salivary secretion to which particles adhere. It then drags this down to the burrow for consumption.

Miss Walshe found that as the oxygen concentration in the water was reduced, the larva spent less time in feeding, and more in irrigating its tube, but at very low concentrations it conserved its energies by not moving at all. The red blood pigment, with its haemoglobin, would hold a store of oxygen equal to nine minutes' consumption by the resting larva. This pigment appears to serve three purposes, in each case by speeding up the rate of exchange of oxygen: the larva can go on filter-feeding when otherwise low oxygen concentration would have forced it to concentrate on respiratory irrigation only; it can continue this irrigation down to a lower oxygen level before having to stop moving altogether; and after a temporary period of oxygen lack the larvae with haemoglobin can recover more quickly.

Although the aquatic larvae of non-biting midges are very much alike in their appearance, they have cultivated a wide variety of habitats. Some remain in their fixed cases all their lives, others have movable cases, and some live free in the mud. A few adventurous types attach themselves to the larvae of mayflies and other aquatic insects and live as 'commensals', which is a polite term for eating the crumbs that fall from the rich man's table. *Symbiocladius* is a true parasite of mayfly nymphs, weaving a silken sac under the wing-pads of the nymph, making a small hole in its skin, and sucking its blood. Some *Endochironomus* live with water-snails, species of *Limnaea* and *Physa*, and may possible parasitize them, and *Chironomus limnaei* is said to complete its whole development within the bodies of snails of the genus *Limnaea*. Others, like *Chironomus varus*, merely attach themselves to the shell.

Many other aquatic animals, especially water-beetles, water-bugs, nymphs of mayflies and dragonflies, and even other dipterous larvae,

63

LIVERPOOL JOHN MOORES UNIVERSITY
LEARNING SERVICES

provide shelter for tube-making larvae of non-biting midges, sharing food with them, or gathering an algal growth upon which the midge-larvae feed. Commensals confer no benefit on their 'host' animal, but Brock [29] has recently described an association between a midge larva and an alga, where the benefits are believed to be mutual. Such an association is called *symbiosis*, though for some reason Brock prefers the ugly word 'mutualism'.

Cricotopus nostoicola and *C. fuscatus* live in the blue-green alga *Nostoc parmelioides*, which grows in pads attached to rocks in a stream. Each pad is tunnelled by one larva, which makes no web or net. The tunnel has no external opening, and there is no current in it. Apparently the larva feeds exclusively on the alga, and excretes only a yellow fluid which is absorbed by the alga. The midge pupates in the tunnel, and the pharate adult bores its way out, and swims to the surface. The mutual benefits claimed are that the midge larva lives a peaceful life protected from being eaten by fish, or exposed by drought, both great hazards of midge-life, while the alga gets the organic excreta back again; rather like taking in each other's washing. If the alga comes adrift in the stream the midge larva is said to sew it back again with a silken net, and the final escape is said to release trichomes of the alga which drift in the stream and found new colonies.

Larvae of a few Chironomid midges mine into the stems of water-plants, and damage to rice-fields has been reported from this source. Finally, the entire sub-family Tanypodinae are carnivorous as larvae, using their efficient mandibles to pierce and devour the larvae of other insects. Besides feeding on mosquito-larvae, they also eat many of the filter-feeding midges and even smaller individuals of their own species. Cannibalism is fairly common among carnivorous insects, though perhaps its significance, as well as its frequency in nature, are exaggerated because of observations made under experimental conditions, when larvae of the same species are unnaturally crowded together. It does not follow that in nature they would actively seek out members of their own species. At least this is true of larvae: cannibalism by adult carnivorous flies is another matter, which we shall come to later.

As we might expect in a family that is so highly adapted to living in water, Chironomidae have colonized waters that are still or torrential, stagnant or highly aerated, warm or icy. The sub-family Diamesinae in particular like the cold waters of the Arctic as well as those of the high mountains, and may be found breeding in water that is running off a glacier, with the adult flies settling on the snow and ice above. The length of larval life is very varied, according to abundance of food and to the temperature. The same species may have several generations a year in warm water, and only one in the cold water at the bottom of a lake.

As a final gesture of defiance, the larva of *Polypedilum vander-planki* can be completely dehydrated, and will enter into a state of 'cryptobiosis' during which all metabolic processes apparently cease. Hinton [135] has studied this phenomenon. The larva consumes no oxygen – it will survive in an atmosphere of pure nitrogen and it can withstand injuries that in normal life would set up a chain of disturbances that would kill the larva. When it is put back into water it quickly swells to normal size, and quietly resumes its breathing and feeding.

Chapter 6

Black-Flies

CERTAIN BITING FLIES of this aquatic group are very different in appearance from normal midges and gnats. Instead of being fragile and elongate, they are small and compact, with short, broad wings which are strengthened by having the strong veins concentrated towards the leading edge, or costa. Even the antennae, although they have nine to eleven segments, have these compacted together, and cannot be called whiplike as is typical of the Sub-Order Nematocera. In general members of this family are more like advanced flies of the Sub-Order Cyclorrhapha, especially some of the Acalypterae.

These are the notorious black-flies, buffalo-gnats, or turkey-gnats, of the family Simuliidae. Only the females suck blood, and not those of every species, while many of the bloodsuckers prefer the blood of birds to that of mammals or of man. They are called 'turkey-gnats' because *Simulium occidentale* transmits a blood parasite of turkeys, and young ducks are attacked by *S. venustum*, but it is chiefly as pests of man and his domestic animals that Simuliidae are important. Note that 'black-flies' are true flies, but 'blackfly' are aphids such as the bean aphis, sap-sucking bugs belonging to the Homoptera.

I shall discuss these flies in their relation to disease in a later chapter, but a great deal of their menace arises from the direct effect of their bites in such large numbers. The combined effect of loss of blood and loss of feeding-time disturbs herds so much that it may often lead to the weakening and death of grazing animals. An extreme instance is the oft-quoted census made in Rumania, Bulgaria, and Yugoslavia in the year 1923, where nearly 20,000 domestic animals, horses, cattle, sheep, and goats, are said to have been killed, along with many wild animals, by the Golubatz fly, *Simulium columbaschense*. D. J. Lewis, in several papers, makes comments that give a vivid impression of the effect of these flies on human activities, especially in the Sudan. He shows photographs of men carrying, and even wearing, masses of smouldering rope to give a protective smoke, and says that they vary their daily work considerably in the 'nimitti', or *Simulium*, season. 'A few people have been known to remain indoors for several weeks on end during a bad nimitti season.'

Black-flies are remarkably persistent in their attacks, and they often surround their victim as a milling cloud. The females often fly round one's legs, and settle at about knee-level, and if one is wearing shorts the knees are soon covered with spots of blood from the bites. Both sexes form a cloud round the head of a person walking along, even though the male flies are not able to bite. Presumably a male swarm is going along with a moving object, in the same way that a cloud of non-biting Chironomid midges will follow a person walking, and if there are two people who separate, each goes his way accompanied by his own share of the flies. Lewis, again, describes how an effective repellent such as dimethylphthalate (DMP) will keep black-flies from actually settling on a person, but holds them off at a distance of only an inch or two, so that he has a sort of aura or sheath of frustrated flies. The very persistent *Simulium griseicolle*, the true nimitti of the Sudan, will settle even when DDT is used, though it will not bite. This fly is mainly a pest of birds, especially turkeys, and of donkeys, both of which it may kill: 'Only a small proportion of nimitti bite man, but this may be a large number when millions are present.' Yet the main complaint against this fly is the intolerable annoyance of having so many crawling about one's face and head.

Although black-flies are thus very active indeed during the daylight hours they sometimes fly at night, and male *Simulium* will come to light traps. These compact, tough little flies are obviously more powerful fliers than the fragile midges of the other families, and they are credited with being able to cover long distances. A much-quoted statement by the late Dr F. W. Edwards that '*Simulium griseicolle* has been found numerous in places as much as 200 miles distant from its home in the river Nile' has been received sceptically by dipterists, because the area concerned is largely uninhabited, and is one in which streams come and go at different seasons. No one could say for certain that there was no breeding-place nearer than the Nile. Yet *Simulium arcticum* is said to have flown 140 miles, aided by the wind, and in all countries where *Simulium* occurs it certainly roams over an area much greater than most flies.

This is a very homogeneous family, and although some other

generic names are used, such as *Prosimulium* and *Eusimulium*, we can safely use *Simulium* as a generic term for any black-flies in this general, biological account. This seems to be an example of a single evolutionary experiment in larval habitat, a family that is perhaps only now beginning to change its way a little, as we shall see in a moment. As a sign of change in adult life, *Cnephia dacotensis* has reduced its mouthparts and ceased to suck blood, thereby reducing its

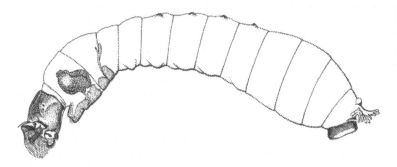

Figure 15. The aquatic larva of *Simulium damnosum*, with its mouth-brushes and posterior sucker, a static larva filtering its food from moving water.

adult life to a very short spell. This is a step comparable to that taken by Chironomidae, but *C. dacotensis* has gone further, and the adults do not disperse, mating within a few yards of their breeding-site.

Black-fly larvae are entirely aquatic, and this is the first family that we have yet met that has gone over completely to living in the water, without any members wandering off into other habitats, or straying back from the water into terrestrial media. They are also among the minority of flies that make no use of rot or decay, but live in clean, clear water, from which they filter off any animal or vegetable particles that are the right size to eat. Although they or their ancestors have been aquatic long enough to have lost all external spiracles, and so to breathe through an entirely closed tracheal system, they have not raised this to the same efficiency as the mud-loving Chironomidae. On the contrary, the most characteristic feature of the biology of black-fly larvae is that they must have their water well aerated. They are usually thought of as living in turbulent, broken water, and are to be looked for in or just below waterfalls and rapids, but they can also breed to pestilential populations in deeper water if it is very weedy, and so full of small air-bubbles from the growing plants.

In either situation there must be a current to bring along a stream of food particles, and the larva leads a mostly sedentary life, attached by its posterior circlet of hooks (figure 15), with its head hanging downstream, and its mouth-brushes trapping food particles in the water. The larva can move easily with a looping motion, bending its body, and using its fore and hind prolegs alternately. This serves for limited local excursions, but for a controlled transit downstream, without being swept helplessly away by the current, the larva spins a silken thread, and pays it out as it goes, like a spider in a breeze. When it alights on a suitable object, the posterior proleg again takes a grip.

Any submerged object will serve as a base, provided it is secure enough to be gripped, and that it lies in a suitable part of the current. Within a stream, as can easily be seen, there are considerable variations of current, places where the flow is accelerated by a constriction between rocks, and others where the current is slack, or even reversed by eddies. Consequently black-fly larvae tend to collect in the most favourable places. This is even more true when the larvae are clinging to aquatic vegetation. Leaves which droop and trail into flowing water may become so encrusted with *Simulium* larvae that when they are lifted out of the water they glisten with what appears to be a continuous gelatinous layer.

Methods of egg-laying vary according to the situation, even within one species. Sometimes the females operate singly, and lay small batches of eggs on vegetation floating on the water, or emerging from it, but always at or above water-level. Sometimes a female drops an egg without actually alighting to do so, or she may alight just long enough to lay one egg. In *S. arcticum* it is believed that the eggs are laid on the surface of the water and sink to the bottom. Muirhead-Thompson [210] has described communal oviposition, in which a dense swarm of female *Simulium* came along just before sunset. They moved along a leaf as a body, laying a carpet of eggs that was estimated to contain as many as 50,000 on one leaf! These leaves were hanging down, dangling in the water, and seemed to be unusually attractive to flies. Muirhead-Thompson speaks of the possibility that 'two or three hundred females must have taken part in the pre-oviposition swarm, and in the communal ovipositing itself', so he evidently believes that these flies must first assemble and then act communally: ie that they are a true swarm, and not just an aggregation (see Chapter 20).

Like the non-biting midges, the larvae of black-flies make use of the other occupants of the water that they live in, but they give the appearance of perhaps only just beginning to do so. The search for the breeding-place of *S. neavei* in East Africa was as exciting as a criminal investigation. Of the two African *Simulium* that are known to carry the disease onchocerciasis (see Chapter 19), *S. damnosum* can be found breeding in a variety of situations, in cascades and rapids, and sometimes in the sluggish lower reaches of rivers. Indeed, the puzzling thing about *S. damnosum* is that its habits seem to be so different in different parts of its range; its vitality as a species, and its danger as a carrier of disease, come from its versatility. Larvae of the other species concerned, *S. neavei*, could not be found in any of the usual situations. Search was extended to boggy ground, and even to dry land, just in case this should be a *Simulium* that was leaving the water, and moving back to a terrestrial life. Then it was discovered that adult *S. neavei* disappeared from a locality in which the streams had been treated with DDT, so clearly the larvae must be somewhere in or close by the streams.

The search for larvae was conducted by the most drastic methods, even to the extent of diverting the streams, damming and blasting, as the dry bed was studied in detail without result. McMahon and his associates who carried out this search argued that, since the larvae were not attached to the bed of the river, and since they could not be living freely in the water, having no power of swimming against the current, they must be attached to some other water-creature, living in what the ecologists call 'phoresy' to indicate that the smaller associate is no more than a passenger.

For three weeks 2,000 water-insects a day were examined – early stages of mayflies, stone-flies, caddis-flies – as well as crabs and fish, but no larvae of *Simulium* were found. Then at last it was found that crabs of one species, *Potamon niloticus*, were carrying larvae and pupae of a *Simulium*. Usually the larvae were attached to the dorsum of the carapace, or the basal segments of the legs, and were very difficult to see against the colour of the crab itself; but when looked for carefully they were present on nearly half of the crabs examined. From these pupae emerged adults of *S. neavei*.

Other black-fly larvae have been found in phoretic association with nymphs of dragonflies and mayflies, and recently (1960) a group of specialists on *Simulium* reported in the magazine *Nature* that they had found a batch of eggs of *Simulium* attached to a dragonfly nymph in Southern Rhodesia. The larvae from these eggs were like those of *S. adersi*, which was also common on rocks in the same stream, and the authors also report a similar occurrence in *S. medusaeformis hargreavesi* in Nigeria. They believe that this is a pointer showing how these larvae may be in process of changing from a sedentary to a nomadic life, a step that has already been taken by *S. neavei*. Corbet [49] suggests that the chief value of such attachments is the protection they give during the pupal stage, reducing the risk of being washed away by the current, or injured by shifting stones.

When the larva is fully fed it spins a silken cocoon and attaches it to the surface on which it rests; if this is a crab or another insect larva, then the pupa is attached to that. The pupa has long, branching respiratory filaments, which usually protrude from the cocoon in species of *Simulium*, but not in *Prosimulium*, nor in *Cnephia*. When they are visible these filaments are often diagnostic of the species, and are much used in classification. There is argument as to whether each species always has the same arrangement of pupal filaments, or whether these also vary with the chemical and physical condition of the water in which the pupa is living: a problem analogous to that of gall-making insects and their galls on different species of plants.

As the adult matures, the cocoon becomes distended with air, and if it is still submerged in water when emergence is due, the adult rises to the surface covered in a glistening air-film held by the surface hairs of the body. It is then able to take off and fly away without being wetted, and without having to wait for its body to harden, as do most insects.

The only remaining large family of these aquatic midges and gnats is that of the mosquitoes, but these are so well known, and so much creatures in their own right, that they must have a chapter to themselves. But before we come to the mosquitoes there are still two or three small families that are of minor importance, but interesting in their natural history.

Thaumaleidae, with the alternative, almost unpronounceable name of Orphnephilidae, are tiny midges about 3 mm long, with their eyes close together in both sexes, and not, as is usual, in the males only. Although about fifty species exist they are enigmatic insects, remote

71

and little known. They occur in high latitudes and in mountainous areas: in Europe chiefly in Lapland, Norway, and Scotland, and in the mountains down to Italy; in the United States and Canada; and in New Zealand and Tasmania.

At one time Thaumaleidae were thought to be land-midges akin to the Mycetophilidae, but after almost a century the larvae were discovered in cold mountain streams where a shallow layer of water runs over rocks. So shallow a layer, in fact, that the larva is not completely submerged, but creeps about in a mere film of water, browsing on vegetable detritus on the stones. According to Vaillant the larva normally touches the rock only at its two ends, and can glide quickly away on the surface film when disturbed. The larva is superficially like those of the biting midges of the family Ceratopogonidae, with a well-formed head, a thoracic proleg, and hooks which serve as a proleg posteriorly. The pupa is found near by in mud or wet moss.

The correct position of this family among the Nematocera is a matter for argument among systematists, who reach different conclusions by considering adult and larval characters. Biologically it is interesting as an experiment towards aquatic life, one which has probably been made a number of times in the evolution of flies. The larvae are amphipneustic, that is they have open spiracles on the thorax and at the tip of the abdomen. In this, in having abdominal plates, and in their habitat, they are remarkably like some Psychodids, particularly those that we have mentioned as flourishing in sewage filter-beds, where thin trickles of water flow over stones. Their present distribution, broadly interrupted in all the warmer areas of the globe, suggests an ancient group in retreat, clinging to its specialized habitat. So they are best considered as an experiment in evolution that did not succeed.

Two similar, but slightly more spectacular experiments are represented by the families Blepharoceridae and Deuterophlebiidae. These 'mountain midges' are associated with torrential streams in mountainous areas; Blepharoceridae world-wide, but rare or local; Deuterophlebiidae known only from mountainous areas of Central Asia and North America. The larvae of both families have developed a bizarre equipment of lobes and suckers along the abdominal segments, with which they cling to the rocks against the rush of the current, or in the perpetual spray from waterfalls. Plate 4 shows the larva of a Blepharocerid, head and thorax fused into one great mass,

9. Two flower-loving flies; above, a hover-fly; below, a much smaller March-fly, family Bibionidae.

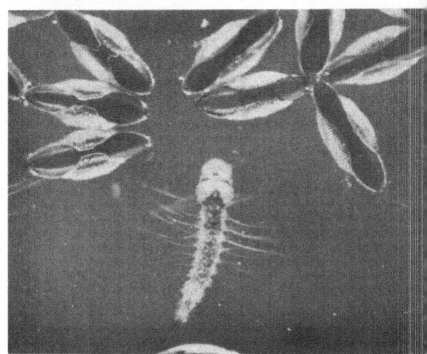

10. Eggs of mosquitoes: (a) laid in a floating raft by a culicine mosquito; (b) laid separately by an anopheline mosquito, each egg equipped with floats.

11. A male mosquito, *Culex molestus*, displaying very clearly the long hairs of the bushy antennae.

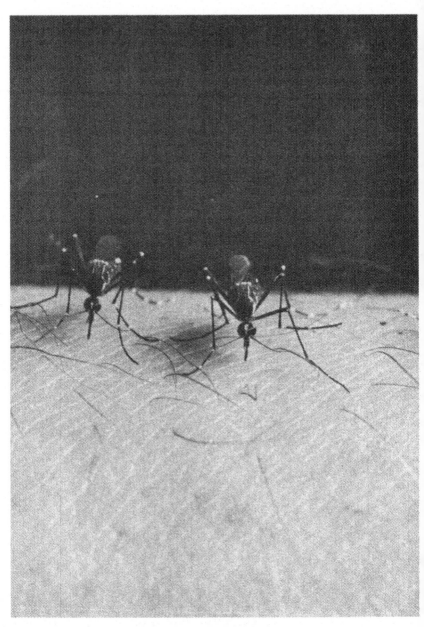

12. Female yellow fever mosquitoes sucking human blood.

whereas the larva of *Deuterophlebia* has three distinct thoracic segments. These two families are deceptively alike, the larvae because of their sucker-discs, and the adult flies because in both families the wings have many folds or creases that look like supernumerary veins. It is not certain whether the two families are really closely related, or whether they may not have reached their present similarity convergently, through adaptation to a similar way of life.

Adult female Blepharoceridae are mostly carnivorous, catching and devouring other insects in flight, like some Ceratopogonidae. Adult Deuterophlebiidae have reduced mouthparts, like the non-biting midges of the family Chironomidae. So in these two families we see a rather mixed collection of habits and structures reminiscent of other, more successful families of flies: rather like earlier, discarded models of current motor-cars.

According to Kellogg [160] the decline of the Blepharoceridae, and possibly also the Deuterophlebiidae, as a consequence of following such a difficult way of life, is illustrated by two examples of how vulnerable they are.

The larva is able to survive only in the most highly oxygenated water, and so has to cling to rocks in such places as the lip of a waterfall, or 'the sides of a pot-hole where the water is ever whirling and boiling'. In this extreme current the floating particles that a *Simulium* larva can capture are whisked away before they can be seized, and the Blepharocerid larva can merely browse on the diatoms covering the rocks. This means that it must move about, yet to release hold, or even to allow part of the body to be seized by the current, is to be swept away to destruction. The larva moves sideways, cautiously releasing three of its suckers and attaching them again before releasing the other three.

The pupa is found in the same place, and is permanently anchored to the rocks, not by detachable suckers, but by three pairs of suction pads. The adult fly emerges from this pupa under water, which must be shallow enough for the adult to reach the surface while still clinging with its hind legs to the anchored pupal skin. In this precarious position, the adult fly can only pause for a few seconds to harden before it either takes flight, or is carried away. If the water at the time of emergence is too deep for air to be reached without letting go of the pupa-case, the adult is trapped in the rushing water without hope of escape. Apparently deeply immersed pupae have some capacity to delay emergence for a limited period, which may be enough for them to wait through one of the brief spates that are characteristic of mountain streams, but there is great loss of life at this critical stage.

Kellogg suggested, very plausibly, that the distinctive network of folds from which these flies get their name of 'net-winged midges' is a consequence of this unseemly haste to get into the air. Other flies take time to expand and harden their wings after they have emerged from the pupa. The net-winged midges must be ready to fly after a second or two, and so the pharate adult, within the pupal skin, must have its wings fully prepared, but tidily folded like an airman's parachute.

It is interesting to compare these with the black-flies, Simuliidae, which have made a much more successful adaptation to life in rather similar conditions. Instead of a precarious locomotion by manipulating sucker-discs, the larva of *Simulium* remains fixed in one place, or makes a positive, purposeful movement with only two suckers to control. Against the risk of being swept away it has its silken thread, and its oxygen requirements are such that it can choose less rigorous places in which to cling, where food can be captured from the flowing water. When the adult black-fly emerges it is provided with a bubble of air, and so can commit itself to the current with impunity, and wait to take flight until a quiet pool is reached.

The net-winged midges and the black-flies give us an example of the two types of adaptation that we can see throughout the animal kingdom. The one meets each problem with a piecemeal adaptation, and thus accepts a progressive restriction of its way of life, until decline leads to extinction. The other responds in some generalized, simple way, which increases rather than decreases the number of things it can do, and so makes life easier instead of more difficult. It thus retains its versatility, and is ready for the demands of any environment. Perhaps each group of animals starts off in this expansive phase, and eventually becomes entangled in a contracting spiral. History suggests that human aggregations may do the same.

Finally among the midges there are two groups of near-mosquitoes. It is probably more correct to take the old-fashioned view that the family Culicidae has three sub-families: Culicinae, the mosquitoes; Chaoborinae; and Dixinae. From the viewpoint of natural history these should certainly be considered together, if we are to see the mosquitoes in their true perspective as nematocerous flies. Yet so much has been written about mosquitoes that their study has reached a different level from that of flies in general, and it would be pedantic to insist upon keeping them in their place as a mere sub-family. It is also inconvenient to have to qualify so many general statements, which would be true of the mosquitoes, but not of the others. So I shall mention the Chaoborinae and Dixinae in this chapter, and start a fresh chapter for the mosquitoes proper.

Dixinae are non-biting flies rather like small crane-flies, but distinguished by not having a V-shaped groove or suture of the thorax. There are only two genera, *Dixa* which is world-wide and has about 200 species, and *Neodixa* which is confined to New Zealand. The larvae are cylindrical, with a pair of spiny prolegs on the first two abdominal segments. These are another group of larvae that are only just aquatic, crawling about on vegetation or rocks at the water's edge, and often having the body curved into a ∩-shape, still kept wet by surface tension, but actually above water-level except at the two extremities. This is a curious adaptation, since it would seem more logical to have the ends exposed with their open spiracles to the air, and the middle wetted to avoid desiccation.

Dixa larvae collect small organisms in the water, using mouth-brushes. Their eggs are laid in a gelatinous mass, rather as in Chironomidae, but on a solid sub-stratum and not floating in the water. The pupae of Dixinae are rather like those of mosquitoes (plate 7), which we shall discuss in the next chapter.

Chaoborinae are mosquito-like flies with hairs and scales. They have a short proboscis, but do not bite, and the male flies have plumose antennae. The adult flies sometimes occur in immense swarms, and are particularly notorious over the lakes of the African Rift Valley; Livingstone spoke of them over Lake Nyasa, where countless millions of *Chaoborus edulis* used to be caught and pressed together to form a sticky mass called 'Kungu Cake' (plate 30). Similar swarms occur on the other lakes, and their empty pupal skins may form a floating island.

The North American *Mochlonyx cinctipes* swarms in clearings in the woods when the light intensity has fallen to a definite level, estimated at 3·5 ft-candles. The swarms have eight males to every female, and are typical mating swarms.

Collectors of pond-life are familiar with the 'phantom larva' of *Chaoborus* (*Corethra*), a sausage-shaped, transparent larva with two glistening, kidney-shaped air-sacs. This carnivorous larva lies horizontally in the water, slowly descending, until its prey comes near; then it makes a convulsive, lashing movement, and darts unpredictably in any direction, to resume its passive waiting. It is completely aquatic, has no open spiracles, breathes entirely by absorption, and rarely comes to the surface except at night; by day it may lie in the mud at the bottom.

The eggs of *Chaoborus* are laid in jelly on the water-surface. Those of *Mochlonyx cinctipes* are deposited with leaves and other debris at the edge of pools, and are exceptional for the duration of the egg stage. According to O'Connor they are laid in May and June, and remain on the leaves throughout summer, autumn, and winter, hatching early the next March. There is thus only one generation per year, of which nine months are spent in the egg, and all the rest of the life-cycle is compressed into three months. The pupae of Chaoborinae swim actively, like those of mosquitoes.

Chapter 7

Mosquitoes

THE TREATMENT of mosquitoes in a work of this kind presents a problem. Like birds, butterflies, and fruit-flies, mosquitoes have attracted to themselves a devoted, nay dedicated band of investigators, some of whom study the group in isolation, making little attempt to relate it to other animals. Consequently in each of these groups there exists a body of detailed knowledge which might profitably be applied in other directions, if only there were more liaison between the butterfly man, the mosquito man, the *Drosophila* man, and those interested in other groups.

In this book we are not primarily interested in mosquitoes for their medical importance, nor with their effect on the doings of man, though these will be touched upon in a later chapter. Just now we are looking at mosquitoes as flies, to try to see how they fit into the evolutionary pattern that we are tracing, and to gain from them ideas that can be applied to other families that have been less intensively studied.

The Spanish word 'mosquito' is merely a diminutive, and means a little fly, but it has long become established in English, and with variants in other languages, for bloodsucking flies of the sub-family Culicinae of the family Culicidae. Entomologically, then mosquitoes are just part of one of almost a hundred families into which flies are divided. We have already briefly mentioned the other two sub-families of Culicidae, Chaoborinae, and Dixinae, and seen that they belong with the group of water-midges and black-flies, in which the larvae are predominantly aquatic, and the adult females of many families supplement their nectar diet by piercing the skin of vertebrates and sucking their blood.

It is these two habits, more highly developed, that give the mosquitoes their importance to man. Because they suck his blood, and moreover because they take a succession of blood-meals from different persons, adult mosquitoes act as carriers of some of the most virulent of human diseases; and because they have learned to exploit such a variety of aquatic larval habitats, permanent and temporary, they are able to maintain an unpleasantly close relationship with man, and to elude his determined efforts to eradicate them.

Adult mosquitoes are typically nematocerous flies, slender, fragile, with antennae, abdomen, and wings all narrow and elongate. They are distinguished from nearly all other related families by having the veins of the wings festooned with scales, and from all other families by the combination of scales with a long, projecting proboscis. Scales are modified hairs, and many of the hairs of the rest of the body are similarly modified, so that the abdomen often has a striking pattern of bands or spots of colour. Bands of scales can often be seen, too, on the legs, and can be used in the identification of particular species.

In 1959 Stone, Knight, and Starcke catalogued 2,426 species of mosquitoes throughout the world.

From the viewpoint of natural history, as well as of structure, mosquitoes fall into three tribes: Anophelini, Culicini, and Megarhinini (Toxorhynchitini). As in other families we have met so far, only females suck blood, and this habit is believed at first to have been universal in the family. Certain members have later abandoned bloodsucking, and feed only from flowers. This is true of the whole tribe Megarhinini, and individual species in the other two tribes have also ceased to bite. As we shall see again with the horse-flies in a later chapter those that have given up their blood-meal have, as it were, stagnated, making little evolutionary advance, and declining in numbers. The bloodsuckers on the other hand, have found in the search for blood an evolutionary challenge, and have multiplied and flourished in their recent evolution. Whether they will continue to do so is another matter.

We will leave aside the non-biting Megarhinini for the moment, and talk about the other two sub-families. The Anophelini are the 'true mosquitoes', which carry human malaria, and are the smaller and more homogeneous tribe. Culicini are a more varied tribe, both in the habits of the adults and in the diseases they transmit. Any reference to 'mosquitoes' in what follows, may be taken as applying to both tribes, unless it is qualified by the adjectives anopheline or culicine.

It is when mosquitoes come to bite that they attract attention, and arouse fear, and naturally this part of their life has been extensively studied. In the last twenty years a great deal of attention has been given to 'biting-cycles', and to 'vertical distribution', but it has to be admitted that these studies inevitably have a human bias. They tell us much about the mosquitoes that seek human victims, and less about those that bite birds, reptiles, and mammals other than man.

Flies in general are lovers of light. They move towards the brightest light, and if they come indoors, unless it is for the purpose of hibernation, they soon appear on the window. Mosquitoes, however, are well-known creatures of the night, and Europeans in the tropics

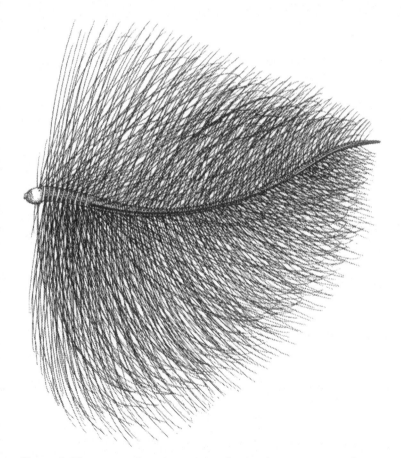

Figure 16. The antenna of a male mosquito, showing the many-segmented flagellum with its whorls of long, sensory hairs.

have long slept under mosquito-nets. Studies of nocturnal mosquitoes have led to the discovery that many other flies are about at night, some rather unexpectedly. It seems reasonable that tiny biting midges should be about in the dark, but it is startling to find that this is true of many large horse-flies.

Mosquitoes in general seem to shun very bright light. *Anopheles* and *Culex*, the best-known biters, are crepuscular, or nocturnal, while the diurnal species of *Aedes* and *Sabethes* are flies of forest or woodland. It seems likely that these diurnal mosquitoes like moderately light and warm conditions such as are found in open woodland in a wide range of latitudes. In the dense tropical rain-forest it is dark, cool, and very humid, and the mosquitoes – indeed all the biting flies, notably the horse-flies of the genus *Chrysops* – generally prefer the high canopy of the tree-tops, where conditions are more like those of the woodland. Much recent work, in which Dr A. J. Haddow and his colleagues in Uganda have taken a significant lead, has shown that these flies feed mainly on monkeys or birds in the tree-tops, or at specific intermediate levels. They may descend to lower levels if they are able to see activity taking place below them: that is, if there is a clearing, or a special type of vegetation such as rubber-trees, without undergrowth, or if some parts of the ground rise to the general level of the canopy. In this way infections such as jungle yellow fever may break out sporadically and unpredictably among people who only occasionally visit the forest. The forest edge where the canopy descends to the ground, constitutes an important zone of contact as a 'fringe habitat'.

Bloodsucking, as we have seen, is a way of getting more animal protein to supplement the stores accumulated during larval life. Thus it is said that *Culex pipiens*, the house-gnat, can never develop eggs without a blood-meal, in contrast to *C. molestus*, which can live for years on only vegetable food. Yet *C. molestus* does bite, and bites man much more readily than does *pipiens*. *C. pipiens* is the dull-brown gnat that commonly hibernates in houses in temperate countries, and which is periodically accused of biting. In fact it takes its blood-meal from birds, and the culprit is usually *C. molestus*, if it is not some other mosquito, or even some other insect altogether.

The understanding of the difference in behaviour between these two species of Culex was reached during the Second World War, when *C. molestus* attacked people taking refuge in deep air-raid shelters. Since then a great deal of investigation has gone on, but the exact limits of *pipiens* and *molestus* in different parts of the world are still not fully clarified. It is likely that the two are not fixed entities, not even sub-species, and that they may belong to a shifting population that has developed, and is still developing, biological races. These races may have different feeding preferences, and a different physiological balance between larval and adult stages, though the adult flies may be almost impossible to distinguish from each other. Mosquitoes, more than most groups, emphasize the fact that flies are not cast in a mould, but are constantly modifying their habits, and adjusting themselves to meet the demands of their environment.

Twohy and Rozeboom [292] compared the food-reserves of lipids, glycogen and nitrogen compounds in *molestus* and *pipiens*, and found that the pupae and newly emerged adults of *molestus* were provided with substantially larger amounts of these reserves, having taken slightly longer to mature in the larval period. By the time that the eggs ought to be ready for laying, adult consumption has used up all the reserves in *pipiens*, leaving nothing to provide a yolk for the eggs. The authors conclude that the *molestus* larvae do not live in an any more favourable environment than those of *pipiens*, but that they make more efficient use of it. By doing so they relieve the adult of the need to search for a blood-meal.

'Biting-cycles' of mosquitoes are widely studied today. Between mating and egg-laying, that is during the period when the eggs are maturing, a female mosquito divides her time between resting and flying in search of a blood-meal, so that for this period, at least, her biting activities give a good picture of her active day. Under natural conditions out of doors there seems to be a principal division into those that have one peak of biting activity in twenty-four hours, and those that have two. Thus diurnal mosquitoes will either have one peak of activity in the middle of the day, or two peaks, morning and afternoon, with a lull at midday, when few or none of them are about. Similarly nocturnal mosquitoes may be active all night, or have a burst of activity soon after sunset and another just before dawn.

Of course, since we are dealing with living things, we should not expect all the individuals of a species to start biting at one stroke of the clock, and all cease to do so at one stroke. A typical graph of a diurnal species, *Aedes nubilis*, shows that it is active from 6 am to 8 pm, and that in that period the average catch per man per hour never falls below five: in the morning it rises to twelve, and in the evening to eighteen, but between 11.30 am and 5.0 pm it stays between five and seven.

Two obvious factors that may influence the activity of mosquitoes are of course temperature and humidity. Light intensity has a limiting value, acting in the opposite sense upon diurnal and nocturnal mosquitoes, by preventing activity unless the light intensity is above or below a certain value; but it does not seem to enter into the distinction between one-peaked and two-peaked activity. Temperature and relative humidity tend to oppose one another: that is, the middle of the day is hottest but driest, the middle of the night coldest and most humid. A possible explanation of the one-peaked and two-peaked types is that all the diurnal species are most active in the highest temperature and all the nocturnal ones in the lowest, thus having basically one peak. Some species, however, are less tolerant of variations of humidity. For the diurnal ones the middle of the day is too dry for them, and for the nocturnal ones the middle of the night may be too humid. In other words, the two-peaked species are the fussy ones who want everything just right.

Closely similar biting behaviour has been described in the forest horse-flies of the genus *Chrysops*, showing that this is a physiological problem of general interest, and not just a peculiarity of mosquitoes. A further indication that flies are not automatons reacting directly to variations of warmth and humidity, is shown by the fact that these cycles of activity may continue, for a time at least, when captured mosquitoes are kept in cages. If the hours of light and darkness are artificially reversed, the mosquitoes will change their ways to suit, like the small nocturnal rodents in zoos that are exhibited in darkened houses with dim red lights.

A great deal of work is going on at the present time on 'biological clocks', inbuilt rhythms that are being detected in a wide range of different organisms, plants as well as animals. We ought not to be surprised at this. Our own life is full of such rhythms, waking and sleeping, which help us to deal with many situations that recur frequently, and thus to avoid the need to waste conscious thought on them. As we go down to animals with less and less conscious thought, the value of such helpful rhythms becomes greater. As Marston Bates points out, such rhythms can be confusing when we are experimenting with mosquitoes, or any other flies, making them react in ways that we cannot understand unless we allow for rhythmical habits that they have acquired previously.

Mosquitoes share with the other midges and with the horse-flies and others (see next chapter), the habit of mating in male swarms that are visited by individual females. Though this is the most spectacular form of mating behaviour, it is not necessarily the most important, and Dr Haddow makes the point that the females seen to take part are so few that they must be an insignificant fraction of the whole population. The species is primarily kept going by, as it were, clandestine matings taking place in or about the resting-places of the mosquitoes.

Resting is just as important an occupation for adult mosquitoes – and for other flies – as is hunting for food. Blood is a highly concentrated food, and we commonly use the expression 'a blood-meal', which implies feeding only on one or two well-defined occasions. A meal may be interrupted if the victim is restless, but ideally a fly probes, penetrates, fills itself to repletion, and goes off to rest until the eggs mature. Taking moisture and sugars from flowers is more intermittent, like eating sweets between meals. In short, mosquitoes behave more like lions, which gorge and then sleep it off, than like zebra which graze all day.

Most mosquitoes rest among vegetation, and are hard to see, though they may be collected by sweeping or beating. In woodland they rest on tree-trunks, and the mosquitoes themselves are often striped or banded with brightly coloured scales. This makes them very conspicuous in a collection, but in the dappled light and shade of their resting-places they merge into the background.

Whether they are diurnal or nocturnal in habit, they rest in shady places, just as we do when we want a quiet nap. Shannon calculated that a light intensity of 1–5 ft-candles was most favoured. So nocturnal mosquitoes, resting in the daytime, need deeper shade than the others, and seek caves, spaces under stones, tree-holes, and of course buildings. The night-biters are the ones that are found indoors during the day, though they do not always bite in the same places in which they are found resting. When they wake up, their blood-sucking instincts may lead them quite differently from their sheltering reflexes. A study of resting-habits is of great value in control, because it is then that the adults of some species are most vulnerable: they are concentrated together, and are stationary, and others will often come to the same shelter, so that a persistent insecticide will have effect over a long period. Many species rest in dispersed places out of doors, and against such species DDT is less effective.

Male mosquitoes, and those females that are able occasionally or permanently to lay eggs without a blood-meal, are assumed to take nectar from flowers. Some females may bite cold-blooded vertebrates, or other insects. Occasionally, males have been noticed among a mass of female mosquitoes congregating on a person, or on the walls of stables and other farm buildings. Some of these have been proved to be gynandromorphs: that is the body is a mosaic of male and female cells, more particularly male on one side and female on the other, with one biting mandible and maxilla instead of a pair of each. Such flies seem to be partly female in instincts as well as in structure. Yet, apart from this explanation, males and non-biting females may be attracted on occasion by sweat, dung, mucus, and all the other media that attract flies.

One remarkable eccentricity must be mentioned. Mosquitoes of the genus *Harpagomyia* (now confusingly called *Malaya*), widely distributed from Java to tropical Africa, have discovered that certain *Cremastogaster* ants, which run up and down tree-trunks, are carrying honey-dew which they have taken from aphids. A mosquito hovers about an inch from the trunk and suddenly alights in front of an ant, not touching it, but vibrating the wings in an apparently hypnotic way and waiting until the ant opens its jaws. The proboscis of the mosquito is pushed into the ant's mouth and the honey-dew is filched from it.

This is not, as it might seem, an occasional report that could be a traveller's tale. It has been filmed in detail, and is apparently the normal method by which all the mosquitoes of this genus feed. The proboscis is suitably modified, and probably takes no other food. That the ant should give up its booty in this docile way is at first surprising, but it must be remembered that ants are very highly evolved social insects whose life involves reciprocal relationships with many other insects. In this instance the ant has only just taken the honey-dew from an aphid, and it is a pleasant irony that its reflexes should be cleverly exploited by the mosquito.

We think of mosquito-larvae as aquatic creatures. After all, we know that ponds, creeks, swamps, water-courses of all kinds breed mosquitoes, and so do temporary pools and puddles. Both larvae and pupae are familiar as 'wrigglers', swimming actively about in deep, clear water. Yet they are air-breathing insects, and only imperfectly able to exist away from the free air: much less so, in fact, than most other larvae of Nematocera of the 'watery' group.

They are really creatures of the surface film, enjoying the advantage of breathing atmospheric air, combined with the protection of the water against desiccation, and the constant supply of microscopic food that the water supplies. The nearest equivalent lies in those Psychodids that live in a surface film over stones and in shallow water, and we may see in this a hint that mosquitoes have evolved in a direction of their own, exploiting fully the peculiar advantages of the water/air boundary.

Only a few mosquito-larvae seem able to live on the bottom for a long time, immobile except for the minimum effort that is needed to circulate water over the body. Those of the genus *Mansonia* are able to extract air from the tissues of water-plants, as we shall see in more detail presently.

Breeding-places divide broadly into permanent water and temporary accumulations, and this makes a basic cleavage through the family. Those which breed in permanent water have life-histories geared to the slow rhythm of the seasons, and the long-term effects of temperature, rainfall, food-supply, and enemies. Those breeding in temporary water have to be able to wait patiently until the time is ripe, and then develop quickly and efficiently before the opportunity has passed.

Mosquito-larvae are unable to live except in water, and they have none of the versatility of say, Stratiomyids which we shall see later can survive in mud, or even leave a dried-up water-hole and move over the ground to another. Hence timing is of paramount importance in the life of an individual mosquito. Marston Bates points out that the egg-stage of mosquitoes provides a 'buffer mechanism', introducing the necessary flexibility into the life-history to meet the demands of its particular environment.

In permanent water the problem is to choose the right amount of light and shade, of salinity, and of movement, and then to keep the eggs afloat, and if necessary together, so that they do not drift away from the best situation. These eggs have some kind of float, more elaborately developed in those that are laid singly, and often strikingly different in different species. *Anopheles* and such *Aedes* as breed in permanent water lay the eggs singly, either by settling on the water-surface for the purpose, or by dropping them during an 'oviposition dance' close to the water. *Culex, Theobaldia, Mansonia,* and related mosquitoes mostly lay their eggs in 'rafts' (plate 10), the eggs being placed on end and bound together with a sticky material.

The female rests on the surface, and carefully places each egg in position.

These egg-laying routines are not rigidly fixed, and may vary in closely related species, and even in the same species under different conditions, eg in captivity. In the South African genus *Trichoprosopon* one species lays eggs singly, while another makes a raft. It is apparent that in this highly important stage of their life-history, on which depends so much the success or failure of the subsequent larval life, mosquitoes are still actively adapting themselves, and are in a flexible stage of evolution.

Once these eggs have been laid they are provided with a permanent habitat for the larvae, and development can go on without delay. Eggs that are laid in permanent water hatch as soon as the embryo has developed sufficiently to begin an independent life as a larva. But in temporary water – ephemeral pools, rot-holes in trees, the water accumulated in cut bamboos, plant-axils, Bromeliads and so on – the hazards are much greater. It is here that the real buffering effect of the egg-stage is seen. If the eggs had to be laid directly in water the mosquito would have to carry them round until just the right moment, and after a storm the multitude of small pools could not be efficiently exploited at short notice.

So the eggs are laid in soil or debris in suitable hollows, or among leaves, or even on the ground. One cannot of course credit the female mosquito with the foresight consciously to select a place that is going to collect water next time it rains. One must believe that selection has favoured the good choosers, and eliminated those which put their eggs higher up, where they were never immersed in water. This is one of the rare instances where simple, Darwinian natural selection could operate decisively.

The eggs having been laid, they must not then develop and hatch forthwith, or the larvae may find themselves with no water to live in. The buffering effect has taken the form of a *diapause*, or enforced delay, which is a common feature of larval life in several Orders of insects. The embryo develops normally until it is ready to hatch, and then stops, and does not actually emerge from the egg-shell until a particular stimulus has been received. Such eggs are resistant to frost and to desiccation, and usually begin to hatch only when they are immersed in water.

A device similar to this is usual in fleas, which have a diapause in the pupal stage. The adult develops within the pupa, but does not break out except under the stimulus of vibration and perhaps warmth. This is an obvious advantage, because if adult fleas were to emerge when no animal was near they might starve before they could find one. Such a device for protecting an insect against the irregularities of life is well named a 'buffer stage'.

Eggs that require a hatching stimulus are not usually able to develop in permanent water, because conditions there do not change abruptly enough to constitute such a stimulus. An exception is a drop in oxygen tension in water that has become stagnant, and therefore well provided with micro-organisms upon which larvae can feed. The hazard that can be avoided by diapause varies according to the climate of the country concerned. In temperate, subarctic and arctic regions, winter is the hazardous time, when frost deprives the mosquito-larvae of the water-surface they need, and air-temperatures are too low to suit the adult flies. Hence many mosquitoes of the colder regions hibernate in the diapause of the egg-stage. Not all do this, however. *Anopheles claviger* hibernates as a larva in mud at the bottom, breathing through its cuticle sufficiently for its limited needs. The water may be frozen on the surface without ill-effect, but the larva itself must not be frozen into the mud. Opinions differ as to whether this is a true diapause, or whether it is merely a state of quiescence: the difference being that diapause, as we have seen, needs a definite stimulus to break it. Sautet says that *A. claviger* is in a true diapause and that the rise in oxygen content of the water provides the necessary stimulus.

A number of mosquitoes hibernate as adults, two well-known examples being *Culex pipiens*, the house gnat, and *Theobaldia annulata*, the banded house mosquito. I get both of these in my house in every winter. *Theobaldia* continues to bite at intervals and is a nuisance in the bedroom during the night. *Culex pipiens* does not bite humans, but has often caused me trouble by hibernating in my dark-room. Aroused by the glow of the enlarger, it flies round my head, and sometimes settles on the photographic paper on the enlarger, to appear, ghost-like, on the print.

In hot countries it is drought that is the hazard, and this is met by a summer diapause called aestivation. The coming of rain is usually the stimulus that breaks this.

The permanent-water habitats of mosquitoes are lakes and flowing streams. The larva (plate 7) is highly adapted to its job of living and feeding at the surface film, but it has only feeble devices for clinging to rocks or plants, and its powers of swimming are energetic rather than co-ordinated. Hence anything that approaches a torrent is out of the question. Even where mosquitoes can be seen breeding in obviously moving water, it will be seen that they are in fact congregated in little patches of still water between the eddies, or under the shelter of stones and plants. 'The adaptation of the stream-breeding larvae seems to be not so much to life in running water as to the avoidance of running water' – Marston Bates [15].

The surface of the water must be clean, at least over a reasonable proportion of its area, or else the larva cannot reach the air without getting its siphons blocked. Since the ability to keep a tube open through the surface film depends on the surface tension, anything which lowers this is a menace to the larvae. In theory at least, any substance in solution reduces the surface tension below that of pure water. Hence mosquito-larvae prefer the cleanest water for breathing purposes, though they may find more to eat in water that is stagnant.

Though many mosquitoes breed readily in tidal marshes and brackish coastal swamps, experimental work suggests that they tolerate salt rather than actively seeking it, and that they would do as well in fresh water. The familiar picture of coastal and estuarine marshes as malarious, mosquito-ridden places comes about partly because such places are more difficult to drain and eliminate than similar areas elsewhere; and partly because certain species or races of mosquitoes, such as *Anopheles labranchiae atroparvus*, have a greater tolerance of salinity than others, and so are able to flourish in the absence of their more fastidious rivals. *Aedes detritus*, a salt-marsh species of northern temperate countries, which can live in brine pools with a salinity much greater than that of sea-water, can regulate both osmotic pressure and salinity of its body fluids to counteract the unfavourable mixture outside. Its cuticle is impermeable to salt, and the balance is maintained by exchange through the gut, the Malpighian tubules, and the anal 'gills'.

The survival of permanent-water mosquitoes depends on the ability of the egg-laying females to choose the right place, where their eggs and larvae will not be washed away. So far as chemical content is concerned, experimental work shows that mosquito-larvae will live and grow in water substantially different from that of their natural habitat. It appears that the 'typical habitat' of many mosquitoes is the one that the females choose to lay in, but not necessarily the only one that would suit the larvae. It is unlikely that the egg-laying female can know much about the chemical content of the water, or even the degree to which it is oxygenated, during the brief reconnaissance that she makes before dropping the eggs. Probably what she does appreciate are light and shade, and a process of selection has narrowed down the choice to one that, on balance, gives the best chance of survival.

It is the mosquitoes that breed in temporary water that show the real ingenuity in making the most of what they can find. The different places in which they have been found can be analysed and classified interminably, according to the point of view of the list-maker. Obviously, someone who is interested in the chemistry of the water will make one classification; someone who is interested in the physical effects of light and shade, and temperature, another list; and the practical man who wants to arrange the most economical spraying campaign will have yet another grouping. From the viewpoint of comparative natural history we may draw a basic distinction between ground pools, and what are called 'container habitats', ie small quantities of water in cavities, natural or artificial, away from the ground.

The first category, 'ground pools', includes those of every size: areas of flood-water left behind; rainwater pools formed in hollows of the ground; cart-ruts; footprints and hoofprints of animals; seepage of foul water; and even pools of urine. In these habitats the problems of egg-laying is simple, and no specially clever behaviour is needed, except the instinct to lay eggs in the sort of cavity that is likely to fill with water sooner or later. Then the egg lies dormant in diapause until immersion gives it the hatching stimulus. Larval life is a race to get through before the pool dries up or drains away, hence such mosquitoes often develop extraordinarily quickly, perhaps in as little as five days. No time here for a long, luxurious childhood: a long sleep in the egg; a quick period of frantic feeding and growing; a short adult life on the wing; and then the cycle begins again.

The problems of container habitats are pretty much the reverse of this. The water usually remains longer, and larvae need not grow so rapidly. On the other hand, it is more difficult for the egg-laying female to find the potential container, and to recognize it when she does find it.

The best-known and most studied of container habitats are rotholes in trees. These occur in every climatic zone, and may be filled with either a spongy mass of rotting vegetable matter or a quantity of water, usually with a high organic content. The eggs are laid just above the water-line, and they settle down to their diapause until the water-level rises enough to give them their hatching stimulus. The evolution of this habit is an evolution of behaviour in the females to select both the time and the exact place to get the maximum chance of beginning larval life within a reasonable period.

Most ordinary rot-holes do not present any special physical problem to the female, but mosquitoes of some genera seem to have gone out of their way to make life difficult for themselves. Thus, while many *Culex* and *Aedes* are content with easily accessible tree-holes, the South American sabethine mosquitoes project their eggs into holes in the shoots of bamboo, to hatch when water accumulates in the hollow space between two nodes. An even greater persistence is shown by *Armigeres*, which lays its eggs on its own legs, and some time later releases them by dipping the legs in the water: an ingenious way of avoiding the waste of eggs that may be laid in places where no water ever comes. This used to be thought a device to allow this mosquito to lay in inaccessible places, but in fact it commonly uses cut bamboos, treeholes and other fairly exposed sites.

These habits have been demonstrated by the modern close-range cine-camera which shows these mosquitoes, and others like *Harpagomyia* (or *Malaya*) and the ant, behaving with an elegant precision that is beautiful to watch. The larvae of *Megarhinus* have been found in bamboo with a hole only 2 mm wide, through which it seems impossible that the large mother could have passed to lay her eggs, or the resulting adult offspring escape. Here is an unsolved mystery.

In tropical forests many leafy plants collect water in the leaf-bases and other cavities, and two groups of tropical plants are celebrated for their aquatic fauna. The Bromeliads of South America, the family to which the pineapple belongs, are often epiphytic plants, that is, they cling for support to other, stronger plants, but do not feed upon them parasitically: the common ivy is now said to do this. In the centre of a group of leaves the Bromeliad collects rainwater, and maintains what is appropriately called an aquarium tenanted by a wide range of aquatic organisms. Mosquitoes of the genus *Wyeomyia* make a speciality of such water, and there is a range of different species of *Wyeomyia* in different plants of this family, demonstrating on a miniature scale the same kind of radiation by isolation that was shown by Darwin's celebrated finches in the Galapagos Islands.

They are not by any means the only mosquitoes of the Bromeliads, from which nearly 300 species have been recorded, from the genera *Culex*, *Aedes*, *Bancroftia*, *Megarhinus*, *Sabethes*, and even a few *Anopheles*.

To balance these, in the Old World tropics of Indo-Malaya and Australia the pitcher-plants, *Nepenthes*, hold great quantities of water in their 'urns', which is rich in enzymes and other dissolved substances by which the trapped adult insects are killed. It is often said that these plants are carnivorous, but the current opinion is that this is a fortuitous effect, and not a regular way of feeding by the plant. But, like the terrestrial larva that occasionally eats another that it happens to meet, we may be seeing the beginning of a change of diet through slow evolution. Meanwhile the nourishing soup often supports larvae of *Tripteroides* mosquitoes, which are preyed upon by larvae of *Megarhinus* (*Toxorhynchites*).

Other plant-reservoirs are found in the North American pitcher-plant, *Sarracenia*, and even in the common Teasel, *Dipsacus*, of temperate countries. Mosquitoes are not a common feature of the fauna of teasels, though Séguy records that he himself has found larvae of *Culex pipiens* in a giant *Dipsacus* in a marshy area.

One thinks of mosquito-larvae as being abundant and highly successful. They appear quickly in any body of water, and whether we are building a reservoir or making a garden pond we must be preoccupied with the risk that it may become infested with mosquitoes. It is therefore surprising when Marston Bates [15] points out that mosquito-larvae are not a 'dominant' element, and that their presence or absence makes little or no difference to the ecological balance.

A little thought shows that they are in fact curiously limited creatures. We have already seen that with few exceptions they live in the surface film, and breathe atmospheric air through posterior spiracles. Anopheline larvae have the two spiracles on the eighth segment, and to keep these exposed to the air the whole body must lie horizontally, supported by tufts of 'float-hairs' on each segment, like a drift-net or an anti-submarine barrage. Culicine larvae have the posterior spiracles at the tip of a conical siphon, and hang vertically downwards like a trapeze artist. It seems astonishing that the effect of surface tension on a system of valves surrounding the spiracles should be enough to support the weight of the larva, though Archimedes' principle simplifies the problem. An oily secretion from perispiracular glands helps to increase the surface tension, and to keep the water out of the spiracles and tracheae. The principle behind the familiar method of destroying mosquito-larvae by oiling the surface to the water is that a light oil is not repelled by the oily spiracles, and thus flows into the tracheae, displacing the air, and asphyxiating the larvae. Modern anti-larval oils are highly toxic, and the quick flow ensures penetration of the insecticide rather than acting primarily by asphyxiation.

I say that the larvae are limited creatures because they can neither exist without the water, nor, as a rule, go into it properly for very long. Stratiomyid larvae, also with a siphon, can lie in mud, move over wet ground, and even be completely dried up, and still recover. Black-fly larvae (*Simulium*) cannot leave the water, but they can lie at considerable depths, in conditions of their own choosing, and ignore what is happening at the surface. Thus in spite of the apparent diversity of larval habitats that we have discussed in this chapter, from the viewpoint of the larvae they all present themselves as merely an air/water interface. A depressing thought, after all the ingenuity that the adult mosquitoes have displayed in finding them such a variety of places to live in!

Many mosquito-larvae, and possibly most, can get enough air at the bottom of a pond if they lie inert, breathing dissolved oxygen through the general body surface. The anal gills that are a prominent feature of both anopheline and culicine larvae have been shown to be used mainly to regulate the salt-content of the body, and consequently to adjust the osmotic pressure between inside and out. Yet D. J. Lewis has pointed out that the gills are in themselves a substantial area of body-surface, and submerged larvae may exploit this by stirring a current of water across them with the mouth-brushes; he considers that 'they should be regarded as tracheal gills in some species'.

In the Northern Hemisphere *Anopheles claviger* passes the winter as a larva, not as an egg, and is able to lie in mud for months on end; and *Psorophora discolor*, which breeds in rainwater pools in sand, is said never to come to the surface for air. These two species are exceptional, for the standard method of existing away from the surface is to tap the stores of oxygen in the roots of underwater plants. Thus the European *Mansonia* (*Taeniorhynchus*) *richardii* makes use of the roots of *Ranunculus*, *Acorus*, *Glyceris*, and *Typha*. The siphon of these larvae is equipped with two sets of teeth, one of which acts as an anchoring device, while the other, like a miniature hacksaw, is moved to and fro in the plant-tissue. The amount of oxygen generated by the plant obviously varies greatly with the season, especially outside the tropics, and *Mansonia* larvae have two large cavities in the tracheal system that are believed to act as an air-store, to offset the fluctuations in the supply. *Ficalbia* and *Aedomyia* have similar habits.

The development of these teeth for piercing plants is one of the few extravagances of mosquito-larvae, which otherwise are astonishingly uniform. Those that live in plant-reservoirs, Bromeliads, *Nepenthes*, arum-spaths and the like may have unusually long and conspicuous hairs on each segment, but these are only a slight exaggeration of the normal vestiture. In general, apart from the distinction between anopheline and culicine, mosquito-larvae the world over are very much alike.

Marston Bates makes the point that all mosquito-larvae are really predaceous, but because most of them catch very small prey – diatoms, Protozoa and so on – they are usually referred to as 'filter-feeding'. Filter-feeding mosquito-larvae have mouthparts of a primitive, chewing type, with mandibles and maxillae that meet horizontally. In addition 'mouth-brushes', which arise from the labrum, are used to sweep floating or suspended particles towards the mouth. Elaborate studies have been made to analyse and classify the ways in which currents can be set up and utilized by the larvae, but we have no space to pursue this here. One significant difference may be noted. Anopheline larvae, which lie along the surface, are confined to filtering from the surface layer; culicine larvae, hanging downwards, normally take food from a greater depth, and in addition can go down and forage on the bottom with what has been called a 'nibbling' technique. Most larvae are also believed to extract a certain amount of dissolved nutriment from the water, especially in organically rich foul water. Rot-holes and plant-cavities, with their quota of dead and decaying organisms, are a rich source of dissolved food.

There is probably a complete sequence from such larvae, through feeding on cast skins and dead insects, to actively attacking other aquatic creatures, including other mosquito-larvae. The best-known predaceous larvae are those of *Megarhinus* (*Toxorhynchites*), where we note that flesh-feeding has been transferred to the larva from the adult, which does not suck blood. In a business-like way, the *Megarhinus* larva has prehensile mouth-brushes, and very strong teeth on the mandibles. It lives in container habitats, feeding mainly on other mosquito-larvae, and there is only one *Megarhinus* larva in each container!

When it is noticed that only one larva of a kind is found in each cavity it is highly significant. We shall note the same thing in rot-hole breeding horse-flies: the carnivorous larvae of *Hinea* occur singly, but the vegetarian *Sphecodemyia* come in batches.

Other larvae that are known or suspected to have carnivorous habits are *Sabethes* in Bromeliads, the Malayan *Zeugnomyia* in the water collected in large fallen leaves, *Armigeres* in bamboo-cavities, *Culex* (*Lutzia*) in ground-pools chiefly, and even *Psorophora* and some *Aedes* when these occur in temporary pools. There is an evident tendency for carnivores to arise in restricted habitats. In temporary pools this might be traceable originally to a shortage of food leading the stronger feeders to eat the others, like castaways in a small boat; but this is not true of rot-holes, where vegetable food is abundant. A more likely explanation is overcrowding, the abundance of animal food offering a temptingly quick, concentrated diet.

Pupae of mosquitoes are characteristic objects, head and thorax rolled into a swollen, disc-like shape, from which the abdomen protrudes like a curved tail. The pupae breathe through ear-like trumpets arising from the prothoracic spiracles and these may be applied to the water-surface, or, in those species whose larvae tap the air-supplies of submerged plants, the pupae may do the same. Just before the adult emerges, the pupa breaks away, leaving the tips of the trumpets still embedded in the plant. A fascinating sidelight on the ways of evolution is presented by the way in which the pupa first manages to push its trumpets into the plant, using the already attached larval skin as a purchase.

Pupae of surface-living mosquitoes are an exception to the rule that most activity that is attributed to the insect pupa is really that of the pharate adult, still in its pupal skin. Mosquito pupae swim actively at any time.

Figure 17. A wingless Empid fly, *Apterodromia evansi*, from New Zealand, showing the simplified thorax of a flightless insect, as well as legs and proboscis suited for predatory habits.

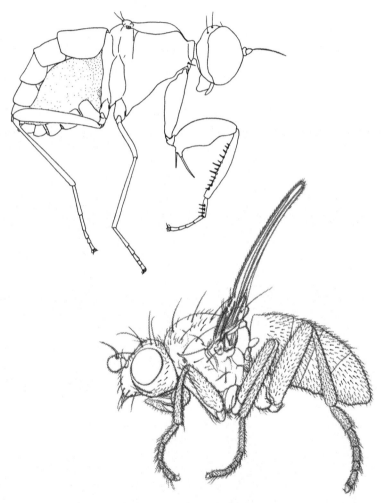

Figure 18. *Anomalopteryx maritima*, an Ephydrid fly with reduced wings that is found on the remote sub-Antarctic island of Kerguelen. Such short-winged or flightless forms often evolve in islands, on mountains, and in other windy places; but even in England we have a Chloropid, an Anthomyzid and a Borborid with reduced wings for no obvious reason.

95

Chapter 8

Horse-Flies

HAVING CONSIDERED all the midges and gnats we are now ready to turn our attention to what I have called the 'flies proper'. We have seen that these are a second main stem of evolution of which the primitive members have many affinities with midges and gnats, while the more advanced families become progressively different, in appearance and in life-history.

At the beginning, the resemblances with the 'watery' section of the midges and gnats are very close. The adult mouthparts are of the same construction, a set of stylets – blade-like mandibles, maxillae and hypopharynx – lying in a grooved labium, with its terminal labella. Indeed Downes has made out a convincing case that both these groups trace back to one bloodsucking ancestor, and that those members of either group that do not now suck blood have lost the habit in the course of their evolution. This fact should be borne in mind all the time that we are considering the Tabanoidea, the group of families centring round the horse-flies. We should compare them with the bloodsucking midges and gnats, as much as with the flies that come after them.

The Tabanidae have many common, though not exactly endearing, names because of the bloodthirstiness of the adult females. Horse-flies, taons, Bremse are names generally used for any member of the family, though they properly apply only to the bigger ones of the genus *Tabanus*. *Haematopota*, called clegs, stouts, breeze-flies, dunflies, whame-flies, burrel-flies and probably many other names in other languages, are ashy grey or brown flies with speckled wings which they hold at rest inclined together like the roof of a house. They arrive silently, and are biting before you have a chance to drive them away. The sinister effect of this silent approach is intensified by the brilliantly coloured eyes, iridescent and shot with zigzag bands of gold, red, and green. The eyes of the males usually meet on top of the head, and often the upper part of each eye in this sex is composed of fewer and larger facets, having a distinct colour-pattern of their own; the lower part of the male eye is exactly like that of the female. This eye-structure possibly has some connection with the swarming habits of the males, and will be discussed again later in that context.

Clegs attack mainly at about waist-height, and often bite the wrist and hands of a standing person. The head and neck are more often the target of the even more brightly coloured female *Chrysops*, which reverse the pattern of *Haematopota*: in place of banded eyes and spotted wings, *Chrysops* has spotted eyes and banded wings, and the body is usually brightly patterned in yellow and black. In North America they are called deer-flies, because they are typically active in the open woodland, with its dappled light and shade, in which deer abound, and are a great nuisance to them. They seem to like to bite man round the forehead and eyebrows, or at the back of the neck, perhaps because the perspiration at these points is attractive to them. They may cause so much swelling above the eyes that it is difficult to see.

Clegs and deer-flies are about 6 to 10 mm long, and altogether more formidable than a single black-fly, or even one mosquito. But the true horse-flies of the genus *Tabanus* start at this size and range upwards to about 25 mm, and are broad and heavy in proportion. When such a fly bites, it makes an appreciable hole, and when it withdraws its stylets a drop of blood usually oozes out. The soft, spongy labella of the proboscis may mop this up, but an animal – or a person – attacked by a number of these big *Tabanus* is soon covered with flecks of blood. Philip estimated that a single grazing animal might lose as much as 100 cc of blood during a long summer's day. It is debatable how much disease is actually carried by horse-flies. Loiasis, a filarial infection of the African rain-forest is certainly transmitted by *Chrysops*, and other horse-flies are suspected of carrying surra, tularaemia and other diseases mechanically; that is, without incubating the organisms before passing them on (see Chapter 19). The practical importance of these flies springs almost entirely from the disturbance, loss of blood, of grazing time and consequently of milk yield, which they produce in pasturing stock.

Stories have come from various parts of the world of the way in which primitive, pastoral people may be forced to modify their habits, and even to move to new areas by the attacks of horse-flies on themselves or on their stock. Usually the latter, because man will put up with almost unlimited discomfort himself, but he is forced to see to the welfare of his herds if he wants to survive. In high latitudes the summer is short, and the biting flies of all kinds are ravenous. Even with twenty-four hours of continuous daylight the flies have their cycles of biting, and some respite can be gained by pasturing animals during what is technically night-time. This practice is not confined to the Arctic Circle. Odysseus, when he came to Telepylus, saw the shepherds who drove their flocks in at night meeting the herdsmen bringing out their cattle to graze. This sentence has been used to take Odysseus through the Pillars of Hercules, and right up to the Arctic Circle, where summer night was non-existent. Professor Pocock has challenged this improbable piece of ancient knowledge. He more convincingly suggests that they were only in Sicily, and that the herdsman had been keeping his cattle away from the horse-flies during the heat of the day.

This might well have been necessary in Sicily in classical times. In the Sudan at the present day Lewis [173] describes how pastoral tribes migrate each year to get their cattle and their camels out of range of the unbearable horse-flies.

The behaviour of *Chrysops*, *Tabanus*, and *Haematopota* has long been known, and is commonly taken as typical of the family Tabanidae, yet in fact these three genera are the *arrivistes* of the horse-fly world. All three have clearly evolved along with the animals upon which they feed, and their history follows closely that of the rise and decline of the hoofed mammals. This is a new feature among the flies we have yet studied, that the direction of evolution should follow adult preferences. Previously the larval feeding habits have been the dominating factor, and they will be so again among the Cyclorrhapha. The increased significance of the adult flies is one of the things that make the Brachycera an attractive group of flies to study.

In particular the clegs of the genus *Haematopota* seem to have followed the fortunes of the Bovidae, the antelopes and the cattle. These are believed to have originated in the Eurasian plains in the Pliocene period, and to have declined there with the glaciations of the Pleistocene, while 'Africa and southern Asia have been havens of refuge from the North Temperate Region' (Romer). Only a few Bovidae crossed to North America, and they never reached either South America or Australia. Since the Pleistocene, the drying up of vast areas, and the rise of man as an enemy have destroyed the bison herds of North America, reduced the wild ungulates of Eurasia, and are now rapidly eliminating the African ones.

Now look at *Haematopota*. The Palaearctic region has about 50 species, the Oriental region about 70, and Africa south of the Sahara about 180. North America has five, only two of which are at all familiar insects; Australia and South America have none, though a species has recently been discovered in Papua, and South America, which has its pampas, has a near relative. The pattern of the two groups, mammals and insects, is very similar, and we can almost feel sorry for the African *Haematopota*. Just when they are blossoming out into a multitude of species, in a great burst of evolution, their food and their *raison d'être*, the game, is vanishing before their eyes.

Tabanus, the very large genus into which most horse-flies are still grouped, is apparently older than *Haematopota*, and more catholic in its tastes. It likes horses as well as cattle, and will also attack such unpromising subjects as elephants and hippopotamus, and reptiles, crocodiles, lizards, and turtles. It should be emphasized that no Tabanid is a specific bloodsucker, confined to sucking the blood of only one kind of victim. These preferences I mention are decided perhaps less by the specific nature of the animal victim than by its habits, the places in which it lives, and the way in which it moves and attracts attention. But it does seem that *Tabanus* has been in evolution longer than *Haematopota*, has spread to all the regions of the world, and has learned to recognize as prey almost any vertebrate animal, except apparently any birds, at least as a regular diet. It would be interesting to know if they ever attack flightless birds such as the ostrich, emu, and cassowary. In quite a different family of flies, one species of the genus *Hippobosca* attacks the ostrich, though all the others of this genus are external parasites of mammals: apparently it treats ostriches as if they were merely feathered relatives of the ungulate mammals with which they roam the African plains.

Chrysops, though its structure shows it to belong to an older branch of the family, is also a genus of recent evolution associated not with open plains, but with broken woodland. 'Deer-flies', as one might expect, are abundant in the countries of the north temperate zone, where Cervidae abound, and have spread into Africa and the Far East, but no further into Australia than the tropical northern tip.

Our knowledge even of these recently evolved horse-flies is mostly derived from the bloodsucking females that come to bite. The males are comparatively little known, because we are still in the group of flies where the males have no mandibles, and can only feed from flowers. Male horse-flies are noted for two habits, for flying high, and for darting down to take water from exposed surfaces. The latter has been described many times, how they swoop down at high speed, strike the surface of the water, and continue on their way up again into the air. No one has yet made a high-speed film to show exactly what happens when they hit the water; presumably they wet the labella and then draw off this moisture during their later flight. They can drink more peacefully by standing on a stone or a stick in the water, but they are then more exposed to attack by enemies, especially by robber-flies (Asilidae) and by *Bembex* wasps. The high-speed drinking act may well be evolved as a protection against such attacks.

The relatively few male horse-flies that are seen in collections were mostly taken while they were sitting on vegetation, fences, posts, and so on. Often, though not invariably, these are newly emerged males hardening off. The best opportunity to get a good series of both sexes of a horse-fly is to be lucky enough to come along just after a mass emergence from the pupa has taken place, usually in the morning. It is then possible to find several scores of specimens, with the sexes in about equal numbers. The flies are sluggish for some time: I once found such an emergence of *Tabanus apricus* near Gavarnie in the Pyrenees, the flies sitting about on umbelliferous plants. When the plants were shaken the flies fell to the ground, struggling helplessly, and unable to take to flight.

Once the flies are fully hardened they scatter in flight. Mating appears to be preceded by hovering of the male, who is presently joined by a female, and the pair flies off to rest on the vegetation. This is a common practice in flies of many families, and the male swarms that we have mentioned appear to be an extension of this, with the advantage that such a swarm is more conspicuous to a female than a single hovering male.

The swarms of male horse-flies proved elusive for a long time. Older authors, such as Brauer, stated that they hovered over forest clearings, and over mountain-tops at dawn, but these statements were not confirmed in living memory until quite recently. Then Haddow and his colleagues [104], investigating the mosquito fauna of the canopy of the equatorial rain-forest of Uganda, found that at their tree-top platforms they caught not only female horse-flies coming to bite their bait animals, but the elusive, long-sought males of some of the commonest flies of the forest floor. Recently they have extended their steel tower even above the tree-tops, and have seen and described the swarms of male horse-flies that occur there, just as they were supposed to.

I shall mention Haddow's work again, later, when we consider the general phenomenon of swarming in flies. It might be mentioned here that he begins to doubt whether swarms exist primarily for mating, as has been assumed. He points out that the number of females to be seen in a swarm is relatively so small that it is difficult to believe that the species can be kept going by this means alone. In his view the swarms seem rather to be a survival of an old habit that may have outlived its usefulness.

Certainly the mass emergence, with equal numbers of males and females sitting about together, seems to offer the best opportunities for pairing. One species of African horse-fly, well represented in collections, is in fact known only from one mass of this kind, which Neave came upon on one occasion. It has never been seen again by him or by anyone else.

Mackerras [188] has recently given a lead in putting these aggressive, modern horse-flies into their true perspective with the rest of the family. It is now clear that almost all horse-flies of both sexes visit flowers for nectar, and specimens in collections often have the head covered with pollen grains. One sub-family, Pangoniinae, have even developed an elongate proboscis with which they suck juices from deep flowers. The best known is the Indian 'Pangonia' (now *Philoliche*) *longirostris*, an insect about an inch long with a proboscis more than twice as long as its body. This magnificent fly hovers over the blossoms of Scrophulariaceae and Labiatae like a humming-bird or a hawk-moth. Yet the female of this fly certainly has powerful mandibles and maxillae and takes a blood-meal. These stylets are very much shorter than the elongate labium, and are ensheathed in it during flower-feeding. To suck blood the long labium is pushed to one side, and the stylets alone enter the skin.

There are all stages of elongation of the labium in this sub-family, which must have developed it as a special line of evolution. Once again, we are mystified to know why they could not get their nectar more easily, especially since horse-flies with a short proboscis have more than once been seen to pick up honeydew from the surface of leaves.

Most female horse-flies have kept both methods of feeding, sugar and blood, but loss of mandibles has occurred rather sporadically throughout the more primitive genera, and in a few of more advanced evolution. This gives an impression that the family might have started out as flower-feeders and later taken to bloodsucking, but in Downes' [67] view all these Brachycera were originally bloodsuckers. In any event Tabanidae that have lost the mandibles have declined on a fully carbohydrate diet, and of course have been unable to resort to bloodsucking because they have no means of piercing the skin. The bloodsucking groups in contrast, have flourished.

One small group of the family has apparently ceased to feed altogether in the adult stage. They are a few small, grey flies, unlike the normal idea of a horse-fly, which are found only on the coast of Brazil, and the eastern seaboard of Africa. Their mouthparts have almost vanished, and if they feed at all it seems to be on the liquid products of the decay of dead animals washed up on the beach – a peculiar and rather macabre diet.

Some 3,000 species of horse-flies are known, yet we know very little about their life-histories, and what is known presents a fairly uniform picture compared with the versatility of, say, the crane-flies. Although they are usually spoken of as an aquatic group they have not moved far from the ancestral home that we have visualized for the flies, the muddy swamp with decaying vegetation. Moreover, they have advanced quite as far towards dry land as towards the water.

Horse-fly larvae are elongate, with a tough, leathery, or 'shagreened' skin, each segment surrounded with a ring of what are called pseudopods or prolegs, though in this family these are rather formless, roughened swellings. The only open spiracles are at the tip of the abdomen, on a breathing-tube or siphon which varies in length in different genera. These larvae are not really aquatic, and have no equipment for breathing continuously under water: if you carry them about in a tube with sand and free water, in constant movement, they will drown.

We know almost nothing about the diet of any except the bigger larvae of *Tabanus* and *Haematopota*, which are carnivorous and very voracious. They will eat any other insects, small Crustacea, worms, snails, tadpoles, and each other. A number kept together in one container will end up as one large larva. Their strong, curved, vertically moving mandibles are very sharp, and can pierce the human skin. They are said on occasion to suck the blood of frogs by rasping the skin, and even once to have injured a toad that misguidedly swallowed one.

The larvae of *Chrysops* do not eat each other, and can be reared easily in a shallow-water aquarium, so that it is inferred that they are vegetarian. Apparently they live on vegetable detritus beneath shallow water, but they do not seem to require organic decay in the sense that many flies do, the decay of the manure-heap, or the foetid water that suits many Psychodids. Crewe says that the larvae of *Chrysop silacea* 'are seldom found in mud beneath more than about two inches of water, and are usually in the top four inches of the mud', and Neave found that the larvae need to come up to the air periodically.

Larvae of *Tabanus* are generally less aquatic than those of *Chrysops*, and occur in the wet mud and sand on the banks of streams, in damp soil away from water, and in rotting wood. The European *Tabanus glaucopis* is particularly associated with chalk downland at an altitude of several hundred feet, and not at all the sort of damp area that one thinks of as typically horse-fly country. Some tropical *Tabanus* live in deeper water, but then use a subterfuge: for example the larvae of *Ancala fasciata nilotica* in Africa, and *Lepiselaga crassipes* in South America have independently learned to support themselves on the broad leaves of the Nile cabbage, *Pistia stratiotes*. The African larvae, and the pupae which live in the same situation, are bright green, with a blood pigment that is carried over into the adult; the South American larvae and pupae are also green, but the adult of *Lepiselaga crassipes* is heavily pigmented with black, and any green pigment there may be is not visible. I have speculated elsewhere about the function of this pigment, and whether it might contain chlorophyll [219]. The larvae of both species are certainly carnivorous, and it seems unlikely that the pigment is merely accidentally accumulated from a vegetable diet. I suggested a possible respiratory function for this pigment, but there is no evidence for this, and a simpler explanation is that it merely gives cryptic protection by making the larvae and pupae less visible against the leaves.

Thus the water-living horse-fly larvae are not at all active swimmers. Those of *Tabanus kingi*, another African species, have particularly well-developed prolegs, and cling to rocks near the surface, but most others are cautious crawlers, like a poor swimmer keeping within his depth, with one foot on the bottom. *Haematopota* larvae are almost terrestrial, and flourish in damp soil. The boggy area at the head of a stream is an ideal breeding-ground for clegs, and so is a typical farm lane, with land-water soaking into the rutted margins.

A number of Tabanines have moved to salt-marshes and to the seashore, and their larvae seem able to withstand brackish water or even salt. The larvae of the Scepsidinae, the tropical, shore-living forms with aborted mouthparts that I mentioned above, live in the sand, and even below high-water mark. Such a larva, found by Dr Callan on the African shore, and which may possibly be the larva of one of this group, has a beautiful disc of branched rays, like a lace collar, round its hind spiracles, which possibly serves the purpose of holding a bubble of air when the larva is submerged by the tide.

A particularly interesting group of Tabanidae is the tribe Rhinomyzini, which extends through Africa, India, and the Malaysian sub-region. They are near relatives of Chrysops, and indeed one genus called *Tabanocella* looks like a large *Chrysops*, and behaves in the same way. But members of this tribe of horse-flies are predominantly breeders in rot-holes in trees. The adults look more like bees or wasps than flies, and mostly stay high up in the trees, seldom coming within the range of collectors on the ground. Many of them are nocturnal. Some of them are vicious biters, but others have no mandibles in either sex; and some of their larvae are vegetarian, and so can be found in numbers in one rot-hole, while others prey voraciously on the other denizens of their tree-hole, and only one larva of the species can be found in one hole. No wonder that the usual adjective for this tribe is 'bizarre'.

The eggs of horse-flies are laid in masses, neatly stacked on end, packed closely together, and sometimes with a second tier on top. They are laid quite close to the larval habitat, attached to plants standing in the water (*Chrysops*), to vegetation over boggy ground, or on the bark beside a rot-hole in a tree. A remarkably isolated piece of behaviour is exhibited by the American *Goniops chrysocoma*, which attaches its eggs to the under-surface of a leaf. The female clings to the leaf with its specially sharp claws inserted into the leaf-tissue, and broods its eggs until they hatch, when the female flies die. Hine says that the vibration of the wings against the leaf makes a rattling noise that can be plainly heard several feet away. Why the females should do this, what advantage is so derived, is not obvious, but there seems to be a parallel with the Rhagionid fly *Atherix*, as we shall see presently.

13. A male adult of the New Zealand glow-worm, *Arachnocampa luminosa*. This is a rare photograph of an elusive fly, but its appearance is typical of land-midges all over the world.

LIVERPOOL JOHN MOORES UNIVERSITY
LEARNING SERVICES

14. Cylinders of mud made by two African horse-flies. The larva cuts off the cylinder by making the spiral groove S, then enters through the hole E. When the adult is ready to emerge, the pupa – or more properly the 'pharate adult' – pushes its way out of hole H.

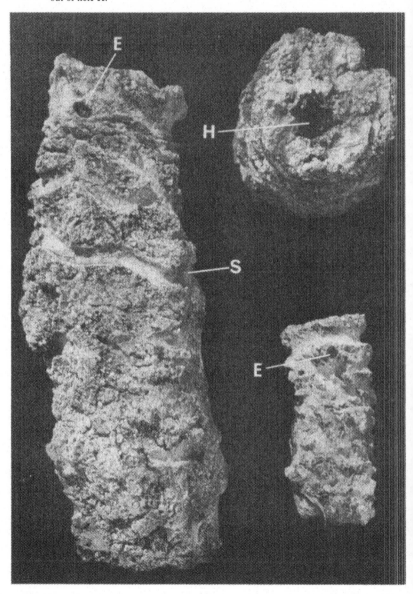

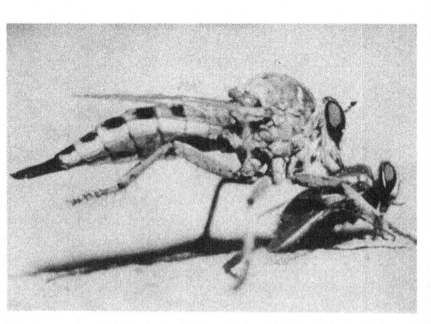

15. Robber-flies: (a) holding a blow-fly and sucking it dry;
(b) newly emerged from the pupa seen on the left. The empty
pupal skin has been removed from the ground to show its powerful
spines and thorns.

16. A female bee-fly on a composite flower.

The length of larval life in Tabanidae is generally one year, but is subject to a very wide fluctuation. It is difficult to know what happens in nature, but it seems as if there is a basic natural cycle. If the larva is not fully fed when the time comes to pupate it may postpone its development for a whole year, and even occasionally for a second time.

The pupae are generally to be found in drier situations, close to the larval habitat, though those of *Ancala* and *Lepiselaga* as we have seen are fully aquatic, and have the thoracic spiracles drawn out into respiratory 'ears', something like those of mosquito pupae. Most horse-fly pupae are to be found in more or less dry soil, round the edges of ponds, or further afield. These terrestrial pupae are very easily drowned, and one of the methods of controlling *Chrysops* when it breeds in isolated ponds is artificially to raise the level and so drown the pupae in the soil at the margins.

A very remarkable adaptation is shown by some tropical species that spend their larval life in mud which is liable to become dry and packed in the long dry season. When the mud is at just the right consistency, the larva descends in a spiral, and cuts off a cylinder of mud from 2 to 5 in long, and 1 to 2 in in diameter (plate 14). The larva then enters the cylinder near the top, and moves down inside it as the surface mud dries out. When the pool is baked hard, and split by deep cracks, the cylinders can be seen as neat circles, each later with a hole in its centre through which the pupa has pushed itself, so that the adult fly could emerge. The remarkable thing about these cylinders is that they are not made by a rare fly of peculiar habit, but by common species of wide distribution. They are already known from two such species, and it is likely that the habit is more widespread. The same flies happily breed in places where the water does not dry up, and in loose, sandy soils where a cylinder of this type could not be formed. Does the larva always try to make a cylinder, and fail except in the rare event that the mud is just right: or is this impulse latent until it is mysteriously called forth? And how did it ever evolve, when by no effort of reasoning can we see it as being essential to the survival of the species?

There is normally one generation a year, and in temperate countries the adult flies may be on the wing for only a few weeks in the spring or summer. In the tropics adult horse-flies may be seen all the year round, yet observation shows that each species has a 'close' season, usually of about two months, during which only a few adults are on the wing. It is not clear what immediate factors cause this seasonal fluctuation, especially in rain-forest areas where superficially conditions seem to be uniform all the year round. Even in rain-forests the levels of streams fluctuate: torrential thunderstorms, especially at the end of the rainy season, put the streams into spate, sweep away the marginal mud in which the larvae live, and probably drown the pupae that are normally above water-level. This may cut the population of adult flies over a period of several weeks.

Although Mackerras laments that the many energetic collectors in Australia surprisingly '. . . wrote so little about what the insects did', yet nearly all our knowledge of the more primitive genera of Tabanidae comes from that region. One reason is that the early spread of the family was apparently a southern one, but not a little of the credit is due to Miss Kathleen English, who has patiently collected and reared some of the most interesting horse-flies in the world. It is significant that none of them is aquatic, thus supporting my earlier contention that horse-flies are inclined as much to the land as to the water.

Larvae of *Scaptia vicina*, a member of a large and primitive genus that flourishes in Australia and in South America, were found in leaf-mould and decaying vegetation under trees in a partly built-up area near a town. *Dasybasis orarius* in New South Wales and the closely related *Dasybasis rubricallosus* over in New Caledonia have spread to the sand of sea-beaches, showing that this step has been taken independently by several different groups of horse-flies: some Mediterranean species of *Tabanus* breed in beach-sand, as well as those peculiar Scepsidinae from the African and Brazilian coasts that I have previously mentioned.

Miss English discovered two life-histories of particular interest to us. *Scaptia muscula*, another member of that large genus, lives as a larva in the pits of ant-lions, to which the horse-fly larva must play the part of 'ant-jackal'. Ant-lion larvae are tremendously voracious creatures, with large jaws, and lie in wait at the bottom of their conical pits of fine sand, to devour any insect that is careless enough to fall in. As we shall see, the family of flies called Rhagionidae has developed a group of which the larvae imitate the pits of ant-lions, but the intrepid horse-fly larvae is to be admired for actually entering the pits of ant-lions and sharing the spoils.

Miss English also discovered horse-fly larvae in rot-holes in trees that formed a thin peripheral canopy over the bed of a creek. These were river oaks, *Casuarina cunninghamiana*, and the cavities were started in the first instance by the attacks of the longhorn beetle *Agrianome spinicollis*. They were not filled with water, but held a moist sticky mixture composed mainly of the chewed wood produced by the beetle-larvae, and extending into the tunnels which led away from the original cavity. The horse-fly larvae were found in the tunnels as well as in the cavity, and once as many as 250 were found together. It is evident that the larvae feed on vegetable detritus and not on each other. Miss English was able to feed them on soft-bodied pupae of the flour-moth, *Ephestia*, but they would go for long periods without food. One pupated after eleven months, and one after twenty-three months, thus supporting the general view that Tabanids can postpone pupation for a whole year if they miss the first chance.

The very few adult flies reared from these larvae proved to belong to a new species, *Chalybosoma casuarinae*, of which no wild adults have ever been seen. The whole story, therefore, is an exact parallel with the African *Sphecodemyia*, which Lamborn found and reared years before a wild adult fly was seen. In each case the adult fly is conspicuous: *Sphecodemyia* is bright orange and brown, like a hunting wasp; *Chalybosoma* is metallic blue, like a bluebottle. It is clear that the adult flies must spend their lives up among the trees, out of sight from the ground. Yet these two genera of horse-flies belong to quite different sections of the family, *Sphecodemyia* to the Rhinomyzini, related to *Chrysops*, and *Chalybosoma* to the Diachlorini, a precursor of *Tabanus*, and so these two must have evolved in a similar way quite independently in different continents.

Arboreal Diachlorini occur in the forests of Central and South America, but as far as I know none have been reared, and no rot-hole breeders have been found, though they most probably exist there; nor have the rot-hole breeders of the Indo-Malayan region been investigated.

Chapter 9

Snipe-Flies and others

NONE OF THE OTHER families of this group has attracted as much attention as the horse-flies. The wood-boring Pantophthalmidae of the Brazilian forest have been mentioned already as an archaic survival. They are very much like huge horse-flies, and the life history of the Australian species that we have just discussed suggest a way in which they may have evolved from some rot-hole-living ancestor. A full account of this family was given by Thorpe [289].

This group of families, like those we have discussed previously, has its evolutionary 'brushwood', a term that is an unintentional pun in this connection, because the flies concerned are to be found among trees, and their larvae, such as are known, live in rotting wood or in leaf-mould. There are about a dozen genera, with one or two species each, that do not fit well into the larger families, and which are now usually lumped together for convenience as the family Coenomyiidae. *Coenomyia* itself is an orange fly, like a moderately big horse-fly, but without biting mouthparts, its carnivorous larva living in leaf-mould and similar debris. It is distributed round the North Temperate Zone, and is clearly an evolutionary relict, though by no means extinct. The others are rare, very local, and specialists' fare, except for the genus *Pelecorrhynchus*, conspicuous, brightly coloured Australian flies that were formerly considered to be primitive horse-flies. For technical reasons they are now removed from that family, though clearly they are a sidebranch of it.

A second group of primitive genera are assembled into the family Xylophagidae, although their wood-living larvae are carnivorous, and do not feed on the wood itself as the name suggests. Biologically they are quite similar to *Coenomyia*, and their interest is as offshoots, 'trial lines' if you like, of an evolutionary stem from which arose the two families Rhagionidae (Leptidae) and Stratiomyidae, which we must now consider.

Rhagionidae are a family of flies that are seldom seen by anyone except a specialist collector. Most of them are solitary and furtive, and those that are gregarious assemble only in certain places, and for a very short period. They are interesting to the student of flies because they stand at the base of the stem of the evolution of all Brachycera, and are comparable in several ways with the family Anisopodidae of the Nematocera. They are called 'snipe-flies' in the books, but I do not believe that any layman knows them by this name.

Like Tabanidae, Rhagionidae are not in origin an aquatic family, though some of their more conspicuous members live near water and have aquatic larvae. Most of them live as larvae in decaying mud, leaf-mould and soil, and are to some extent carnivorous. *Chrysopilus* is a pretty, delicate little fly, with clear, iridescent wings and often golden or silvery scales on some part of the body. The adults flit about among damp vegetation of the dense, shady type that grows on the edges of woods and the margins of streams. *Rhagio* itself, the 'downlooker fly', is usually to be found sitting on tree trunks or posts, often head-downwards, as if it were looking for a victim. These flies have often been suspected of preying on other insects, but in fact no one quite knows what they feed on. Certainly some genera related to *Rhagio* do suck blood, with a mechanism similar to that of the horse-flies, and with much the same technique.

Bloodsucking by females of *Spaniopsis* in Australia has been observed too often to be doubted, and has been reported to occur also in *Austroleptis*, in the South American *Dasyomma*, and in North America in *Symphoromyia* and possibly some *Atherix*. Professor G. C. Varley told me that he himself had been bitten in California by *Symphoromyia sackeni*. Desportes reported that a species of *Atrichops* sucks the blood of green frogs, and may even transmit a parasitic worm, *Icosiella neglecta*.

It seems, therefore, that in this family, or at least within the sub-family Rhagioninae, there is no clear division between those genera or species that can suck blood, and those that cannot. It is possible that even *Rhagio scolopacea* itself, the furtive 'downlooker fly' may occasionally do so, as was reported long ago by three Frenchmen, Mm. Heim, Leprévost et Gazagnaire. G. H. Verrall, the leading British dipterist of their day, rejected their evidence on the somewhat haughty ground that he had never heard of them, but Séguy reaffirms it, for the equally sound reason that he has heard of M.

Gazagnaire. Whatever the credentials of these shadowy figures, it appears that bloodsucking is a declining practice in this family. If Downes' view is to be accepted, then bloodsucking is a primitive habit, and its loss an evolutionary advance, and this is supported by the sporadic way in which it persists, remaining as a regular habit only in Australian flies that probably represent a primitive line, as they do in Tabanidae.

It does appear that Rhagionidae have not blossomed into any modern, voracious, bloodsucking lines to compare with *Chrysops*, *Tabanus*, and *Haematopota*. Furthermore, in this family we see the last of the old ancestral habit of piercing and sucking blood with the help of stylet-like mandibles and maxillae. When we meet blood-sucking flies again, they will have taken up the habit afresh, and, having lost their mandibles and maxillae, will have had to adapt their remaining mouthparts for piercing the skin. The new styles of blood-sucking will be practised equally by both sexes, not by females alone.

No one really knows to what extent Rhagionidae use insect food in place of a blood-meal, and how far the adults feed on nectar and pollen. Certainly a whole sub-family, the Vermileoninae, have gone in for flower-feeding, and have developed a long, slender, probing type of proboscis such as is found in many flower-feeding insects. In particular their long labium, ending in labellar lobes with branching tubular pseudotracheae, is very reminiscent of horse-flies of the sub-family Pangoniinae.

Turning now to the larvae of Rhagionidae, comparatively little is known about them, but what is known supports the view that the family is primitively terrestrial. The larvae of *Rhagio* and of *Chryso-pilus* live in damp soil, leaf-mould, rotting wood, those media that harbour so many dipterous larvae. The larvae have a small head, but well-developed mouthparts, and are said to be partly or wholly carnivorous. The posterior end of the body is drawn out into lobes that close over the hind spiracles like a flower: a very similar structure to that developed by the soil-living larvae of Empididae and Dolichopodidae, as we shall see later. The distance apart of these families, and the lack of any other direct affinities, suggest that this is an adaptive device that in some way protects the spiracles, though how it may do so is not known.

From this central, soil-living type of larva, the Rhagionidae have gone to two extremes. The extreme of terrestrial life is seen in the remarkable larvae of the sub-family Vermileoninae. We have already met the adult flies, with their long, flower-feeding proboscis, which is accompanied by a general elongation of body and legs to a shape very much like certain robber-flies, the genus *Leptogaster* and its allies. Each of these groups has been rejected at times from its family by enthusiastic reclassifiers, and it has even been suggested that they might be placed together, but their similarity of form is clearly no more than convergent.

The curious name Vermileoninae has nothing to do with colour, but is derived from the generic name *Vermileo*, which means 'worm-lion'. This is a clumsy parallel with the 'ant-lions', the genus *Myrmeleo* of the Order Neuroptera, the larvae of which live in pits in sand or fine dust, and there trap ants and other small animals. The maggot-like larvae of the flies *Vermileo*, *Vermitigris*, *Lampromyia*, and other related genera have a very similar habit to the ant-lions, and they must have evolved it quite independently: a remarkable example of convergent evolution.

The pit is conical, the sides steep, and any small animal that once ventures on to this minuscule scree starts to slip, and sends down sand grains which warn the predatory larvae below. The *descensus Averno* is speeded by flinging sand towards the disturbance, and helping to start an avalanche.

Apart from the interesting behaviour of these larvae there are two points concerning their physiology that call for comment. One is that they must be particularly well protected against loss of moisture, or the hot, dry sand would quickly desiccate them. The other is that this entirely carnivorous larval diet is developed in a sub-family in which the adults have turned over completely to taking carbohydrates from flowers. An interesting speculation is to wonder which adaptation came first, and required the other in compensation.

An equally interesting development in the opposite direction has taken place in several genera of Rhagionidae, which have moved towards the water. These aquatic flies are also socially inclined, and several species of the genus *Atherix* – notably *A. ibis* in Europe, and *A. variegata* in North America – practise a communal egg-laying that is spectacular. They lay their eggs in masses on leaves overhanging a running stream, and each female remains clinging to the leaf while others come and cluster on top of her. Thus eventually a mass of immobile, dead or dying female flies takes the shape and appearance, though not the numbers of a swarm of bees. It is on record that the North American Indians sometimes ate these flies, catching them by first building a dam of stones a little way downstream, and then beating the branch until the flies fell into the water, to be stranded on the stones. They were then scooped out, cooked and eaten.

The larvae of *Atherix* are conspicuously adapted for aquatic life, with eight well-developed prolegs on the abdominal segments, and a pair of pointed processes, with long hairs, at the tip of the body.

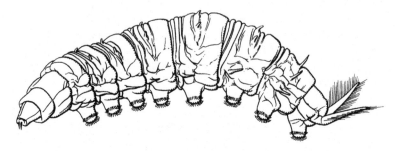

Figure 19. Larva of the Rhagionid fly *Atherix*, another type of underwater crawler.

These larvae live in rotting wood, either floating or submerged, and in the mosses growing on fixed structures in the water. They are clearly adapted for at least a limited amount of locomotion under water, and breathe by means of a single pair of posterior spiracles concealed in a cavity beneath the terminal processes of the body. Of course they are air-breathing, but probably the bubble of air trapped in this hairy region round the spiracles will last them quite a long time when submerged, using it as an underwater lung in the manner we have met before.

The clustering habit of ovipositing females is developed only in certain species, while others closely related, such as *Atherix marginata* in Europe, live a similar life with solitary or only mildly gregarious egg-laying. Recent work by Dr Pleskot has shown that, at least in *Atherix marginata*, contrary to earlier belief, the very young larva may continue to live on the leaf, and fall into the water only when it is ready to change to the second-stage larva. The first period of larval life is always a particularly precarious time for any insect, and it may be that this 'delayed drop' as it were has reduced wastage at this early stage. We may perhaps see the clustering of females as a further evolution in the same direction, giving the very young larvae maternal protection to the death – a noble, if unconscious sacrifice. Stuckenberg [277] mentions various degrees of association in other species of *Atherix*, *Suragina*, and his new and structurally very remarkable genus *Trichacantha*. He calls them 'subsocial', a delightful term that might be applied to some human beings.

In a series of papers in the journal *Mushi*, from 1958–62, Nagatomi has given a very detailed account of the biology and early stages of the aquatic snipe-flies of Japan, giving much precise information for the first time [213].

The fact that the clustering habit may be highly developed in one species, and scarcely or not at all in the next makes one wonder if the behavioural impulse is latent in some species, and may not always be called forth, as we have already seen to be the case in the formation of mud-cylinders by certain African horse-flies.

Finally in this group come the Stratiomyidae, a large but enigmatic family. In English books they are called 'soldier flies', but this can hardly be called a common name, since the flies are seldom noticed except by the collector. The name is sometimes said to derive from the way in which the thorax is often armed with spines, and sometimes from the brilliant colour and pattern of the bigger species, but this is rationalizing after the event.

The adult flies do not bite, nor prey on other insects. They are said to be flower-feeders, and a few like the genus *Nemotelus* do have the face drawn out into a peculiar little snout like a tapir's head. Since the antennae are carried well forward on this snout it is doubtful whether it gives any help in probing into blossoms. In fact the activity most characteristic of Stratiomyidae is just resting, if that is not a contradiction in terms. They are seldom seen in flight, and give one the impression that their adult life consists of little beyond mating and laying eggs. We have seen that this is true of some other flies, notably non-biting midges, but these are soft-bodied, fragile flies, dull in colour, 'utility models', as it were, with no ornament that is not strictly necessary for their limited purpose. Stratiomyidae go far beyond this.

The biggest adults are about 20 mm long, oval in outline and with a hard, shining skin that is often highly coloured. Some are patterned in yellow and black stripes; some with the pale parts green; yet others metallic blue, green, or purple. In addition, the scutellum, the little shield forming the posterior part of the thorax, may be armed with strong spines, and others may be on the thorax proper. Finally, the veins of the wing are characteristically clustered towards the leading edge, in a way that would seem to make this part of the wing very rigid, and leave the rest of the wing more than usually flexible.

If the adult flies do nothing, why have they evolved such an elaborate structure? This is a recurrent question in zoology, and there is no complete answer. A similar type of colour and pattern will be seen again in adult hover-flies of the family Syrphidae, and there it does seem to have a mimetic explanation. Many of the flies of that family, which look like bees and wasps, live in association with these dangerous insects, and would benefit from the deception. But Stratiomyidae have no such ties.

An observation of my own may be pertinent here, as an example of how careful one should be when speculating about the significance of colour and pattern from dead specimens in museums. *Platyna hastata*, an African Stratiomyid has a particularly broad, banjo-shaped body, down the middle of which passes a broad stripe of silvery pile. When the head of the insect faces the light the silvery pile is conspicuous; when the light comes from the rear the pile ceases to reflect it. I saw a number of these flies dancing in patches of sunlight along a forest path in West Africa, and as they twisted and turned, so the twinkling specks of silvery light came and went in the air. I caught some, and the others settled – and vanished from sight. When they came to rest they closed the wings over each other like a pair of scissors, and in that position the wings exactly covered the silvery stripe, so that the whole fly became dark, and quite inconspicuous against a leaf.

Now this scissor-like closing of the wings is general in Stratiomyidae, and it may well have a survival value in concealing the more conspicuous elements of the abdominal pattern, thus attracting less attention. If that is so, then the pattern is probably not mimetic, but it may help in recognizing other members of the same species. Dancing flights of males are common in Stratiomyidae, at least among the smaller members of the family, only a few millimetres long. I have seen the males of *Beris morrisii* dancing in a little swarm under a tree in a country lane. Perhaps the brilliant colouring of these flies is a help during mating flights, and is covered up by the wings when at rest.

There are about 1,500 species of Stratiomyidae, distributed throughout the world in an inconspicuous way. The distribution is remarkably uniform and when the relative areas are taken into account the Stratiomyidae of the tropics do not seem to be either more numerous or more striking than those of temperate regions. From this we may infer that the family has existed for a long time, long enough to spread itself uniformly around the world, and consistent with this view are two small aspects of distribution that do come to mind. The islands of 'Wallacea', the Far Eastern archipelago where Wallace collected and drew his celebrated deductions about geographical distribution, have a great variety of genera and species of the tribe Pachygasterini, which must have had time to evolve there in geographical isolation: and in Australia there are some of the most primitive of all Stratiomyidae, indeed of all Brachycera, elongate, often wingless, belonging to the tribe Chiromyzini. Lastly, Stratiomyidae are notable for having a large number of genera, with only a few species in each, and there are no huge genera comparable with *Tabanus*, *Haematopota*, or *Chrysops*, each with its multitude of still-evolving species. Stratiomyidae seem to have made their great effort some time ago, and now to be marking time, if not actively declining: possibly in competition with the more advanced hover-flies of the family Syrphidae.

All this is relevant to the question, perhaps the most interesting question about Stratiomyidae – where do they fit in? It is customary to list them as the first family of the Sub-Order Brachycera, which gives them a special prominence comparable to that of Tipulidae among Nematocera, but they do not sustain a key position so well. In adult structure they are peculiar in several ways, and even their larvae have followed a line of their own, different from that of other Diptera.

Stratiomyidae, like Tabanidae, are usually thought of as having aquatic larvae, because this is the habit of their largest and most conspicuous members, *Stratiomys* and *Odontomyia* (*Eulalia*). These larvae are elongate, torpedo-shaped, with a very small head that cannot be fully retracted, and posteriorly with a long narrow siphon. There are spiracles both on the thorax and at the tip of the siphon, but the latter pair are the effective ones when the larva is submerged. The siphon reaches to the surface, and has a rosette of fine hairs which help to form a depression in the water-surface, connecting the spiracles with the air. When the larva submerges completely the rosette of hairs folds over and encloses an air-bubble, the familiar external lung that we have seen before.

These larvae are carnivorous, having powerful mandibles and feeding on worms, crustacea, and small insects. They thus compare with the carnivorous aquatic larvae of the Nematocera, but for the fact that the mandibles work up and down instead of cross-wise. This is a small difference, yet it is characteristic of the larvae of the two lines of evolution of flies. It is interesting to note that this horse-fly group of families is biologically linked with the preceding Nematocera, yet in structure it clearly belongs to the other line of Brachycera/Cyclorrhapha.

The larvae of most Stratiomyidae are neither aquatic nor carnivorous. They live in decaying vegetable matter, from wet moss on the margins of streams to soil, leaf-mould, and rot-hole debris in trees. Some, no doubt, eat the vegetable fragments, and perhaps incidentally devour any small organisms that they may chance to meet, in the manner that we have seen in some midge-larvae, and shall see again in those of the robber-flies. There is also a widespread habit of scavenging animal debris, and Buxton said that the Stratiomyidae he met in Samoa devoured the remains of plants and animals after the maggots of house-flies and blow-flies had ceased to be attracted, and thereby '. . . compete with mites and the larvae of beetles, rather than with the larvae of most other Diptera'. This at once brings to mind the family Phoridae, in a connection that we shall see in a minute.

One of the species concerned was the almost cosmopolitan *Hermetia illucens*, a big fly with a cylindrical abdomen which bears a curious paired 'window' of translucent cuticle from which it gets the name '*illucens*'. Its larva is sometimes accidentally swallowed with contaminated food, and so is an agent of *myiasis* (as discussed in Chapter 19). The larvae of *Potamida ephippium* penetrate into the nests of the ant *Formica fuliginosa*, and scavenge there. They are exceptionally long-lived, and are said to spend four years as larvae. Longevity is an asset if it can be invoked under adverse conditions, but if it is habitual it greatly limits the possible rate of evolution by adaption. The insects that evolve quickly are those, like *Drosophila*, that can rush through a generation in days rather than weeks, months, or years.

A remarkable feature of many, perhaps of most, larvae of Stratiomyidae is the toughened skin, which is 'shagreened', that is broken into a roughened surface like sharkskin. This effect is produced by the deposition of calcium carbonate in the cuticle. I do not know that anyone has suggested any possible function for this. It is most heavily deposited in the larger Stratiomyidae, those torpedo-shaped larvae that live round the margins of ponds and streams. It may possibly be an asset if the larva dries up, and these larvae can withstand a high degree of desiccation. I have received for examination *Stratiomys* larvae that have been in the mails for several days, and were stiff, brittle, and apparently dead. They looked as if they could be broken in two easily. Yet, put back into water, they became soft, and soon began to move sluggishly. It is possible that this calcareous layer is a sort of built-in cocoon with the aid of which the larva, during its long life, can retire from adverse conditions in the world outside.

This deposit of calcium carbonate can be dissolved out in dilute hydrochloric acid, and one immediately wonders what can be the effect of acid water on it, and whether the larva is thereby restricted in its choice of water to live in. Apparently not. Brues found Stratiomyid larvae in waters down to pH 5·7, as well as in hot springs which usually contain various acids of sulphur.

Stratiomyidae are unique among Brachycera in pupating within the last larval skin. Before pupation the larva often moves out of the water, or other wet surroundings, into a drier place, soil, or leaf-mould, but some aquatic larvae remain in the water and form a pupa there. It is smaller than the tough larval skin, which thus has an air-cavity, causing it to float to the surface. Remaining in the water avoids the risk of desiccation, and the consequent difficulty of splitting a very tough, dry skin; but on the other hand, exposes it to the many predators that look for floating food, and also to the hazards of the current, to the danger of being carried helplessly along, and possibly trapped in a place where it is difficult for the adult to emerge.

The trick of using the last larval skin as a puparium, or shelter for the pupa, is characteristic of the Sub-Order Cyclorrhapha, and a few dipterists have tried to visualise the Stratiomyidae, not as primitive Brachycera, but as advanced ones moving towards the Cyclorrhapha. I do not think that this view can be substantiated. This use of the last larval skin occurs in the wood-living *Solva* (*Xylomyia*), which is quite clearly a link with other primitive Brachycera, and the habit even occurs sporadically in some Nematocera, as we have seen. Indeed, there is no reason why the same habit should not have arisen a number of times during the evolution of Diptera. We have seen that changes in larval and pupal form are often brought about simply by a shift of timing, and it would seem a relatively simple step to fail to break out of the last larval skin, if this results in protection, and gives a better chance of survival.

Chapter 10

Robber-Flies

AT THIS POINT in our progress from the primitive flies to the most advanced, we come to a sharp break in continuity. So long as we were discussing horse-flies and allied families we were constantly looking back, making comparisons with the midges and gnats, and especially with those that supplement the diet of the adult females with the blood of vertebrate animals.

The next group of families, which make up the rest of the Sub-Order Brachycera, owe nothing to the midges. Nor, with one exception, do they have much in common with the flies that come after them. They stand sturdily on their own, flies that are mostly robust, highly varied and individual in appearance, a group in which the adult is at least as interesting as the larva, and often more so. Perhaps the most conspicuous feature they have in common is that either as a larva or as an adult they obtain animal protein by preying upon other insects, or upon spiders, or by parasitizing them, or at least by scavenging among their leavings.

In the normal way this group of flies do not suck the blood of vertebrates, though some of the biggest members have been known to pierce the human skin in self-defence. Many of them regularly feed by piercing the integument of insects. They cannot do this by using the mandibles and maxillae, for they have lost at least the mandibles and sometimes both pairs of stylets. The commonest weapon is the hypopharynx (figure 21), the single, unpaired stylet that carries the salivary duct, and they may push this in like a hypodermic needle. Two families at least rely upon this equipment for their principal food-supply, the body-fluids of other insects.

This group is dominated by two large families, the Asilidae or robber-flies, and the Bombyliidae or bee-flies. Robber-flies are carnivorous and rapacious as adults, and their larvae are believed to be mainly vegetarian; bee-flies take their animal protein as larvae, by preying upon or parasitizing other insects, and then mature into harmless, nectar-feeding adult flies. These two families set the fashion, as it were, for the group, and the other families can be similarly divided. The group as a whole is a terrestrial one, and its members have little or nothing to do with water until at the very end the two families, Empididae and Dolichopodidae, have both evolved aquatic. members; but we shall come to them later.

Robber-flies are among the most fascinating of all flies, except presumably to other small insects that are about to be snapped up by them. I commented early in this book on the way that some flies,

Figure 20. A hind leg of a robber-fly of the genus *Lagodias*, with long scaly fringes on the tarsal segments. The function of these is not known, but possibly they may be used as an air-brake in a darting flight after prey.

generally those of the most ancient families, are strikingly adapted to the demands of the life they lead; whereas others, including almost all of the more recently evolved families, like the house-flies and blow-flies, have a generally utilitarian shape that is suitable for any way of life. Robber-flies are specialized hunting machines *par excellence*. Look at plate 15 and figure 21, and see how much could be learned of the ways of the fly by studying them feature by feature.

Note first of all that the structure is virtually the same in both sexes, except for the organs that are adapted for mating in the male, and also for egg-laying in the females. A few robber-flies have the two sexes differently coloured in the abdomen and perhaps the wings, and a very small minority are so different that male and female might be taken to be different species. As a general rule, however, both sexes look alike, and behave alike, have the same eyes for finding prey, and the same legs for grasping it. They live by chasing and intercepting other insects, usually in flight, grasping them in their powerful legs, and sucking them dry before dropping the empty skin and looking round for another victim.

The eyes are huge, bulging forward and upward, and separated on top of the head by a deep trough that is a feature of robber-flies. The head, made up in this way principally of the eyes, is relatively short and broad, and the facets that point forward are often fewer and bigger than those round the rim of the eye. To all appearances the fly has the advantage not only of binocular vision both forwards and upwards, but also of a degree of stereoscopic vision too, because the eyes are always far enough apart to indicate distance according to which facets of each eye can see the prey.

Experimental work such as the classic observations of Melin [203] suggests that robber-flies detect moving objects and estimate their distance, but that they are not very good at recognizing another insect in the sense in which we would use this term. It seems likely that they are attracted by the movement rather than by the detailed appearance of the prey, and will chase anything that moves in a way that they associate with food. We ourselves recognise people by their movements while they are still too far off for us to see their features: ''tis Cinna, I know him by his gait'.

Melin found that at close quarters the flies made capture darts with more confidence, and sometimes corrected an earlier mistake, deciding to ignore an unsuitable object such as thistledown. This does not necessarily mean that they see a clear image even at that distance, but that the movement covers more facets of the eye, and is more easily recognised as lifelike or not.

Robber-flies are excessively bristly, and there can be little doubt that these bristles are at once a weapon and a necessary defence. This is especially true of the head, where the narrow strips that are not occupied by the eyes are generally covered with a forest of hairs, interspersed with strong bristles. Like the Viking pirates of old, robber-flies have both moustache and beard (figure 21). The middle of the face, the strip along the rim of the mouth and below the antennae, is nearly always covered with a bristly mass, which extends beside and beneath the base of the proboscis. These are not merely decorative, but serve the very real purpose of keeping the struggling victim, impaled upon the proboscis, away from the vulnerable eyes of the attacking fly. This protection is enhanced in many robber-flies by having the moustache pushed out on to a raised protuberance standing out between the eyes.

The proboscis itself is generally simple, straight, and only moderately long, never developed into a flower-probing type like that of the pangoniine horse-flies, or of the bee-flies that we shall meet later. In a few genera it is heavily sclerotized, triangular in cross-section, and even curved upwards like a scimitar, with the sharp-pointed hypopharynx sliding in and out at the tip in a wickedly effective fashion. Such flies as these last tend to attack heavily protected insects like beetles, and no doubt need all the strength that they can muster.

When a robber-fly intercepts a flying victim this is pounced upon in the air. The legs of the robber-fly are always well equipped with long and particularly strong bristles, and in flight the first two pairs of legs hang down with their bristles overlapping, making a trap which closes round the prey and holds it close to the proboscis. The

Figure 21. Head of the robber-fly *Proagonistes athletes*. Here the dense bristles probably protect the face and eyes when living prey is impaled on the powerful proboscis. Compare figures 2, 4.

hind legs are nearly always stronger than the others, and the femora and tibiae often close on to each other like a pair of nutcrackers, making an effective clamp for further controlling a struggling victim. The hind legs of certain Asilidae, notably *Leptogaster* and *Dioctria*, are long and slender, and more helpful in holding on to a support during feeding than on to the victim itself.

Most robber-flies are powerful fliers, as they need to be. Generally they sit about at some point of vantage – *Dioctria*, for instance, near the tip of a twig on a low shrub – turning the head from side to side, moving the legs, and even raising the body as if peering round for a victim, as indeed they are. Some make exploratory flights. A tropical *Promachus* that I watched for a long time made regular patrols, and each individual had its 'beat' along the secondary vegetation on a bushtrack. The fly would settle for a few minutes, usually out of reach among the branches of a shrub, and would dart out if anything happened to pass by. If nothing came near, after a short time the robber-fly would take wing and fly in a straight line along the edge of the path, and then settle again. If I stayed in one place it would return periodically in its regular patrol.

On the other hand the small robber-flies seem to have less surplus energy, and rely more on concealment than on active search. This is particularly true of the fascinating little *Stichopogon*, tiny grey flies, distinguished by the unusually broad-saddle-shaped gap between the eyes. Presumably this well-marked structural feature must be correlated with some optical peculiarity, perhaps an enhanced stereoscopic effect, or, from their behaviour, perhaps an ability to see in very intense light. These flies habitually sit out in full sunshine, on dry rocks in a stream-bed, or on the open sand, where they are quite invisible in the brilliant glare, as long as they keep still. The only way to collect them under these conditions is to wait until the fly makes a capture-dart, and then put an inverted net over the spot where the fly is seen to alight, only to disappear at once from view. Even then the fly is often cunning and will 'go to ground', refusing to rise into the net.

The slim, elegant *Leptogaster* is unusual in reversing the general practice. Instead of sitting and waiting for a moving prey it makes scouting flights, and pounces on stationary victims including such an unlikely target as a spider sitting on a stem of grass. In this case the victim cannot be recognised by any characteristic flight, and it seems as if *Leptogaster*, at least, must have a more accurate vision than we have so far allowed to robber-flies, with the ability not only to recognise an immobile creature, but to pick it out from a confusing background. This is really remarkable, because if we photographed a mass of vegetation at close range, with a spider or a plant-bug sitting on it, we should have difficulty in making the animal stand out from its surroundings, either in black-and-white or in colour. It is possible that the victims do not, in fact, keep still, that, alarmed by the approach of the robber-fly they make quite small movements that are invisible to us, but are seen by the robber-fly at close range. In all questions of camouflage and deceptive coloration we have to try to see the effect, not from our own distance, but with the short-range vision of the predator.

Since robber-flies and their victims are often seen by naturalists, it is not surprising that attempts have been made to classify the prey, and to enquire whether certain robber-flies specialize in certain kinds of victim. Certain tendencies can be detected, but must not be exaggerated. For instance, robber-flies of the distinct genus *Dioctria* of the North Temperate Zone show a preference for catching parasitic Hymenoptera, especially ichneumons. The robber-flies themselves look rather like ichneumons, with an elongate body and legs, and a shining yellow and brown coloration. They also live in the sort of surroundings where ichneumons flourish, grassland with low bushes, and sometimes more definite woodland. There seems to be some degree of mimetic resemblance, perhaps only to the extent that it gives the robber-flies a few seconds' lead before the prey takes alarm.

Hobby [138] records Diptera as well as Hymenoptera among the captures of this genus, and *Diotria atricapilla* took more Diptera than Hymenoptera. So clearly they are not fussy eaters, and do not refuse other food if they can see it and capture it.

Besides being in the right place, potential prey must also in a broad sense suit the size and equipment of the attacking Asilid. On the whole, a big fly finds it easier to catch bigger prey, though there are surprising exceptions to this, usually in the sense of a robber-fly gallantly tackling a big grasshopper, or a cicada, or some other insect bigger than itself. The shape and strength of the proboscis are more severely limiting factors, because a small, weak hypopharynx cannot penetrate the armoured cuticle of a large beetle; and equally, a tropical *Proagonistes* 5 cm overall would find it difficult to catch an aphis, or to pierce it with a strongly curved proboscis itself 9 mm long.

Many years ago, Dr S. A. Neave made a great collection of robber-flies and prey from Mt Mlanje in Nyasaland. A detailed analysis of this collection is still impossible because so many of the flies cannot be identified to the species, but even superficial examination shows that the flies will tackle any prey that they can catch. For example, a single box of the collection contains over sixty robber-flies of the species *Promachus negligens*, together with the following insects that they had seized as prey: five moths, five ants, an ichneumon, four wasps, two bees, four beetles, a dragonfly, an ant-lion, a termite, one large and one small horse-fly, and several other flies, including a smaller robber-fly. This list of prey seems to have no common factor except opportunity, containing as it does examples of large and small, robust and very thin, scaly and bare, active and sluggish, timid and aggressive.

As a striking contrast to this general lack of precision, the genus *Hyperechia* provides one of the most persuasive examples of mimicry in action that is exhibited by any insect. In India and in Africa are to be found robber-flies that are close mimics of the big carpenter-bees of the genus *Xylocopa*. The bees make their cells in burrows in wood, old trees, dry stems, and the structural timbers of fences and buildings. The larvae of the flies live in smaller burrows in the same wood. The adult flies bear an uncanny resemblance to the bees, not only in general shape and size, but in details of colour and pattern, so that each species of *Hyperechia* can be compared with a particular species of *Xylocopa*. The resemblance must have some biological value. It is impossible that two unrelated insects living in such close proximity should have become so much alike by mere chance. Convergent evolution might account for some degree of resemblance, but why should they follow each other in the extravagances of specific colour and pattern ?

The theory of mimicry is one aspect of evolution that has aroused, and still arouses, most furious argument. The simplest picture would be to assume that the *Hyperechia* prey upon the larvae of the *Xylocopa*, and that the extraordinary resemblances between flies and bees enable the flies to come near without being detected, to lay their eggs on the wood. For a long time it was disputed whether in fact the burrows of the fly-larvae ever penetrated into the larger burrows of the bees, and even if they did, whether the one larva attacked the other. It was held that the fly-larvae merely fed on wood, and that they only met the larvae or pupae of the bees by chance, as is still believed to be the story in other genera of Laphriinae. Then it was suggested that the robber-fly larvae merely scavenged, seeking out pupae that had died and begun to decay. Finally, one or two apparently authentic examples were quoted in which a burrow was opened and a bee-larva was found to have been sucked dry by an Asilid.

A similar doubt attaches to the adults. *Hyperechia* adults do take adult *Xylocopa*, but they do not specialize upon the species that they mimic – very inconvenient for mimetic theory! And they certainly take other insects as prey. So it seems probable that the detailed resemblance to the particular *Xylocopa* whose log they share gives the robber-fly some protection against the bee, or against some common enemy, for the selection of pattern to have taken place. Once established there, the robber-fly both as larva and as adult, probably takes any prey that it can find, without caring whether it is the model or not.

The whole arrangement is not essentially different from the habits of Laphriinae in general, the tribe to which *Hyperechia* belongs. These are all rather wasp-like, or more usually bee-like, and undoubtedly gain a practical benefit from looking like such formidable insects. Specific mimicry is not usually as well marked as in *Hyperechia*, but in South and Central America there are the furry, exceedingly bee-like *Mallophora*, and in North America *Dasyllis* and *Bombomima* are close mimics of bumblebees. Some of these do in fact prey on bees, including hive-bees, and so they may be a nuisance to farmers who rely on the bees to pollinate their crops such as alfalfa; but other robber-flies join in the bee-hunt which have little or no likeness to their prey. In all regions of the world the genus *Laphria* has representatives, some of which are just like bees, others that are 'bee-like' only in the most general way. See also Adamovic [317].

Yet the last word has not been said on this matter of mimicry, and unexpected resemblances are always being noticed, often without any known link between the model and mimic.

In conformity with their vigorous and diversified adult life, female robber-flies have exploited a wide range of places in which to lay their eggs, and have developed effective ovipositors with which to do so. Those that just scatter the eggs on the ground are in the majority, and they have only a simple opening for the purpose, sometimes with a downturned flap which serves to deflect the egg and project it more positively downwards, perhaps increasing its chance of lodging in a crack or crevice.

The next step is to cover up the eggs with soil or sand, and some have bristles which help in doing this. *Philonicus*, a common littoral robber-fly of temperate countries, lays its eggs in soft sand, and uses the two long bristles at the tip of the abdomen like a broom, sweeping over the site of the egg. A great many genera of robber-flies have structures called by G. H. Hardy acanthophorites, or spine-bearing plates at the tip of the abdomen. These bear a crown of spines like the collar that used to be seen protecting the neck of a valuable dog. They are used to push aside sand or soft earth, and so to allow the tip of the abdomen to be inserted while the eggs are being laid. When the abdomen is withdrawn, the soil or sand falls back and covers the eggs.

The acanthophorites are found in a number of genera and tribes of robber-flies in such a way as to suggest that these structures evolved only once, and that the flies possessing them are nearly related. If that is so, the method and site of egg-laying must also have been an ancestral one, and so these flies must be one group of earth-breeding robber-flies. Moreover, a very similar structure occurs in several related families of flies, in Mydaidae, Apioceridae and Therevidae, and indicates that these are closely related to the robber-flies, as we also believe from other evidence, to which we shall return later.

Other genera of robber-flies have evolved peculiar forms of ovipositor to suit their various egg-laying habits, but all of these seem to be of more recent origin than the acanthophorites. They are particularly common in the tribe Asilini, the most advanced of robber-flies, which commonly lay in crevices in growing plants. *Machimus*, with a very simple, flexible ovipositor composed of loosely articulated small plates, pushes its eggs into flower-heads of an open and easily accessible type. *Neoitamus*, with loosely jointed segments flattened into the shape of a ribbon, pushes its eggs into deeply recessed crevices of the flowers of trees, and so is a fly of the woodland rather than of the meadow. In *Dysmachus* the last pair of plates is partly recessed into those preceding, and so forms a flat blade, which is used to push aside spikelets of grasses. And the last stage is reached in *Eutolmus*, where the recessed segments are hardened and flattened into a miniature knife, which is actually used to slit a hole in plant-stems, so that a neat row of eggs can be placed within the structure of the plant (figures 8, 9).

These elaborate ways of hiding the eggs have apparently no relation to the requirements of the larvae, but serve merely to protect the egg against being seen and eaten by predators. The newly hatched larvae are very tiny and vulnerable, especially to drying up, so that a long exposed journey would be fatal to them. Fortunately the great majority need only crawl out on to the surface of the plant and let themselves fall to the ground. Most robber-fly larvae live in soil, and are believed to be vegetable feeders, though some authors have claimed evidence that they are occasionally carnivorous. It seems probable that if they meet a larva or a pupa of some other insect in the soil they will munch their way through it if it keeps still long enough, or if it is newly dead.

This is certainly true of the larvae of the tribe Laphriini, which live in tree-stumps and other rotting wood. The eggs are laid on the rough surface, or in crevices, and the larvae have little trouble in making their way inside. They may move under the bark, or follow the burrows of beetles, but they can also make tunnels of their own, like the *Hyperechia* we have already seen. They turn and twist, and scrape with their mandibles, thus hacking out a hole rather as miners used to hack their way through coal seams far below ground. The mandibles may become quite worn during the life of the larva.

An odd thing about the classification of robber-flies is that the genus *Asilus*, typical of the whole family, contains very few species. The big, handsome yellow and black *Asilus crabroniformis* can be seen in European meadows, laying its eggs in the droppings of animals, especially in cow-pats, and nothing is known with certainty about the diet of its larvae. Having regard to their size, over 30 mm long, their powerful mouthparts, the abundance of other insect larvae in the dung, and the common tendency of many fly-larvae to become carnivorous as their appetite grows with size, it seems likely that the larva of *Asilus* is at least partly predaceous. The larva of *Asilus barbarus* of the Mediterranean basin is said to prey on larvae of the Dynastid beetle *Phyllognathus silenus* in the sand or soil.

When an Asilid larva has finished feeding, it transforms into a pupa which shows externally the legs and wings of the adult, but in addition is heavily armed with spines and bristles on the abdomen, and a few, very robust horn-like excrescences on the head (plate 15). This kind of pupae is characteristic of the terrestrial Asiloidea, and when the adult fly is 'pharate', active, but still confined within its pupal skin, the weird creature levers its way to the surface until the front end projects enough to allow the adult to break out and escape. The pupae of the log-living Laphriini can be seen thus, projecting from the surface of logs like miniature periscopes.

Taking the robber-flies as a pattern we can quickly look over four families that are their close relatives. Mydaidae have much the same general appearance as Asilidae, but with a number of structural peculiarities that systematists would regard as primitive. They are like those robber-flies that resemble bees and wasps, and indeed Mydaids fly in a wasp-like manner, darting to and fro close to clumps of vegetation. Surprisingly little is known about the biology of Mydaidae in any respect. The adults have been reported to hunt like robber-flies, but it is suspected that some field observers have confused the two. The proboscis of an adult Mydaid seems suited to sucking but not to piercing, and it may be used to feed from flowers.

The little that is known about their behaviour would be consistent with this.

There is more evidence that the larvae are carnivorous. Larvae have been reported from old tree-trunks in the manner of Laphriine robber-flies, and have been said to feed on the pupae of Lepidoptera, and on beetle-larvae. A South American species is said to be associated with the parasol ant, *Atta*.

What can be said with confidence is that Mydaidae are an old family, on their way to extinction. There are still about 200 species scattered through the world, in the tropics, sub-tropics, and countries with a Mediterranean type of climate. Hennig points out that this even but thin distribution, with the 200 species spread among about 25 genera, is evidence that they are an older family that flourished in its day, and is now relentlessly being run down and extinguished. Why, we do not know, but it has the classic sign of evolutionary senility – a crop of huge species in the South American forests. *Mydas heros* is a giant among flies, reaching a length of 60 mm, with wings broad to match, and magnificently coloured in red and black.

Apioceridae show the other classic sign of antiquity in evolution. They are few and rare, found only in the southern tips of the continental masses, and in South-Western North America. They have the look of an early attempt at a robber-fly. The larval life of *Apiocera maritima* in Australia has been revealed by the patient work of Miss Kathleen English [79]. The larva lives in sand, and behaves much like an Asilid, but it is definitely carnivorous. The larvae had to be kept in separate jars, and could be induced to feed on pieces of earthworm, though they were not successfully reared to maturity on this diet.

The larva of *A. maritima* is unusually long and narrow, with some degree of subdivision of the segments, and seems to form a link with the peculiar larvae of the other two families of this group. Therevidae and Scenopinidae share the honours of having the most unlikely larvae of all Brachycera, and their long, smooth, pale larvae are often mistaken for worms. Therevid larvae are found in soil, and those of Scenopinidae are most often found in carpets! Since carpets can scarcely be their natural habitat we must look for some natural medium of a similarly unpromising kind. It is thought that the debris of birds' nests is the most likely choice, and these larvae are certainly very resistant to desiccation as well as to lack of oxygen.

It seems likely that the Therevidae are the most primitive family of this group, and Paramonov thinks that the Australian genera are actively evolving at the present day, though in other regions they give an impression of being more dormant. Their larvae are probably still vegetable-feeders, which, like those of robber-flies, attack inert or sluggish insects, including the larvae and pupae of moths. Adult Therevids are fairly like robber-flies and have been said to prey on other insects, but this is apparently mere inference. The lack of evidence from observation, contrasted with the ease with which robber-flies can be seen hunting, suggests a shift of emphasis on protein feeding from the adult back to the larva. This is certainly true of Scenopinidae, where the larvae feed on the larvae and pupae of clothes-moths, beetles, and fleas – plentiful both in birds' nests and in carpets! – while the adults have become harmless 'window-flies', attracted to the light, but otherwise entirely unnoticed.

Chapter 11

Bee-Flies

STILL WITHIN the super-family Asiloidea, a contrasting group of families centres round the Bombyliidae, or bee-flies. It is curious how the bees and wasps have served as models for so many other insects, whether metaphorically, or literally in a true mimetic association. Some families, like the robber-flies we have just seen, have a certain number of bee-like members, but at least four complete families of flies are so much like bees and wasps that in flight they are often mistaken for them, even by entomologists with some experience.

Probably the most widely known bee-fly is *Bombylius major*, the typical species of the whole family. This is a furry, bee-like insect, with a long proboscis, and is seen in the north temperate countries from March to May. On one of those delightful spring days, when winter has at last gone, and the primroses and wild violets have suddenly appeared, you may see this fly busily visiting one flower after another, moving along, usually less than a foot from the ground, and often keeping on in one direction as if it were following a set course. When it settles, the single pair of wings is spread at an angle, and shows a diagonal pattern of brown and white. Every year someone or other sends me one of these flies, pathetically crushed in a matchbox, and says that he has lived twenty or thirty or fifty years in the same village and never seen this remarkable insect before.

Yet *Bombylius major* occurs all round the North Temperate Zone from Europe to Japan in both directions. One reason why so few people notice it is that it flies so close to the ground, and moves along, so that a chance encounter is very brief. Adult life is short, and by the end of May this beautiful fly has disappeared for another year.

Bee-flies are lovers of sunshine, and, surprisingly perhaps in view of the example we have just taken, they are essentially flies of hot countries. Dry, stony, scrubby, parched regions are ideal for bee-flies. This they have in common with robber-flies, and anyone who

likes the sun himself can have a happy time observing, photograph-
ing or collecting adults of these two families of flies. Yet there is a
sharp difference of emphasis between them. Adult bee-flies are not
at all aggressive or carnivorous, they are quite unable to pierce, and
their mouthparts are entirely adapted to sucking from flowers.

The bee-flies are divided into two groups with rather unpronounceable Greek names: Homeophthalmae and Tomophthalmae.
Broadly speaking, the first group are like the ones we have described,
and have the body furry and a long proboscis for sucking the nectar
of flowers; the second group have the hair modified into scales and
in life are beautifully patterned, with areas of overlapping, brightly
coloured scales. Unfortunately these scales are very easily damaged,
and specimens that are seen in collections are often only a poor
shadow of the living fly. This second group have a short proboscis,
and I do not think anyone knows exactly what food they take.
Possibly nectar and honeydew, but very likely, like other flies, they
will take organic liquids from wet and rotting vegetation, stagnant
pools, perspiration, perhaps even liquids from dead animals. They
are not known to take blood, even from running wounds, in the way
that some other non-biting flies do.

In the bee-flies protein feeding has shifted to the larval stage, and
all their larvae are parasitic or carnivorous, or at least scavenge
among debris rich in animal protein. A great many of them infest
the nests of solitary bees, which they find easily in the dry, scrubby
areas that the adult bee-flies frequent. Thus the name 'bee-flies' has
a double relevance, to the appearance and to the habit.

When bee-flies are seen hovering close to the ground they are not
always looking for flowers, but are often female flies actively egg-
laying. The egg is dropped near the burrow of a solitary bee or wasp,
and it has been suggested that the fly flips it into the opening, with
the technique used by low-flying bombers when attacking railway
tunnels. This is probably a picturesque flight of the imagination. In
any event, the larva has to be very active at first in order to make its
way into the recesses of the nest, and then, having once reached its
goal, its remaining larval life is inactive, living on the food that is
available close at hand. With the versatility that has already led flies
to exploit the larval stage so well, bee-flies have gone one better, and
evolved two different kinds of larvae to meet these successive requirements.

The first-stage larva is a minute, wormlike creature, which moves actively over the rough ground and seeks out the cell, into which it penetrates. This larva may feed on pollen or honey stored for the use of the bee-larva, but after moulting it becomes more maggot-like and after a time attacks the larva or pupa of the bee. The rate of feeding is apparently such that the bee-larva remains alive as long as possible, thus providing fresh animal protein, without the complications of putrefaction.

The evolution of more than one type of larvae to meet successive problems is called *hypermetamorphosis*, and occurs sporadically in different Orders of insects. It is not generally a reliable guide to relationships, but in the present instance it does seem significant that the other two closely allied Nemestrinidae and Cyrtidae, also have this flexible device.

Bee-flies do not confine their unwelcome attentions to bees and wasps. In the second, scaly half of the family, they have become more versatile, and may be internal parasites of caterpillars – for example of both larvae and pupae of the other 'army-worm', the moth *Laphygma exempta*, in South Africa – or attack the already parasitic larvae of ichneumons and of Tachinid flies; these last two groups thereby becoming *hyperparasites*. In the genus *Thyridanthrax* certain African species are internal parasites of the pupae of tsetse-flies. The tsetse-fly, as we shall see later, is viviparous, dropping a mature larva which immediately forms a pupa in the ground. This is immobile, and therefore vulnerable to being eaten by the larva of *Thyridanthrax*, and a number of people have studied the association in the hope of encouraging the bee-fly against the tsetse. So far not much progress has been made and a similar lack of success attends the use of various species of *Systoechus*, a furry bee-fly closely related to Bombylius, to eat the buried egg-capsules of locusts.

One of the giants of the family is a huge black bee-fly which was bred from the pupa of a goat-moth in Malaya by Mr H. T. Pagden. In spite of its great bulk and sinister appearance this fly has tiny mouthparts, and does not appear to feed as an adult. I gave it the name of *Oestranthrax pagdeni*, putting it into a small genus of bee-flies without functional mouthparts, although none of those previously known were nearly so big. Recently Dr A. J. Hesse in Cape Town has described a similar fly and created for it a new genus, *Marleyimyia*, to which undoubtedly the Malayan specimen belongs. It is strange that such magnificent insects should exist, of which we know nothing until an occasional specimen is bred. Where do the adults live? The black colouring suggests that they may be noc-turnal. A huge, rare, peculiar creature like this is usually considered to be on the verge of extinction, yet it seems to linger on in two widely separated continents, and perhaps may eventually be found else-where.

Bee-flies are one of the biggest families, with about 3,000 species that have been given names. Some of these no doubt will prove to have been duplicated, but on the other hand there are certainly many yet to be discovered. The world-wide genus *Exoprosopa* has a vast number of species, some of them the biggest and handsomest of their kind, with a wing-span up to 25 mm, yet little or nothing is known of their life-history. Another group of which we know almost nothing is composed of flies that are quite different in shape from the rest. *Systropus* is a very narrow-bodied fly, with long abdomen and legs, looking almost exactly like a Pompilid wasp. It flies like one, too, darting to and fro in the air, with the hind legs dangling, in the same way that a hunting wasp does. Again we do not know why, or should I say to what end, it looks and behaves like this.

The smaller Bombyliidae are no less interesting than the big ones. *Toxophora* spends its larval life in the nests of solitary wasps, and is a small, hump-backed fly of very distinctive shape. In this case there is no great resemblance to the involuntary host, and it is difficult to say why *Toxophora* should look like this. Many of the smaller bee-flies are hump-backed, and some are extremely tiny, some *Phthiria* being little black flies only one or two millimetres long. Most flies of these genera have a tiny proboscis, and the adults apparently feed from flowers, especially Compositae, on which at times they may be numerous, though locally distributed.

It is odd that the tiniest bee-flies should have elaborate equipment for feeding in their miniature world, while the giants I mentioned above have almost no mouthparts.

Linked with Bombyliidae are flies belonging to two enigmatic families. Nemestrinidae are bee-like flies that hover in front of flowers, making a most beautiful picture, as plate 16 shows. Mac-kerras says of them: 'When feeding, the insect grasps the flower, or the stem just below it, with its legs, and keeps its wings vibrating rapidly, but not as rapidly as when in flight, the note dropping markedly in pitch each time the fly touches a flower. I have never seen [the flies] actively hovering when feeding, the plant in all cases being grasped by the legs. It is very rare, however, to see the wings come to rest.' Other observers have described the fly as remaining motionless, but poised for instant flight at the smallest alarm.

Nevertheless, in spite of their beauty, Nemestrinidae have no common name, perhaps because they occur only in warmer countries where showy insects abound. They are just among the hoverers at flowers, and are not recognized by the layman as being a distinct family. To the collector they are unique among flies in the elaborate strengthening of the wing by what appear to be extra veins, especially by the so-called 'diagonal vein', which runs obliquely across the wing. In fact this is no more than alignment of parts of a number of normal veins in an unfamiliar pattern. Some of these flies, especially in the genus *Nemestrinus* itself (figure 22), have the tip of the wing

Figure 22. Wing of the fly *Nemestrinus*, with the network of cross-veins that is sometimes claimed to be the *archedictyon* of primitive insects.

reinforced with a number of extra cross-veins, making a network of veins that I mentioned in an earlier chapter. I do not think it is the 'archedictyon' of primitive insects, but a special modification that this family has evolved in conjunction with its exquisite powers of hovering flight.

Mackerras stresses the fact that the males hover with a loud, high-pitched note, and when they are hovering thus in the air like a Syrphid fly, they will similarly dart away if disturbed, and then return to the same spot in the air; but if one is disturbed when it is feeding at a flower it will not return.

About 250 species of Nemestrinidae are known, but the life-histories of very few have been worked out: not surprisingly, perhaps, since this is a minute fraction of the world population of insects, and these few species are widely scattered, and occur in locally restricted habitats. The larvae of *Hirmoneura* have been found in tunnels of Cerambycid and Lamellicorn beetles in much the same way as the larvae of Laphriine Asilidae. *Trichopsidea* was bred by Miss Mary Fuller, from *Chortoicetes terminifera*, the Australian Plague Locust; she found them in one grasshopper in twenty, of four different species. 'The larvae are thus fairly abundant, in striking contrast to the scarcity of the adult, which is seldom seen in the field, and is comparatively rare in collections.' *Neorhynchocephalus* and *Trichopsidea* have also been bred from grasshoppers in the New World, and *Symmictus costatus* from *Locusta pardalina* and *Schistocerca gregaria* in the Old World.

These examples compare with Asilidae and Bombyliidae respectively, while the South American *Hirmoneura exotica* links with both families by laying its eggs in the nest of a carpenter-bee, *Xylocopa*. It does not, however, mimic the bee like the robber-fly *Hyperechia* that we have met previously, and it flies quite differently: so much for the evolutionary importance of mimicry!

The first-stage larva of a Nemestrinid is actively mobile, and moves to a suitable feeding-place before becoming more sedentary. That of *Hirmoneura* is said to hatch from eggs that are thrust into beetle-infested wood, and then to leave the wood and fall to the ground, where it finds and attacks the larvae and pupae of the root-feeding beetle *Rhizotrogus*. Greathead's account [93] of the life history of *Symmictus flavopilosus* brings out the following points of particular interest: (a) the larva in its second to fourth instars is astonishingly like the maggot of Cyclorrhapha, with mouth-hooks working vertically, and breathing through posterior spiracles; and (b) the spiracles are connected by means of a respiratory tube to the hole in the host's skin through which the larva originally entered. This tube is short in the second instar, but has become elongate in the third and fourth. The pupa on the other hand, is typically of the Brachycera and resembles that of a Bombyliid. This history illustrated the value of hypermetamorphosis, or a multiplicity of larval forms. Just as the complete change that takes place between larva and adult allows the flies to live two different lives in succession, so a succession of two or more larval types allows the eggs to be laid in a safe place, relying upon the active first-stage larva to find its own way to its food-supply.

This happens again in the family Cyrtidae. This is one of the family names that are bedevilled by problems of nomenclature, and these flies are sometimes called Acroceridae, and sometimes Oncodidae; some who in the interests of pedantry want to call them Ogcodidae suffer from a cold in the head as well as lacking any sense of euphony. Cyrtidae, too, have peculiar wing-venation, but their veins are reduced either in number or in strength, and the energy of the very rapid movement of the wings is devoted to keeping the fly just airborne, in what is appropriately called a 'balloon-like' flight. Cyrtids are very clumsy-looking flies, with a tiny head almost lost beneath a great, bulging, arched thorax, and inflated abdomen. They have sometimes been called 'fat-flies', not without reason.

Cyrtids are often numerous in one place, but the 250 species are even more widely dispersed than those of Nemestrinidae, since they occur far into temperate latitudes, as well as in the tropics. Adult flies

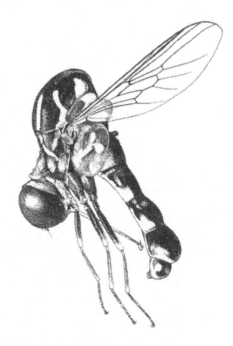

Figure 23. *Megalybus crassus*, a hump-backed Oncodid fly from Chile, where it was found on flowers of *Chrysanthemum* and *Berberis*. From Edwards.

are scarce in collections, perhaps because they do not get about much, and perhaps because they have a short life as an individual. There are all stages from having a long slender proboscis and feeding at flowers, to having no visible mouthparts at all. This is one of the families in which all the protein food is taken by the larvae, and all known larvae are internal parasites of spiders.

The eggs are laid in masses, usually at the tips of twigs, though in captivity the female will drop the eggs anywhere. Cole saw one individual lay 800 eggs, and another 900, and not surprisingly they were 'discharged rapidly from the ovipositor'. Possibly this huge number

137

of eggs may be necessary to keep the species going with such small numbers of adults, though it is not obvious why the adult flies should be so few. Heaven knows there are plenty of spiders everywhere, and Cyrtids are apparently not very discriminating. The young larvae hatch when the foliage is moist, and then wait for a spider to come near. They can leap considerable distances (5 or 6 mm), by bending and suddenly straightening the body, and can move along the thread of a spider's web with a movement like that of a looper caterpillar.

When they attach themselves to the body of a spider they grip by the head, and often stick out at right angles from the body, like bristles. The spider may kill some of them but others penetrate to the interior of the spider and devour it slowly. It is said that the larva may breathe by applying its posterior spiracles to the lung-hooks of the spider instead of keeping them open to the outside air.

Only a few larva of Cyrtidae are known, but such facts of natural history as are known suggest that this family is a relatively consistent one, in habits as well as in appearance.

We bring this series of families to a close with two who begin to mark a change-over to the 'higher flies' of the Sub-Order Cyclorrhapha. Empididae are sometimes called 'dance flies' and Dolichopodidae 'long-legged flies' or 'long-headed flies', though the last is a mistranslation from the Greek. As I have suggested about some others, these two families really have no common names, because the ordinary person just does not notice them, and the vernacular names are invented by the specialist. *Redolet lucerna*.

Empididae invite comparison at all stages with the robber-flies of the family Asilidae. The adult flies look like robber-flies, with the same spare, bristly, alert appearance, the same powerful grasping legs, and the same prominent, often dagger-like proboscis. They pounce upon their prey and suck it dry in much the same manner, and it is easy at first glance to mistake the two families. Empididae have a more rounded head, with the eyes not bulging like those of robber-flies, and not separated by a trough at the top of the head. They are thus deprived of any stereoscopic effect that the bulging eyes may confer, but on the other hand, they have a big area of eye-surface, which is enhanced by having the eyes meeting, or nearly so, in many males, and some females, and extending a little round the antennae in both sexes. It looks as if Empids may have the better vision in poor light, but less accurate estimation of distance than Asilidae. In the males the upper facets of the eyes are often enlarged, and sometimes the forward-facing ones as well, thus neatly linking both with the robber-flies and with primitive families such as the horse-flies and many Nematocera.

About 2,000 Empids are known, and their distribution is worldwide, but they are more characteristically insects of cooler climates, and so they flourish best in the temperate zones of both Northern and Southern Hemispheres. In the tropics they are fewer, except at altitudes where conditions are cooler and moister. Their distribution thus complements that of Asilidae: the typical picture of a robber-fly would show it hunting over scrubby vegetation on a hot, stony hillside or plain, while that of an Empid would show it flitting silently among leafy vegetation in the pale sunlight of a temperate country. Colyer and Hammond [47] use the expression 'leisurely hops' which perfectly describes the progress of *Empis tessellata*.

Like many of the family, this fly has a proboscis elongate beyond anything that is necessary, or even desirable, when attacking living insects, and uses it to probe flowers. The carnivorous habit is by no means the only way in which Empids feed, and Melander, in his account of them in the *Genera Insectorum* series, says: 'Many anthophilous females have never been seen to imbibe insect juices except at the moment of copulation.'

Indeed in this family, feeding and mating are linked together in a curious way. The name 'dance-flies' that they are given in textbooks refers particularly to the small black *Hilara*, which twist and turn in aerobatic manoeuvre close to the surface of still or slow-moving water. They are a very familiar sight in summer in temperate countries, as much a part of pond-life as the skaters and water-measurers that move on the surface film.

Pairing takes place during this darting, dancing flight, and when a pair is captured the female is usually seen to be holding a tiny ball of silk which may or may not envelop a small insect victim. The female may drop the silken ball when she is caught, and it should always be looked for in the collecting net or tube. The silk is spun from special glands carried in the enlarged tarsal segments which make the front legs of the male flies look large and clumsy. It is well established that the male offers the silken ball to the female as a part of courtship, but the sequence and significance of events have been much debated.

It is clear that the spinning of the silk is an embellishment in *Hilara*, and that in other genera the titbit that is offered may be unwrapped, as it were, or enclosed in a frothy 'balloon' of anal secretion as in *Empis aerobatica*. The simplest explanation is that the female Empid – like the female of the praying mantis – is apt to seize and eat her mate, and that he diverts her compulsive eating from himself by offering her some other food. Though this may have been the origin of the habit, observation suggests that in many genera the female has come to associate the two events, and that the offer of food woos the female into a mood responsive to mating. *Hilara* represents a final stage in evolution of behaviour, when the element of love-play has become the dominant one, and the prey may not be actually eaten by the female.

Sometimes, indeed, the male offers an empty ball of silk, or even some other object picked up for the purpose. An ironical sequence, like wooing a lady with a diamond necklace, then putting the neck-lace into a showy box, and finally wooing her with the empty box! Those who want to know more about these charming activities of 'balloon flies' than there is space for here will find the papers of Kessel [162, 163] interesting and instructive.

These spectacular, aerial species are the best-known Empids, but are only a small section of a varied family. Many Empids run rather than fly. *Tachypeza* (*Sicodus*) run about on smooth tree-trunks, *Tachydromia* on leaves and low vegetation. Others, like the curious little hump-backed *Hybos*, *Ocydromia*, and *Hemerodromia* sit about in damp, shady places, flying only sluggishly, if at all. In fact the ending – *dromia* that is so common in names of Empids comes from the Greek *dromos* and refers to running, as in hippodrome. These flies run about swiftly, and seize their prey with their legs, often having the femur and tibia of at least one pair spiny, and shaped to close upon each other like the jaws of a wrench. There is a rich fauna of small insects, mites, and spiders running about on the surface of vegetation, and these insect thugs have a good time of it.

This is a direction in which the robber-flies of the family Asilidae have not evolved, much as they resemble Empids in other ways. Robber-flies are slow and ponderous on their feet, and their powerful legs are used as an aerial basket, 'fishing' as it were through the air.

Empids have also gone much further than Asilids in exploiting the water. Adult flies of the genus *Clinocera* have found a new source of food in the insects that are trapped in surface films, either where water flows over stones, or on the alga-covered surface of deeper water. Laurence [171] has studied the modifications of the mouthparts that are associated with this change of food, resulting in a proboscis that is softer and much less conspicuous than in those flies that pierce living prey. This hyena-like habit has arisen surprisingly rarely among flies, considering how they have exploited almost every other possible source of food. *Hydrophorus* and *Campsicnemus* in the next family, Dolichopodidae, also do this, and after that we have to take a big jump to the acalyptrate family Ephydridae before we find any others. Meanwhile, back among the Empids, Laurence has also described how *Microphorus crassipes* steals its food from the prey caught in spiders' webs.

Empids also join the miscellaneous flies of the seashore, most of which are modified in some way for the rigours of their life. *Chersodromia cursitans*, as its name suggests, runs about, and is able to fly little or not at all, because its wings are reduced. This takes us back to the Chironomids, and further back still to the littoral Tipulids that we met long ago. Later on we shall see still other flies of the seashore, of which some cannot fly, like the Borborid *Apterina*, but others like the seaweed fly *Coelopa frigida* make constructive use of their powers of flight in a way that is useful in their peculiar environment.

Linked, no doubt, with the varied habits of adult Empids is the variety of shape and spacing of their eyes. It is common all through the Diptera for the eyes of the males to be enlarged until they almost or quite meet above the antennae. Empids show their versatility by having not only females in which the eyes meet, but also many genera in which the eyes meet below the antennae as well as above. All these modifications must have some optical value in the struggle to survive, but no one knows what it may be.

Before we look at the larvae of Empids, which are rather dull, let us consider first the adults of the next family Dolichopodidae. The name, from the Greek, means 'long-legged flies' and not 'long-headed' as sometimes it is mistakenly rendered. Most of them are small, shining metallic, bristly flies, which seem to have more kinship with bluebottles and greenbottles than with horse-flies and robber-flies. Some specialists share this impression, but the weight of evidence seems to be against it. Laurence has shown similarities in mouthparts with those of the Empid *Clinocera*, and in several ways the Dolichopodidae remain tied to the Empididae, and through them to the other families we have been discussing.

These, too, are predatory flies, but they do not openly pierce and suck. In place of the lancet-like labrum and hypopharynx, they have developed the sponge-like labella of the labium, and they catch their prey much in the manner of the Roman *retiarius*. Small, soft-bodied insects, adults or larvae, are enveloped in this spongy bag, and burst open by tearing their skin with hard, sclerotized 'teeth', which are

Figure 24. *Thambemyia pagdeni*, a Dolichopodid fly from Malaya, with hairy eyes and a grotesque proboscis. Though obviously predatory, no one knows what it feeds on.

not really teeth, but are adapted from convenient hard parts surrounding the mouth-opening. In Dolichopodidae they come from the pseudotracheae, the open channels that are so familiar in the photographs of the proboscis of house-flies. Mr Jobling recently told me that in his view some use a very much reduced lypopharynx to pierce, and are thus structurally akin to Empids and Asilids.

We have already noted how some Empids are aerial insects like the robber-flies, and how others run about and catch their prey on foot. Dolichopodidae also are footpads, as their name suggests, and although they have well-developed wings and can fly competently, they spend nearly all their time walking about, or just standing in a watchful attitude. The stance of a Dolichopodid is very characteristic. The length and proportions of the legs cause the front of the body to be raised, and the head is held in a 'chin-up', questing sort of way. The head is long, with large eyes, often bright green, and it is significant that the eyes are nearly always separated in both sexes. The touching eyes of many male flies are clearly associated with swarming flight, and aerial mating dances, and not to the mundane capture of prey, whether on foot or in the air. Asilidae, which catch aerial prey, and nearly all Dolichopodidae which catch sedentary or cursorial prey, each have the eyes separated.

Dolichopodidae are to be seen in great numbers on waterside vegetation, more particularly in temperate countries. They occur throughout the tropics, but, like the Empids, they tend to be elbowed out in warmer countries by more efficient flies. Empids are largely displaced by Asilids, and Dolichopodidae by the curious stalk-eyed Diopsidae, which we shall meet much later.

On the whole, Dolichopodidae are rather uniform in general appearance, though they are rich in bizarre details. Their males often sport long, strap-like antennae, and an array of ornamentations of the legs, discs, or fans of black or white scales which can be seen with the naked eye (figure 25). Courtship, like hunting, does not take place in the air, and those males that have these devices wave them near to the eyes of a female, much as one passes a hand to and fro in front of someone who is day-dreaming, to attract his attention. In other species the males have conspicuously marked wings, and raise and lower them rhythmically in a way that is very eye-catching. This is a common habit among flies of the section Acalypterae, which we shall come to in due course, and is another of the tantalizing 'previews' of later evolution that we can find in Dolichopodidae.

The courtship of Dolichopodidae is intriguing to watch, and can be followed comfortably because it takes place in slow motion, on an exposed leaf or tree-trunk, and the participants are not easily disturbed. It is noteworthy that the male does not offer any gift to the female, either to woo her, or to protect himself from attack. He is in little danger, because the mouthparts of Dolichopodidae are only suited to taking prey small enough to be at least partially engulfed in the sponge-like labella, and when wooing is necessary he carries his bait permanently attached to his person.

The largest Dolichopodidae are seldom as large as a house-fly and the smallest are almost as small as midges such as *Sciara*. One extreme of shape is that of *Argyra*, cylindrical and tapering, the males with the abdomen covered with a pile of silvery hairs that shine from one direction only. These males fly more than most members of this

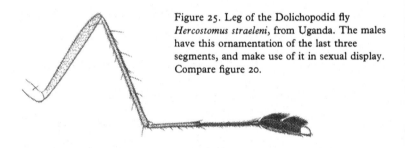

Figure 25. Leg of the Dolichopodid fly *Hercostomus straeleni*, from Uganda. The males have this ornamentation of the last three segments, and make use of it in sexual display. Compare figure 20.

family, and the hairs then sparkle and glitter as the light catches them. At the other extreme, the tiny *Campsicnemus* run about in dark places, such as ditches, and their males have curiously distorted legs. *Medeterus* hunts over the surface of tree-trunks, spending most of its time immobile, but occasionally making a short flight. Laurence watched them catching thrips, small insects of the Order Thysanoptera, which they followed and picked up in the labella, demolishing the wretched insect by working the two labella against each other.

The biological similarities between these two families are even more marked in the larvae than in the adults, and the larvae of the two families are so much alike that it is very difficult to distinguish them. The head is equipped with a set of internal rods, and again gives us a preview of what to expect in the third great group of flies, the Cyclorrhapha.

The larvae of both families are essentially carnivorous, though they make do at times with carrion food, such as dead insects, and even with wholly vegetable material. The larva of the Dolichopodid

Thrypticus bores into monocotyledenous plants, such as the reed *Phragmites*, and is believed to be vegetable-feeding. Though these two families are essentially terrestrial, and are properly linked with the terrestrial Asiloidea, some members of both families have begun to stray towards the water. This may not be unexpected, since the larvae look for animal food, and have not found it in the parasitic or semi-parasitic ways of the Bombyliidae, Nemestrinidae, and Cyrtidae. We have seen that water has much to offer to carnivores.

Larvae of Empididae and Dolichopodidae have spread out on to the mud flats, the surface of the water, and even to the seashore, and between tidemarks. They have not really gone into the water properly. The Empid *Hemerodromia* is perhaps the most completely aquatic, and has developed abdominal prolegs, with hooked crochets, in the way familiar in larvae that crawl under water. Its pupa, too, is aquatic, and is found on submerged stones. On the other hand, Bertrand [21] says of the larva of *Clinocera*, which is found in streams, particularly in mountainous areas, throughout the north temperate regions: '*Les larves sont plutôt hygropétriques que franchement aquatiques, et la nymphose [ie pupation] à l'inverse des* Hemerodromia, *a lieu hors de l'eau.*'

Certain Empids of the genera *Roederoides* and *Wiedemannia* prey on the pupa of black-flies (Simuliidae), and having eaten the pupa, themselves pupate either in the cocoons, or behind crevices of the stones. They are thus beginning to follow the same path as the Bombyliidae, in moving towards a form of group-parasitism.

Miss Smith [266] gives the following summary of the ways of Dolichopodid larva, which may also serve as a general pattern for Empids.

Marine Habitats

Matting of filamentous algae covering rocks, Pacific coast, USA – *Aphrosylus praedator*; muddy pools with algae, Hawaii – *Cymatopus acrostichalis*; beach sand, England – *Machaerium maritimae*; under decaying vegetation, Hawaii – *Asyndetus carcinophilus*; beneath decaying seaweed, Europe – *Hygroceleuthus* sp.

Freshwater Habitats

Shallow water and mud – *Argyra albicans*; *Hydrophorus* spp.; wet rocks near a spring, with diatoms and other plant life – *Eurynogaster*; muddy streams – *Dolichopus*; wet moss – *Hercostomus*; sand at edge of stream – *Leucostola*; sand at edge of pond – *Porphyrops*.

Terrestrial Habitats

Rich garden soil – *Dolichopus*; beech humus – *Sciapus, Neurigona*; beech fungus, rotten wood, exuding sap – *Systenus*; decaying vegetation – *Dolichopus, Campsicnemus, Asyndetus*; stem-borers – *Thrypticus*; under bark and in burrows of wood-boring beetles – *Medeterus*.

Add many records from dung and we have surely a pattern of versatility!

A well-marked divergence between the two families is the liking of Dolichopodidae for protecting their pupae with a cocoon, a habit that is typical of them except for the stem-boring *Thrypticus*. Aquatic species use sand or mud glued together with a tough gelatinous substance. Even *Medeterus* in its tunnels makes a cocoon of tough silk. The pupa has respiratory 'horns', which sometimes protrude through a small hole. In *Hypocharasus* the opening is big enough for the adult to emerge through it, but in others the pupa escapes from its cocoon, and then the adult from the pupal skin.

Now at last we have reached the end of the flies of the Sub-Order Brachycera, and are ready to move on to the 'higher flies', and report the acme of evolution in the fly-world.

Chapter 12

Coffin-Flies

THE NAME 'COFFIN-FLY' has been applied to *Conicera tibialis* of the family Phoridae, because this fly is able to maintain itself through many successive generations in coffined bodies that have been interred for a year or more. The fact has long been known, but even today no one really knows how the fly gets there. It seems unlikely, for a variety of reasons, that the eggs are laid on the body before burial, and so either the adult fly or the larva must find its way down through several feet of soil. This seems a difficult feat when we think of its being performed by each larva or each adult fly, but, as in many problems to do with animals, time and continuity make most things possible.

My colleague Mr R. L. Coe studied adults of this species associated with the corpse of a dog, buried about 3 ft down, and found them coming and going freely between this and the surface. Flies emerging from the pupa underground made their way to the surface, mated there, and returned down through the soil to lay eggs on the carrion. Though this observation shows what is possible, enough circumstantial evidence has been accumulated by those who have studied this grisly subject to show that all stages of the fly are able to remain far below ground, where they live a complete life-cycle in the dark, getting enough air to breathe from natural cavities in the soil. This permits the existence of a subterranean population of flies, and once this has become established there is no problem about access to the surface, at least as long as the larval food-supply holds out.

I start my account of the family Phoridae on this macabre note, because this is a macabre family, and the life-history of *Conicera tibialis* is no more than a spectacular extreme of habits that are widespread in this family. As Phorids have no common name, 'coffin-flies' is a not unjustified invention for them. The furtive habits, the liking for dark, secretive damp and mouldy places, penetrating far into small orifices, the larvae feeding on organic matter that has begun to dry up and mummify – these are characteristics that run through the Phoridae, though they are exploited differently by various members of the family.

Phorids are tiny flies, the biggest of them only about 6 mm long, and the smallest minute, perhaps $\frac{1}{2}$ mm. They are among the most distinct of all flies, especially the winged individuals, because the veins are arranged in a curious way: the longitudinal veins are strong,

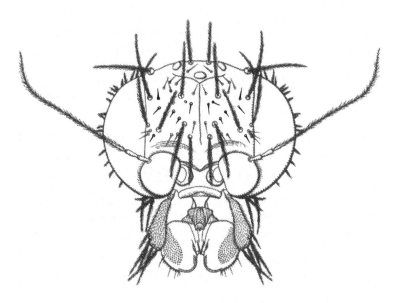

Figure 26. Head of the Phorid fly *Megaselia meconicera*. Note the characteristic arrangement of bristles, many of which are microscopically fringed.

but short, compressed into the first half of the leading edge, while a series of fainter veins arise at an angle, and run parallel to each other to the hind margin. It is difficult to fit this pattern into the classification of flies, and the rest of their structure does not help very much. The head is usually very bristly, and the bristles, their position and inclination, and the plates on which they stand, are different from those of other flies. The antennae, again, are peculiar, with the second segment fitting into a cavity of the globular third segment, like a ball-and-socket joint. The proboscis is short and fleshy, and the palpi are unduly large, often with strong bristles and obvious sense-organs.

Earlier in this book I expressed the opinion that Phoridae are out on a limb of the evolutionary tree, and this seems to be the truth. They have clearly found a highly successful niche, and have been going their own way for some considerable time. No one really knows where they come from, though ingenious attempts have been made to derive them from various families of midges. It is not even certain whether they belong to Brachycera or Cyclorrhapha; older authors usually put them at the very end of Brachycera, but I think they are best seen as a collateral branch of the Cyclorrhapha. The peculiarities that we have just noted are also to be found among Hippoboscidae, a family of bloodsucking parasites that we shall meet much later on. It is tempting to try to make something out of these resemblances, but probably they are no more than convergent adaptations to a similarly furtive life.

The winged Phorids of more or less free-living habit are very uniform in appearance and are classified, with difficulty, by details of the arrangement of bristles, especially those on the legs. Those flies that have taken up an unusually secretive existence, and especially those living in the nests of ants and termites, have evolved into bizarre little insects. Reduction or loss of the wings has been accompanied in many cases by modifications of the thorax and abdomen. With the reduction of the flight muscles the plates of the thorax become thinner, and either reduced in area or fused together, and the females of some have settled for a life of inactivity, with the abdomen enlarged into a membranous bag. This again is a common feature of the parasitic Hippoboscidae.

Phorid larvae and pupae are distinctly of the type that we associate with the 'higher flies' of the Cyclorrhapha. The larvae are maggot-like, with a pointed front end and a truncate rear, the head reduced to vanishing point, and the mouthparts to a set of mouth-hooks. They can generally be distinguished from other families by having rows of tiny, spine-like processes along the abdomen, and occasionally, as in *Megaselia*, these may be modified into fringed processes. This is an adaptation to living in organic semifluid media of a certain consistency, and is most conspicuous in the lesser house-fly, *Fannia*, belonging to quite a different family.

Phorids pupate inside the last larval skin, and the puparium is curiously flattened and shield-like at the head-end, with the pro-thoracic spiracles held out at the tip of slender 'horns'. These belong to the pupa, and are pushed forcibly through two spots on the puparium that have remained soft for this purpose. Keilin [159] describes in detail the precision of this operation by the pupa, which is nearly always successful; when, rarely, the pupa fails to find the right places it dies of asphyxiation.

Fully winged, active Phorids are common everywhere. When they are seen on windows, or are collected with other insects in a net or on occasion make a nuisance of themselves by swarming indoors, they can be recognized by their peculiar scrabbling sort of movement. This is probably forced upon them by the relative lengths of the legs, as it is in Dolichopodidae. Usually they are not noticed by the lay-man until they mass in his house. The late Dr Hugh Scott recorded a number of 'swarms' of *Megaselia meconicera* in his own and other houses, and was satisfied that they originated outside, and did not breed in the house. This behaviour, and whether it can properly be called swarming, I hope to discuss in a later chapter of this book.

Adult Phorids are particularly typical of caves and grottoes, and are regularly collected by people who investigate the fauna of such places. While some of them breed from the corpses of small mam-mals and birds, a great many others breed from fungi. The conditions of mushroom cultivation are just what Phorids like, and one species in particular, *Megaselia halterata*, has been a pest of cultivated mush-rooms, often enough to be known to the trade as the 'mushroom fly'. Hundreds of larvae may occur in a handful of mushroom com-post.

Perhaps the most favoured food material for larvae of the free-living Phorids is dead snails. These are always plentiful, and the interior of the shell provides them with a supply of drying, highly nitrogenous food in a still, moist atmosphere. The pupae are to be found stuck to the empty shell, and plate 30 illustrates a curious by-product of this way of life. People will leave their empty milk bottles lying about in the garden, instead of returning them promptly to the dairy. If the bottles are still wet with milk they attract small acalyptrate flies of the genus *Drosophila* whose larvae feed actively in the curd that forms. If the bottle is washed clean before being put out, then it often gives refuge to larvae of Phorids, looking for a place to pupate.

This is particularly a habit of *Paraspiniphora bergenstammi*, a little fly whose formidable name is matched by the impressive rows of bristly papillae that are carried by both larvae and pupae. The pupae stick firmly to the glass of milk-bottles, and may escape the washing plant, to be first noticed by the customer to whom the refilled bottle is delivered.

Not surprisingly, perhaps, people sometimes accidentally swallow larvae of Phorids, through eating fruit, vegetables or meat contaminated with eggs or larvae. The tropical species *Megaselia scalaris* and *M. rufipes* are the principal offenders, especially the former. It is reliably claimed that a European in Burma passed larvae, puparia and adult flies over a period of a year, under conditions that indicated a colony of the flies to be living and mating in his intestine. This is another example of the extreme resistance of these flies at all stages to asphyxiation and to chemical action. Larvae will also infest wounds, and the exudations from sores, and will live in preserved material of all kinds, even it is said, specimens preserved in formalin.

Of greater interest, perhaps, in this family are those which live in association with other animals. The giant snails of the tropics, mainly of the genus *Achatina*, are fascinating creatures, ranging up to 3 or 4 in in length, and sometimes more. They feed mainly at night, and eat an enormous quantity of vegetable food, so that once they are transported away from the lush tropical forests that are their home they become a serious pest. The recent war saw them spread across the Pacific, and the US authorities have had to take the most stringent precautions to avoid their becoming a plague, even on the American mainland.

If you catch one of these snails and watch it moving about in daylight or artificial light, you will probably see small white specks in the slime covering the upper surface of the foot. Under the microscope these are seen to be wingless females of one of the two or three species of the Phorid fly *Wandolleckia*. The thorax is heavily cut away from behind, and obviously much reduced from that normal in a winged fly. Halteres as well as wings have been lost. The head is a typical Phorid head, with its spiny bristles and its complex antennal joints, but the eye has been reduced to about twenty-five to thirty facets. Since they are not so tightly packed, they have reverted to a circular shape. The legs are fairly normal, since the insect is an active runner, but the abdomen is almost entirely membraneous; there are one or two plates, and the rest is a bag of thin membrane, which in life is packed with about twenty large eggs. It is the white abdomen that can be seen with the naked eye.

For a long time *Wandolleckia* was thought to be parthenogenetic, and early attempts to associate males with the white-bodied females were mistaken. In 1953 Professor Baer described how he had reared a male as well as a female from eggs that he had found in excreta of a giant snail, and on the ground near by. There are three larval instars, but larval life is comparatively short, and less than three weeks after egg-laying there hatch stenogastric individuals: that is, with narrow abdomen, not inflated in the familiar way. The swelling into a physo-gastric form comes later, but Schmitz points out that it is more than a mere stretching of the intersegmental membrane as the eggs grow. It is an irreversible process of growth, which is carried even further in some members of the family, as we shall see in a moment.

These larvae and adults are living on external secretions, but some Phorid larvae have become parasites by attacking living tissues. Earthworms and beetle-larvae are recorded as having nourished Phorid larvae, and *Pseudacteon formicarum* is an internal parasite of ants, including the common *Lasius flavus*. The fly lays its eggs be-tween two segments of the ant's body, and the larva penetrates the integument. The tropical Phorids *Plastophora* and *Apocephalus* lay eggs on the ants *Solenopsis* and *Camponotus* respectively and their larvae enter the head of the ant and destroy its tissues, ending in decapitation.

Many Phorid larvae live as scavengers or commensals in the nests of ants, bees, wasps and termites. *Metopina pachycondylae* lays its eggs on the larva of the ant *Pachycondyla harpax*, and its larva en-circles the neck of the ant-larva. Established there, it steals the food brought by the worker ant to feed the larva, and behaves like a cuckoo in the nest. Other larvae raid the food-stores in nests, includ-ing the fungus-gardens of termites, and even devour the eggs.

Certain of the Phoridae associated with termites have become so highly modified for their secretive life that the adults are scarcely recognizable. They are often separated into one, or perhaps two dis-tinct families, but Schmitz has described *Alamira termitoxenizans* from a termite's nest near Nairobi in Kenya, which he considers to be the missing link between the sub-family Termitoxeniinae and the rest of the Phoridae.

Termitoxeniinae have a first-stage larva that moults immediately after hatching from the egg, and a second-stage larva that takes no food; the third-stage has been suppressed, and the whole larval life lasts only from a few minutes to one hour, according to species. This

17. Hover-flies in flight.

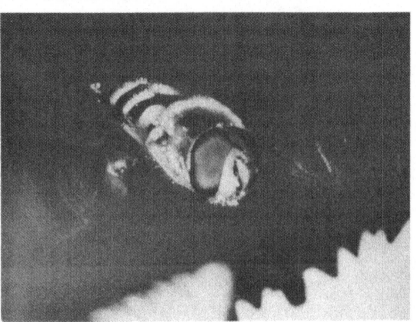

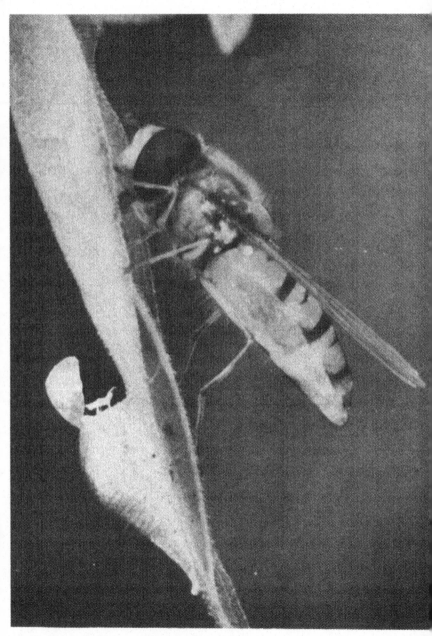

18. A hover-fly newly emerged from its puparium, seen below it on the left.

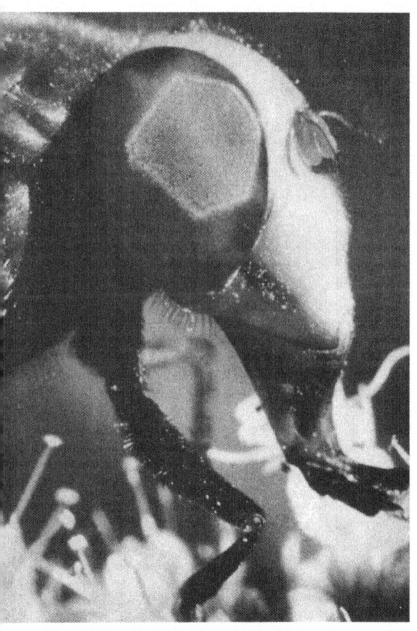

19. Head of the hover-fly *Volucella zonaria*, showing the nose-like prolongation of the face, an adaptation to feeding from flowers.

20. A male of the Mediterranean fruit-fly, *Ceratitis capitata*, with characteristic knobbed bristles on the head. The elaborate pattern and vivid colour of this fly seem to go far beyond the demands of mimicry or mutual recognition.

is a further extension of the reduction to insignificance of larval life, which we have already seen foreshadowed in *Wandolleckia*. We noted early in this volume how the existence of a larva that was quite different from the adult enabled flies to evolve into two different worlds simultaneously. These flies have 'voluntarily' (if that is the right word) confined themselves to one world: *Wandolleckia* to life inside the shell of a snail; Termitoxeniinae to the interior of a termite's nest. With a remarkable economy and adaptability, they have begun to dispense with the complication of larvae and larval moults, concentrating all their energies into the highly adapted, flightless, adult fly. We shall see later how a similar decisive step has been taken by the parasitic families Hippoboscidae, Nycteribiidae, and Streblidae, as well as by the bloodsucking tsetse-flies.

But Termitoxeniinae have gone further. From the puparium hatches a small adult fly of the 'male phase', which has both ripe sperm and rudimentary ovaries. This undergoes a 'postmetabolic change', comparable to the irreversible change that we have seen in *Wandolleckia*, to become a swollen, physogastric 'female phase', with the ovaries fully developed. As Schmitz says, so many associated changes occur that it is remarkable that so much can be achieved without a moult. It is the exception that proves the rule that adult insects are fixed and immutable.

The late Father Schmitz, S.J. was the world's leading student of this family, and he rightly said of them: '*Es gibt am Körper dieser kleinen Fliegen inwendig and auswendig kaum ein Organ, das nicht gelegentlich bei dieser oder jener Subfamilie, Gattung oder Art ganz exorbitante Werge der Entwicklung einschlüge.*'

Two things the Phoridae have not managed to do are to take to the water, and to suck blood. It is true that Neveu-Lemaire, in his textbook of medical zoology, had a footnote: '*A. Roger a découvert, au nord de Bornéo, une espèce du genre* Aphiochaeta *occupée à sucer le sang d'un moustique*', but he does not explain what he means by this: whether the Phorid was robbing the abdomen in the manner of the Ceratopogonid we mentioned in an earlier chapter, or was taking the body-fluid of the mosquito itself. On the data it seems no more than a single observation, perhaps misinterpreted.

Placed awkwardly next to the Phoridae is a small family of soft-bodied, yellow flies, 5 mm or less in length, the Lonchopteridae. It is misleading to call this family a rare one, since individuals are fairly easy to find, and sometimes occur in numbers, among leafy vegetation in damp places. There are only about twenty species in the world, and there is a curiously static air about them, as if they were on the way to extinction.

The adult flies are remarkable for their curiously pointed, lanceolate wings, with a venation that is different from that of any other flies, and which even differs between male and female. Some link can be seen with Phoridae, both in wing-venation and in the bristles of the head and legs: there is also some similarity with Dolichopodidae, thus reinforcing the feeling of a gradual transition from Brachycera to Cyclorrhapha, without showing a clear line of evolution.

The larvae of Lonchopteridae are unique among flies. They are flattened, scale-like creatures, with the abdominal plates clearly separated, and bordered with a sort of striped band. The small head is almost hidden under a shield-like prothorax, and there are long filamentous processes fore and aft. Beyond the fact that the larvae is found among dead leaves, and in decaying vegetable matter, nothing is known of its habits. Mandibles are said to be absent, so presumably the larva is not carnivorous, nor would its shape suggest predatory habits. It is tempting to suppose that the larva must have some exotic method of feeding, sufficient to ensure that it survives, but in small numbers, but this is purely speculation.

A further oddity of this odd family is that, while most species are known from both sexes, *Lonchoptera furcata* is known almost exclusively from females, which on examination are found to have the spermathecae, the sperm-storing receptacles of the female, empty. It seems evident that parthenogenesis is regular in this species, though not in the spectacular manner of the Termitoxeniinae. Stalker [270] found that the American *Lonchoptera dubia* is a mixture of four chromosomal races, and he suggests that the scattered distribution of this fly within quite recent times has encouraged selection for parthenogenesis in small, semi-isolated populations where males happened by chance to be few. This device may for a time stave off extinction in *dubia* and *furcata*, and may be resorted to in turn by the other species that are at present bisexual.

Chapter 13

Hover-Flies

HOVERING, THE ABILITY to keep the body motionless in mid-air, is probably the most characteristic ability of flies and is practised by some of the most primitive, as well as by some of the most advanced. We have already seen it in many families, particularly as a pre-nuptial flight, where the male is usually the one to hover in a conspicuous place, alone or in a swarm, while females come and go.

The flies of the family Syrphidae are known as 'hover-flies' *par excellence*, because of the perfection to which they have brought the art of hovering flight, together with their conspicuous and attractive appearance. Some tiny flies, notably Pipunculidae and Drosophilidae, hover equally competently, but they are so tiny that they are seldom noticed by anyone who is not specially looking for them. Males of some of the larger hover-flies use the trick of hovering for its original purpose of attracting the attention of females, but in most of the smaller species both sexes hover indiscriminately. We have all seen these flies, the wings beating so quickly that they are invisible; the body motionless in the air until we approach, then darting rapidly off in any direction, forwards or backwards, sideways, up or down.

If it were not to risk being branded as anthropomorphic and sentimental, it would be pleasant to think that Syrphids enjoy their hovering as much as we who watch them. After all, flies have to do something all their waking hours. Carnivores have to hunt, and if their prey is hard to find, hunting may take up most of their time. Flower-feeders have an easier time. They can feed at any time, and unlike bees they do not have to collect food for a never-ending succession of larvae in a hive, or, worse still, to make honey for an insatiable bee-keeper. Many flower-feeding flies bask in the sunshine, while others strut about in an absurd way. Syrphidae spend much of their time, weather permitting, hovering in the air, with every appearance of enjoying it: or if you prefer to express it differently, the state of hovering seems to be one of content, from which they are disinclined to depart.

Though their hovering is conspicuous, they also feed at flowers, and after the bees are the most important pollinators. Sometimes they

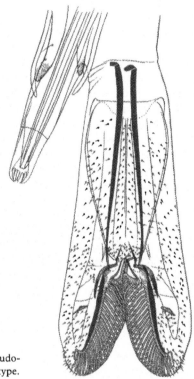

Figure 27. Proboscis of the hover-fly *Eristalis*, showing the sponge-like pseudo-tracheae at the tip. A flower-feeding type.

go too far, like the ones recorded by Banks [11] which collected in a greenhouse, and pollinated cucumber flowers. The resulting fruit was coarse and unmarketable.

A peculiarity of Syrphidae, not entirely confined to them, but most often observed here, is to go on making a noise after they have settled and ceased to move the wings. It has been likened to the 'singing' of a tea-kettle, and is apparently produced by rhythmical contractions of the thorax, possibly accompanied by vibrations of the halteres. Although Aubin's [6] paper provides some basis for consideration, little or nothing is known with certainty about either its mechanism or its significance. It will be noted that the Syrphids are among those flies, not all closely related, which have large *squamae*, or lobes at the bases of the wings, covering the halteres.

Collectively, this is a family of wasp-like or bee-like, flower-haunting flies, almost the only family of flies that everybody knows, and everybody likes. The adults have totally abandoned any carnivorous diet that their ancestors may have had, and have no piercing or crushing mouthparts. They now live entirely on nectar and honey-dew, which they suck up through the sponge-like labella of the labium (figure 27). In this family protein-feeding has been relegated to the larval stage, and almost all larvae either prey on other insects, or scavenge organic debris, or feed in foul water rich in organic decomposition.

This kind of life (of the adults, that is, not of the larvae!) would be described in human terms as carefree, if not arcadian. It is quite different from the busy, purposeful life of a robber-fly, on the one hand, ever finding and striking its prey; and on the other hand from that of the bluebottle searching endlessly for a place to lay its eggs. The life of an adult hover-fly is possible only if the larva has accumulated an ample store of protein, so that the adult fly need not hunt. On the other hand, the active hovering, so different from the lazy rise and fall of Chironomid midges, consumes a great deal of energy, which must be provided by a supply of carbohydrate food.

The flower-feeding habits of adult Syrphidae are thus explained, but it is curious that in the course of evolution, they have not produced more flies with an elongate proboscis, in the manner of Bombyliidae, and Nemestrinidae, of butterflies and hawk-moths. *Lycastris*, an Oriental genus, has a very long proboscis of this type, and also has the leading edge of the wing strengthened with a number of extra cross-veins, like some Nemestrinidae (figure 22). There is unlikely to be any direct link between these two families: rather does this illustrate how flies of similar habit may evolve along convergent lines.

The proboscis of almost all other Syrphids remains short or at least inconspicuous. In an earlier chapter I said that it seemed a lot of trouble to have evolved a specially long proboscis just in order to reach down into the depths of tubular flowers, when the majority of flowers are easily accessible. The Syrphids have reached the same conclusion! They do, however, develop the face into a nose-like structure, and many genera have a characteristic profile. *Rhingia* has a conical snout about as long as the head itself. Some of these facial swellings may help in pushing a way into a flower, and to that extent perhaps they take the place of an elongate proboscis. No doubt some of them have developed further than was strictly necessary, as is the way of things in evolution.

My colleague Mr K. G. V. Smith points out to me that the membraneous area at the base of the proboscis is capable of great expansion in some Syrphidae, presumably under the pressure of blood. Mr Smith has seen *Rhingia* extend its proboscis by this means to penetrate into deep flowers such as bluebells. Perhaps the ability to do this makes it unnecessary for Syrphidae to have a long visible proboscis.

The appearance of adult Syrphidae is simply described as being like bees or wasps, with which they can easily be momentarily confused. Among flies they can be recognized by that way in which at least two of the veins of the wing curve forward on to the preceding vein, making closed cells which leave the trailing edge of the wing without support; there is also a longitudinal 'spurious vein', not part of the normal pattern.

Adult hover-flies are aesthetically attractive, but lack variety when compared with many other families. As I have said, higher flies cease to show by their shape much about their way of life, and Syrphid adults are an early example of this. There are plump and hairy ones; plump bare ones, with banded patterns; slender ones with a near-cylindrical abdomen, and wasp-waisted ones. Although a widespread, cosmopolitan family, they look much the same from any country of the world. Some of them seem obviously to mimic Hymenoptera, sometimes as in the genus *Volucella*, those Hymenoptera with whom they are associated, but a great many that look like Hymenoptera seem to have no obvious reason for doing so.

They are popular objects of study, especially among novice dipterists, because the almost infinite variations in pattern that are possible make identification a direct and often an easy process. The behaviour of the adult flies is probably also highly individual, within the rather narrow limits permitted itself by the family, and deserves more study than it has yet received. Such colourful insects are attractive to the collector, who learns enough about them to know where to look for them, but only rarely spends much time in observing the living flies.

No student of Syrphidae can avoid giving some thought to the question of mimicry. The familiar drone fly, *Eristalis tenax*, is common in everyone's garden in temperate countries, and has become spread throughout the world, yet many people will tell you that they never saw it before: they have dismissed the flies as bees. Many tropical Syrphids look very much like bumblebees: e.g. *Lycastris austeni*, one of those with a long proboscis, mimics the bumblebee *Bombus orientalis*, and *Graptomyza* resembles the stingless Meliponine bees.

Wasps too, are imitated. *Volucella inanis* looks like a wasp, with the same yellow-banded abdomen, and actually breeds in the nests of wasps and hornets. So does the bigger *Volucella zonaria*, which is often mistaken for a hornet. This is one of a number of flies that are not very firmly established in the British Isles. Until the nineteen-forties it had not been seen in Britain within living memory, and old records of it were suspect. Then it suddenly appeared in the extreme south-east, and is now accepted not only as an annual migrant, but also as a permanent, if local, resident. It appears even in London, and is often killed by alarmed householders and keepers of greengrocers' shops, who think it is a dangerous wasp or hornet. *Volucella bombylans* exists in two well-marked forms, one of which looks like *Bombus terrestris*, and the other like *Bombus lapidarius*, but little use seems to be made of this resemblance in practice.

Chrysidimyia and *Mutillimyia* closely resemble the Chrysidid and Mutillid wasps respectively, and the whole tribe or sub-family of the Cerioidinae are extremely wasp-like in appearance. Two Australian species fold the wings longitudinally along a deep groove in the membrane, in a way that increases their resemblance to the yellow and black wasps with which they are found.

In the family Syrphidae evolutionary originality is displayed in the larval stage rather than by the adult. Syrphid larvae are conveniently recognized by having the posterior spiracles placed close together, and more or less fused into a cylindrical tube of varying length: the first-stage larva of one species of *Mesogramma* is said to be the proverbial exception. Each segment bears twelve short spines in a definite arrangement, except in the slug-like larvae of *Microdon*, where the segments themselves are indistinguishable.

Microdon is the exception to so many rules that it may be well to dispose of it here. A small, oval object, flattened underneath, and without visible segments, the larva of *Microdon* looks like a little slug, and indeed was first described as a mollusc. It is found in ants' nests, and the details of its association with various ants, which unfortunately we cannot spare the space to describe here, are set out in Donisthorpe's book *The Guests of British Ants* (1927). It is generally regarded as a scavenger, but feeds principally on pellets dropped by the ants, and said to be ejected from infrabuccal pockets.

This scavenging habit is not exceptional. Other Syrphids have gone into the nests of bees and wasps, and flies of other families, notably Phoridae, have established themselves in the nests of ants. Ants are the most highly evolved of social Hymenoptera, and life in their nests is elaborately organized, with much interplay between the inhabitants. Wheeler gave the name trophallaxis to the mutual stimulation, and exchange of excretory or food material between members of such a community. Living in an ants' nest must be rather like being a castaway among a tribe of head-hunting cannibals. There is a chance of survival if one can avoid attracting attention by fitting into the ritual of everyday life, but there is always the risk that after an incautious move one may be killed and eaten.

This is apparently the situation of *Microdon*. As long as the larva crawls about right way up, and keeps quiet, the ants tolerate it. If it turns over on its back it immediately becomes a strange object to the ants, and is attacked. Perhaps the shape of the *Microdon* larva, with its reticulated upper surface, without segments or sutures, and its lower surface closely pressed to the sub-stratum, makes it inconspicuous and also difficult to attack. The ants try to drive away the female when she is laying eggs in the nest, but do not destroy the eggs. The pupa also remains in the nest, inside the hardened last larval skin, and with the horns of the thoracic spiracles protruding.

It is interesting to note that adult *Microdon* have the eyes separated in both sexes, suggesting that the males do not hover for mating as most Syrphids do.

Syrphid larvae, including *Microdon*, can be seen as having evolved in various directions from a central saprophagous type, that is one that feeds on rotting compost-like materials, with a mixture of vegetable and animal protein. This is where we always start with the Diptera, as we have seen before, and once again the most obvious modifications are in those larvae that go into wetter habitats, or become fully aquatic. Since Syrphid larvae already have the hind spiracles united, it is a simple matter to elongate them into a siphon of whatever length is suitable to the medium in which the larvae choose to live. Thus we have 'short-tailed' larvae and 'rat-tailed' larvae, the latter with an extensible siphon a great deal longer than the rest of the body.

According to Hull, the most primitive Syrphidae are probably some Cheilosiinae, the larvae of which live in dung, or in compost-like materials. *Rhingia campestris* the hover-fly with the snout that we have mentioned earlier, breeds in large numbers in cow-dung, but many genera of this group have moved to the pulpy matter in rotting wood (*Myolepta*), rot-holes in trees (*Xylota*), or the sap that runs down from wounds in trees (*Brachyopa*, *Ferdinandea*). The bulb-flies *Merodon* and *Eumerus* are two independent lines of evolution from this type of habitat, finding in bulbs a concentrated store of food material, where they can soon create a pulpy mass for their own use: incidentally ruining the bulb for us. *Merodon* is the large bulb fly, or narcissus fly, which attacks bulbs of the lily family by burrowing through the basal plate. It spends the winter in the bulb as a larva, and then pupates in the soil, and has only one generation per year. *Eumerus*, the lesser bulb-fly, usually enters round the damp neck of the bulb, and enlarges the damp area to make a pulpy internal cavity for itself (plate 32). This fly has two generations per year.

Hartley [323] supports the view that the primitive hover-flies were feeders on decaying matter, and that the plant-feeders, the carnivores and the scavengers came later. He, however, suggests the rot-hole-feeding Milesiini as an ancestral group.

The aquatic Syrphid larvae are reached by stages of greater bogginess of the food material, and the extension of the posterior spiracles into a full 'rat-tail' generally follows. The elongate siphon is so obviously adaptive that it has almost certainly arisen more than once in different groups. The well-known rat-tailed larvae belong to *Eristalis* and related genera, and probably arose from the rot-hole breeders, adapting themselves to live in cavities with a more or less aquatic content.

The drone fly, *Eristalis tenax*, has been mentioned already. Its rat-tailed maggot has an exceptionally long, extensible siphon, and the larva is built for crawling about on the bottom by means of its ventral prolegs, with its siphon reaching to the surface: like an entomological trolley-bus (figure 10). The successful spread of this fly is explained by its versatility. These larvae can be seen on the one hand in deep-water habitats, such as garden pools, water-butts, and roof-tanks; and on the other hand in heaps of sodden vegetable debris, not the fine manure of a compost-heap, but such places as a heap of cabbage-stalks, or the rich brown ooze at the edge of a farm-yard manure-heap. I have recently heard of attempts to use these larvae to purify accumulations of poultry manure, in a way reminiscent of the *Psychoda* larvae of sewage works (see Chapter 5).

These are different from any aquatic larvae that we have yet met, in that they are neither predators with powerfully developed mandibles for seizing their prey, nor yet filter-feeders with mouth-fans to gather food from the water. *Eristalis* and possibly some others have an internal filter-apparatus. In fact, they are underwater scavengers, swallowing organic fragments that they find, and probably also absorbing some dissolved nutriment from the foetid water in which they choose to live, in the manner of mosquito larvae. Since they eat protozoa and tiny invertebrates they technically rank as carnivores, in part at least. There is really no clear distinction between these and the land-scavengers such as *Volucella*, whose larvae are similarly 'snappers-up of unconsidered trifles', in the nests of bees and wasps.

The rat-tailed larvae that live in boggy or peaty soil, for example *Sericomyia* and *Tropidia*, are possibly a different line of development from the breeders in rotten wood. The adults of some of these are particularly inhabitants of boggy places, and of flats near the coast. Systematically they are grouped with *Pocota*, which breeds in rotten wood, and with *Eumerus* which, as we have seen, makes its own 'container habitat' in a bulb.

In temperate regions two genera are known to have made more extensive adaptations to larval life in water. *Callicera* larvae live in water-filled holes in pine stumps, and instead of elongating the siphon, have retained a short 'tail', and evolved a conspicuous group of anal gills. According to my colleague, Mr R. L. Coe, who made a study of *C. rufa* in Scotland, these well-tracheated gills do have a respiratory function – unlike many anal gills – and enable the larva to live submerged, with only infrequent visits to the surface for air. Perhaps because they have slowed down the tempo of development by this means, they may take up to five years to pupate, at least in captivity.

The larvae of marshy habitats have also produced an enterprising form in the genus *Chrysogaster*, which, instead of striving to keep contact with the outside air, has hardened and sharpened the posterior spiracles, and uses them to pierce the stems of aquatic grasses. Varley [294] describes how the larva of *Chrysogaster hirtella* pierces the stems of *Glyceria aquatica*. We have met this habit already in the *Mansonia* (*Taeniorhynchus*) mosquitoes, and shall meet it again in the genus *Notiphila* of the acalyptrate family Ephydridae. Aquatic insects of other Orders, notably the Donaciine beetles, have evolved a similar habit.

These examples have been taken from hover-flies of the north temperate region, because these are the ones that have been most studied. Other regions of the world, and more particularly the tropics, provide a wide range of habitats in rot-holes, fallen trees, and other media rich in organic decay, in a hot, humid atmosphere. Many hover-fly larvae must live in such places, and many interesting life-histories remain to be discovered, perhaps with larvae that have developed a line of their own, as *Callicera* has. A few have been recorded in rot-holes, along with the predaceous larvae of Tabanidae and Stratiomyidae, but the life-histories of a great many remain unknown.

Turning now to those hover-flies whose larvae have become terrestrial instead of aquatic, we have already mentioned *Microdon*, highly modified to survive in an ants' nest, and *Volucella* getting by in nests of bees and wasps without any special aid. A very large number of larvae have gone further than this, and become actively predatory, feeding openly upon the masses of aphids that encrust the vegetation of many countries.

These aphidivorous Syrphids, as they are called, have all the marks of a vigorous group of fairly recent origin, actively evolving at the present time. The adults are small, active, very numerous both in individuals and in species, and showing endless permutations of a simple structure and pattern. Many of the little hover-flies of the garden of the genera *Syrphus*, *Pipiza*, *Paragus*, and so on belong to the aphidivorous group, and therefore should be looked on as good friends of the gardener.

The feeding of these larvae has been studied and several times filmed. The maggot-like larvae hatch from eggs laid under a leaf, and are blind, of course: that is to say, though they may detect light and darkness by a general sensitivity of the skin, they have no eyes, and no power of forming an image of their surroundings. They could not actively chase an elusive prey – no maggot can – but the aphids on a stem present what is literally a sitting target. In fact what impresses me most when I see films of aphids and ladybird larvae feeding among greenfly is the way in which the aphids show no reaction at all while the one next to them is being devoured. I suppose it is what one would expect, but the effect is macabre in the extreme.

The Syrphid larva finds its victim by raising the front end of the body and swinging it from side to side until it touches and seizes an aphid. When a victim has been seized, it is held up, away from the surface, and sucked dry, while the fly-larva clings to the plant by the grip of its abdominal segments. Though at first this would seem a difficult way to tackle prey, it has the advantage that the aphid cannot escape by running away. The technique is certainly efficient: starting with a modest three or four a day, one Syrphid larva is said to work up to a daily consumption of fifty or sixty.

Aphid-feeding Syrphid larvae become green as they grow older, with a pale stripe – an attractive pattern, similar to that of some larvae of Stratiomyidae. The presence of green pigments in some flies, either adults or larvae, is something of a problem. In some instances, like the present, it is tempting to think that the pigment may be derived from the food, but this is certainly not always true. The parasitic Hippoboscidae that are green in life live entirely on red blood! In the Syrphid larvae, and in some others such as the green horse-fly larvae that live on the Nile cabbage, it is possibly an aid to concealment, but on the other hand the ladybird larvae that also prey on aphids are dark-coloured and conspicuously patterned.

The diet of these larvae is not entirely confined to aphids. They occasionally attack other aphid-feeding larvae, such as those of flies of the family Chamaemyidae, and some attack larvae and pupae of moths. This has been recorded for the very common *Scaeva pyrastri* and *Syrphus luniger*, indicating once again that choice of prey depends as much on opportunity and ability as on preference. An old record by Giard claimed that larvae of *Melanostoma mellinum* even attacked and fed upon adult house-flies and other Muscidae that were incredibly large, tough, and bristly compared with the larvae; these flies were in a torpid state at the time.

Melanostoma and *Platychirus* are interesting in another way, in that their larvae have been reared entirely on rotting vegetable or fungal matter. *Mesogramma*, on the other hand, is completely vegetarian. The larvae of *M. polita*, when young, feed on the pollen of maize, and become yellow in colour as a result. In the second instar they begin to scrape the tissues of the plant with their mouth-hooks, and to feed on the escaping juices, and in the third instar they make quite serious lacerated patches on the plant.

Hull [145] speaks of a transition from the aphid-feeding group, through *Melanostoma* and *Platychirus*, to the wholly vegetarian *Mesogramma*, but is the reverse not the more likely? Just as we see the bulb-flies as diverging from the mush-feeders, producing their own pulp from healthy bulbs, so we can see *Mesogramma* as a last remnant of a stage in which pulp was produced by scraping at healthy stems. On aphid-encrusted stems some of the aphids would inevitably be devoured, just as the soil-living larvae of Asilidae and other families eat their way through any small animals they may meet. The more concentrated diet resulting from this may have led eventually to giving up attacking the plant, and feeding exclusively on the encrusting aphids.

Allied to, and always mentioned along with the Syrphidae, and perhaps justifiably included in this chapter about 'hover-flies', are two other very much smaller families. Pipunculidae are remarkable for their almost spherical head in both sexes, and for their magnificent powers of hovering, two items that may be related to each other, as we have seen. Verrall noted that a Pipunculid could hover in the small space within a folded collecting-net, and they commonly hover in among the branches of the vegetation. Perhaps the all-round vision provided by their near-spherical eyes allows them to maintain station by watching all the surrounding objects simultaneously.

The spherical head is often huge compared with the rest of the body, which is only a few millimetres long. About 400 species are known, principally from the countries where flies have been most thoroughly collected. There must be very many species still hovering away among the tangled tropical vegetation, that have still to fall into a collector's net. Certainly the larval food-material is abundant enough. They live as internal parasites of plant-bugs of the Order Homoptera, particularly frog-hoppers (Cercopidae), leaf-hoppers (Jassidae) and lantern-flies (Fulgoridae). The egg is laid on the host, and the young larva penetrates into the abdomen, where it feeds and grows until it fills the cavity. It is said to breathe through its skin during the first larval instar, and only acquires working tracheae in later life. The fully fed larva breaks out between two abdominal segments and pupates in the ground.

The other family is the Platypezidae, also small, obscure flies, and recognized by the flattened feet and by the distinctive wing-venation. The males are often sombre in colour, the females more brilliant. Both sexes are even smaller than Pipunculidae, with a strongly rounded head in which the eyes touch in males, and in the females of some species. This is consistent with the behaviour of the adults, for swarming in the air is characteristic of the males of most species and of the females of some.

Males of the genus *Microsania* are the 'smoke-flies', which mysteriously appear in the smoke of wood fires out of doors. It is one of the oddities of nature that these flies, so seldom caught by the methods of ordinary collecting, can be conjured up, as it were, by the demoniacal process of lighting a fire.

They do swarm elsewhere. Russell found that swarms were quite common near the upper branches of a sycamore tree, and that individual flies could be taken nearby by sweeping, and Fonseca found the *females* of three species swarming together, while the males rested on foliage. Collin noted that specimens could only be taken by sweeping for a brief period after a shower, suggesting that they may spend most of their time on the wing. There may be a parallel here between these flies and some of the canopy-dwelling flies of the tropical forest, such as *Chrysops silacea*, which also assemble in the smoke of fires. It is thought that the movement of the swirling smoke may attract attention. In the case of biting flies it suggests the nearness of a potential victim; in the case of hovering flies it may imply convection currents, like the 'thermals' to which glider pilots steer. Recently, however, Kessel has produced evidence that the smell of wood-smoke is attractive to the flies, in the absence of movement.

In a later chapter on swarming in flies, I shall have more to say about the way in which flies escape notice when they are scattered, and become a pest and an 'infestation' if anything causes them to assemble in one place.

The larvae of Platypezidae live in various fungi, and are peculiarly developed, presumably in adaptation to the soft, glutinous nature of the medium when decay has set in.

The links between these two families are obscure and speculative. The life-history of Pipunculidae seems as it were an apotheosis of Syrphid life, with even more highly developed hovering, and the predatory larva having become an internal parasite. On the other hand Platypezidae form a link back to the Brachycera. They have some features recalling Dolichopodidae, and a small genus of flies

Figure 28. 'The most wonderful fly in the world' – *Sciadocera rufomaculata* was so-called by Bezzi because it combines features of Empids, Dolichopodids, Phorids, and Platypezids, at one of the most obscure points in the evolutionary plan of the flies.

called *Sciadocera* (which Mario Bezzi called 'the most wonderful fly in the world') with one species in Tasmania and one in Patagonia, is said to bridge the gap between Dolichopodidae, Platypezidae, and Phoridae. While these do not fall into a neat pattern, they do at least seem to show signs of a common ancestry (figure 28).

Conopidae are another family that, like the Phoridae, have followed a divergent line of evolution, leading them away from close relationships with any other living flies. Hennig [123] gives them a separate section, the *Archischiza*. We may take advantage of the fact that this is a work of natural history, and consider them alongside the hover-flies, rather than in their true systematic position.

Conopidae look like wasps, with a nipped-in waist, and sometimes with an elongate, club-shaped abdomen. Even the colours are wasp-like. Larval life is passed as an internal parasite of Hymenoptera, and so once again we are faced with the question of mimicry: do the adult flies profit from their resemblance to Hymenoptera, and have they perfected it by natural selection?

It would seem not. There is no specific mimicry, such as that between the robber-fly *Hyperechia* and the carpenter-bee *Xylocopa*, and there is not even a more general protective resemblance. *Conops* looks like a wasp, and does indeed parasitize wasps, but these range from *Vespula*, the common, plump wasps of the jam-pot to *Ammophila*, the slender, emaciated sand-wasp. *Conops* also parasitizes the bumblebees, which it does not resemble at all. Similarly, *Physocephala* looks like a sand-wasp, but only goes for bees. And in South America *Stylogaster* has been seen to fly over the columns of driver ants and drop its eggs, which the ants pick up and take to the nest. An exactly similar habit has evolved independently in the Muscid fly *Stomoxys ochrosoma*, which has quite a different appearance from either *Stylogaster* or the ant.

One cannot say that the appearance of Conopids is never advantageous, but in the main the coincidence that wasp-like flies should parasitize Hymenoptera tells against the theory of mimicry rather than supports it. It seems more likely that we are seeing the result in Syrphidae and Conopidae of two independent lines of evolution from closely related ancestors, with certain ancestral traits persisting. Some Conopidae even have traces of a spurious vein like that characteristic of Syrphidae. That the end product in each family should happen to be like a wasp is interesting, but why should wasps have a monopoly of that particular appearance?

Conopid larvae are found in the abdomen of the host-insect, and ultimately grow to occupy the whole of it. They pupate there inside the last larval skin, and emerge by breaking open the skin of the dead host. Their parasitism is thus of the destructive kind to which Wheeler applied the name *parasitoid*. The larvae are said to be able to alter their shape to an unusual extent, to conform to the space available in the host. This is not hypermetamorphosis as we have seen it in, say, Cyrtidae, because there is no succession of definite shapes adapted to different ways of life: in particular there is no actively mobile first stage. At first the larva breathes through its skin, but soon the hind spiracles are able to draw air through the tracheal system of the host. It is thought that the very young larva, which has penetrated through the intersegmental membrane of the abdomen of the host, lives on the blood of the body-cavity, and only later begins to destroy the tissues.

Adult Conopids have no common name, except that they are unjustly called 'thick-headed flies' in some books. Although about five hundred species are known from all over the world, they are nowhere familiar insects. Some are over an inch long, but they are much less conspicuous than hover-flies because they spend most of their time sitting on flowers and basking in the sun. As is general in acalyptrate flies, the eyes do not meet even in the males, and hovering for mating purposes is not one of their habits. They pursue worker bees and wasps on the wing, and it is believed that they lay eggs on the abdomen of the host. According to Cumber [54] the workers are able to forage until the larvae are grown quite big, and then they either fall by the wayside, or return to die in the nest.

Chapter 14

Compost- and Dung-Flies

THERE REMAIN NOW a large and varied assortment of what are called the 'higher flies', the Muscaria, or flies that more or less resemble the house-fly, *Musca domestica*. They are considered to be one great evolutionary branch, having in common the possession of a ptilinum, the sac in the head that is blown out when the adult fly is escaping from its puparium, and forcing its way out of the soil.

Figure 29. Another sponge-like proboscis, that of the Sciomyzid *Elgiva albiseta*, used for mopping up liquid food.

This great group comprises about two-fifths of all known flies, which may be as many as 30,000 species, and it is always split into two sections, the calyptrates and the acalyptrates. These take their names from the presence or absence respectively of the *calypter*, or squama, a lobe at the base of the wing that covers the halteres when it is well developed. As usual, there are exceptions in each group, but the division of the families into these two sections is nevertheless a natural and convenient one.

The calyptrates are the very bristly, buzzy flies most like the house-fly and the bluebottle. They are not really the equivalent of the whole of the acalyptrates, but rather are an early offshoot that has gone ahead in evolution at a greater rate than the others. We shall come to them in due course, but first we must take a look at the more varied range of the acalyptrates, because these derive more directly from the flies that we have already studied.

This is the most difficult group of flies to write about. Very few of them have a common name, and even their scientific names are little known except to specialists, who disagree among themselves about how the acalyptrates should be classified. They – the flies, not the specialists – are an example of the evolutionary 'brushwood' that I have mentioned, a mass of diverging lines that defy attempts to reduce them to order. Fortunately this is a work of natural history, and so I feel entitled to disregard the problems of the systematist, and to talk about these flies according to their habits and life-histories.

I have thought it convenient to deal with the acalyptrates according to the food-materials of the larvae, and to arrange these in four groups: (a) compost-like vegetable debris, including rotten wood, leaf-mould, and peaty soil; (b) dung, carrion, and similar materials of higher protein content, leading on to carnivorous larvae; (c) the tissues of living plants, stems, roots, living fruit, and berries, including leaf-mining and gall-forming larvae; (d) the products of fermentation, rotting fruit, and the fermenting sap from wounded trees. The first two will be the subject of this chapter, and the other two of the next.

It must be noted that this is a classification of food-materials, and not of the flies themselves. As I stressed in the Introduction, there have been many evolutionary experiments in the past. The acalyptrates are pre-eminently a group in which this experimenting is going on, and we find that a family that is primarily compost-feeding has produced a few leaf-miners, while a vegetable-feeding family may have a few carnivorous members.

The first of the four types of food-stuffs is that basic compost-like material from which all groups of flies seem to have started, and the other three are media that are more concentrated in nutritive materials. There is a notable absence of filter-feeding aquatic larvae among acalyptrate flies, which are essentially a terrestrial group. Those that have taken to the water are relatively few, and either prey on other insects or feed on the stems of aquatic plants. Neither of these can be called a truly aquatic habit, in the sense of being a way of life that could not be practised on dry land. Rather are they terrestrial practices carried over into water.

An example of a basic acalyptrate family is the Lauxaniidae or Sapromyzidae, the larvae of which breed in decaying vegetable matter, leaf-mould and so on. Those of *Halidayella* (or *Calliopum*) *aenea* mine in the leaves of clover, and make a gall in the flowers of *Viola*. The adults are mostly soft-bodied, plump little flies, only a few millimetres long, sometimes pale yellow, coppery, or black. A few of them have patterned wings and coloured eyes, but in general they avoid all structural extravagances. In the tropics of the Old World, however, they have an offshoot in the 'beetle-flies' of the family Celyphidae, where the scutellum of the thorax is so much enlarged that it not only conceals the abdomen, but also acts as a fixed cover for the wings in the resting position (figure 30).

Dryomyzidae are similar flies, more robust perhaps, and fond of rather more actively putrefying materials. *Neuroctena anilis*, a handsome brownish yellow fly with spotted wings, has a liking for the stinkhorn fungus, *Phallus impudicus*. Helomyzidae are a more numerous family of small, soft-bodied flies, often with a grey thorax and a yellow abdomen, which for various reasons look as if they had an evolutionary future before them. They have, as it were, begun to broaden their taste in larval food material, and besides compost-like substances some of them now breed in dung, dead animals, fungi, and such things as rotting potatoes, which culture a variety of moulds. These habits take both larvae and adults into the burrows of animals, including human habitations, and an odd specimen of *Tephrochlamys* is often to be seen on windows indoors. This association with burrowing and cave-dwelling animals is a promising evolutionary line, which in the past may have been the origin of such families as Mormotomyiidae (which we shall return to later) – not to mention our own friend, the house-fly. An immediate advantage is that the fly is protected from extremes of weather, and lives in a relatively equable temperature all the year round.

Helomyzidae are as yet a generalized family: that is they are not obviously committed to any particular way of life, and that is a further reason for looking on them as a 'promising' line. So are the little yellow Chiromyiidae, only 1 or 2 mm long, that breed in burrows, nests, caves, and such places. They are so obscure that they

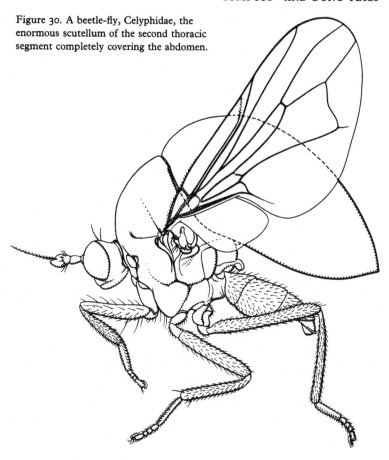

Figure 30. A beetle-fly, Celyphidae, the enormous scutellum of the second thoracic segment completely covering the abdomen.

are seldom noticed except on windows indoors, a characteristic so marked that one common species is called *Chiromyia oppidana*.

Other flies immediately related to these have already committed themselves. The seaweed flies, or kelp flies of the family Coelopidae, for instance, have exploited the banks of wrack, or stranded seaweed that lies and rots on the shore above high-tide level. They are a fascinating family. The rotting seaweed maintains a warm, compost-like environment, and provides a nourishing slime upon which the larvae feed. The flies normally pass their whole lives on the weed thus, and the adults are flattened, leathery insects, greasy of surface, which possess the useful ability of rising to the surface and taking off from the water.

Seaweed flies are abundant in the wrack all the year round, and indeed for climatic and meteorological reasons they are often most abundant in winter. Like all flies that keep out of the way of man, their numbers are not appreciated until something makes them come out into the open, or move into fresh surroundings. At times they fly in a band, a little above the ground, and occasionally they stray far inland. One species, *Coelopa frigida*, seems to be the restless one, and since this species is also strongly attracted by certain odours, particularly that of trichlorethylene, it can be a great nuisance when large numbers of the flies congregate in shops, garages, and particularly dry-cleaners'. These habits will be described in more detail later, when we come to discuss swarming and other aggregations of flies, but one example of their activity deserves mention here as a part of the natural history of the family.

In the autumn of 1954 a great many *Coelopa frigida* were reported from inland localities in southern England, between the English Channel and a line just north of London. After a close study of the records I was satisfied that these flies had come from the south coast, that they were generally distributed through the intervening countryside, and that the reports came from places where the odour of trichlorethylene and kindred substances causes the flies to assemble. In spite of rumours to the contrary there was never any evidence that they were able to breed anywhere away from the seawrack, and the flies quickly disappeared from inland localities when the weather turned colder. The invasion was clearly linked with an abnormally long spell of southerly winds, and has probably occurred from time to time in the past, but the scattered flies have not been detected, and the chemicals that cause them to assemble were not previously in general use. Paramonov has recently revived the suggestion that some Australian *Coelopa* may breed inland, but no direct evidence is yet forthcoming.

Coelopa frigida is the largest and best-known species of the family, and is now being bred as a laboratory animal at King's College at Newcastle. It can easily be reared in glass jars, like *Drosophila*, using selected cut pieces of seaweed stem as a medium, and the bred flies show an interesting range of intraspecific variation, both structural and cytological.

The hind spiracles of Coelopid larvae are prominent, and elaborately fringed, in a way that helps to keep them from flooding when the larva is submerged. A rather less effective device is seen in the larvae of *Helcomyza ustulata*, a large, Dryomyzid-like fly that also haunts the seashore, its larval habit less closely specialized to the kelp than is that of *Coelopa*. These flies, and others that frequent the seashore as part of a wider distribution, are discussed at length by Ardö [5].

The families of acalyptrates so far discussed are known mostly from north temperate regions, perhaps because they have not been collected in tropical countries, but perhaps also because it is in the winters of temperate climates that their particular trick of exploiting the self-heating wrack beds has most value. They also occur in southern temperate latitudes, and it is surprising how closely the northern and southern *Coelopa* resemble each other.

A small step from this basic group has been taken by a large number of families which augment their diet with animal protein, either by living as a larva in dung or carrion, or by being carnivorous as a larva, or as an adult, or in both stages.

The basic family of this next series is perhaps the Sepsidae, a world-wide group of dung-breeding flies of a characteristic appearance: round-headed, elongate, smooth, shining, ant-like in general shape, and sometimes with a dark spot at the tip of the wing. Often the males have the forelegs curiously distorted in shape, and equipped with spines and bristles, apparently used for seizing the female in copulation. These are very active flies, which run about restlessly on the foliage and wave the wings up and down, or shake them with a peculiar tremor. This is a common habit among certain families of acalyptrate flies, and undoubtedly is a part of courtship behaviour. Those Sepsidae that have a spot at the wing-tip make great play with it on these occasions.

Although not many species of Sepsidae are known – Séguy says 150 for the whole world – as individuals they are often numerous. *Themira putris* breeds in the sludge of sewage works, and occasionally a huge mass emergence releases vast swarms that infest neighbouring buildings. The little *Sepsis* with spotted wings are sometimes abundant among heather and low, scrubby vegetation. Occasionally the males can be seen in a mating flight under the shade of low bushes, and it is said that the females emit an odour that can be detected by humans. It is curious that these little flies seem to need so many aids to success in copulation – spotted wings and male display, aphrodisiac odour by the females, special male clasping organs – when other flies manage without any of these.

Attractive though the adults are to watch at their display and to study under the microscope, the habits of the larvae are unpleasant to the human way of thinking. They breed in excrement and carrion, particularly if it is in a semi-liquid state.

Closely related to *Sepsis* is *Piophila*, given a separate family because it avoids the extravagances of the Sepsids both in structure and behaviour. Male *Piophila* indulge in a queer little mating dance round a stationary female, but have none of the wing-spots or leg-structures to help them. The larvae of *Piophila* feed in highly nutritious animal materials of a drier kind, particularly in preserved meats, ham, bacon, dried fish, and cheese – and corpses.

The larva of *Piophila casei* is well known as the cheese skipper, because any larva that finds itself in an exposed situation will leap suddenly into the air, rising perhaps 2 or 3 in, and covering a horizontal distance of several inches. This leap is achieved by bending the body into an arc until the mouth-hooks can grip the posterior tip of the abdomen, where *Piophila* has two small papillae to catch hold of. The abdominal muscles contract, and when the grip is suddenly released, the larva is flung into the air.

Larvae of several other acalyptrate families can do this, notably the Clusiidae – a family of breeders in rotting wood – and some leaf-mining larvae of the Agromyzidae. The leap is obviously a way of getting quickly out of an unfavourable situation, and avoiding a tedious crawl over an exposed surface. Of course such a leap in the dark – literally as well as metaphorically, since the larva is blind – may end in even more dangerous surroundings. Once again, what puzzles us about such a habit is not why it should have evolved at all, but why these particular flies should have evolved it when so many others have not. This is especially pertinent because *Piophila casei* is notorious for eating out a cavity in, say, a large ham, without showing any trace on the surface: what use is the power to leap inside a ham?

Cheese skippers are often common in cheese, and are sometimes eaten in a mistaken epicurean fervour. It is unwise to do so, as the larvae are extraordinarily resistant, and can survive in the human intestine, and lacerate its walls with their mouth-hooks. As a pest this larva is most often reported from provision shops, where it breeds in scraps of ham, bacon and so on around the slicing machines, and in tanneries and fur stores, since it feeds happily on hides of all kinds. Other species of *Piophila* live in carrion, but at a later, more mummified stage than is favoured by the larvae of Sepsids. In particular they will clean bones of the last fragments of tendon and connective tissue. When a series of seventeen elephant skulls was exposed to the weather beneath my window at the museum a fine collection of *Piophila* was a small compensation for the smell. Corpses left lying in the open air may nourish great clusters of *Piophila* larvae (plate 25).

Sepsidae and Piophilidae are world-wide, and the fact that they are best known from temperate countries may be partly because these areas are more urbanized, and better collected. A mainly tropical relative is the curious family Diopsidae, the stalk-eyed flies, in which the small eyes are borne at the end of stalk-like prolongations of the head-capsule. In the way they sit about they remind one of the Sepsidae back home, but recent discoveries about their breeding-habit place them with the plant-breeders of the next chapter.

Micropezidae, also called Calobatidae or Tylidae, are an allied family of abnormally long-legged flies that prey on other insects, aphids, and small flies, which they stalk through the vegetation. Their larvae develop in rotting wood, vegetable matter, or fruit, often secondarily infesting a burrow made by other insects. The flies are world-wide though more spectacular in the tropics. With dark, blue-black wings, and brightly coloured head, some tropical Micropezidae are 'mimetic' in the sense that they look like anything but a fly. Some of the males are known to 'fight', or at least to measure up to each other in a miniature version of the behaviour of rival male birds.

Coming back from these more showy adult flies to the inconspicuously successful, the Borboridae or Sphaeroceridae claim a leading place. These little black flies are easily recognized by the way in which the first of the small tarsal segments of the hind leg is enlarged into a broad pad. This is found in both sexes, and I do not think anyone knows why. Borboridae are universal dung-feeders, and occur everywhere, sometimes in very large numbers. The association in this habitat between larvae of Psychodidae, a primitive nematocerous family, Sepsidae, and Borboridae is shown by the work of Laurence [170]. Out of 18,860 larvae that he collected from cow-pats, 8,269 were Psychodidae, 2,267 Sepsidae, and 4,268

Borboridae, a total of 14,804, leaving only 4,056 to be divided between 26 other groups of animals. Crisp and Lloyd [53] bred no
Borboridae from their patch of woodland mud, but found great

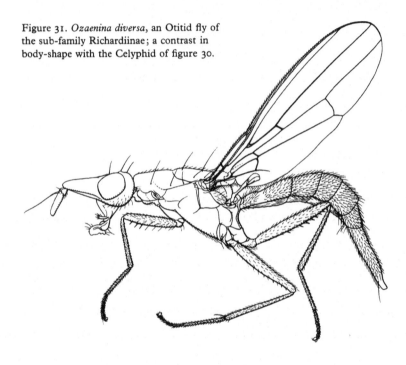

Figure 31. *Ozaenina diversa*, an Otitid fly of
the sub-family Richardiinae; a contrast in
body-shape with the Celyphid of figure 30.

numbers of larvae of two species of this family in fouler mud in a
different locality.

Individual species have their preferences, for birds' nests or the
burrows of small animals, or litter on the seashore. *Thoracochaeta
zosterae* is a permanent inhabitant of the beds of seaweed, along with
the other seaweed flies that we have mentioned. Some even enter
into the water that accumulates in bromeliaceous plants.

Perhaps the most interesting aspect of the natural history of
Borboridae is the readiness with which they have evolved short-
winged, or entirely wingless forms on islands, high mountains, and
other remote places. Like other flies that have exploited sheltered

larval habitats, such as dung, they are able to breed through the winter of temperate or even sub-Arctic regions, and loss of flight is sometimes an advantage in keeping them from straying away from their protected habitat. Byers [38] makes the interesting suggestion that in very cold climates adult flies are not able to warm up enough to fly, and that there is no selective advantage to favour the possession of unusable wings. Professor O. W. Richards has made a speciality of the study of these short-winged Borborids.

In the familiar pursuit of still more nutritious larval food, the remaining families that I have included in this group have all turned to a more directly carnivorous diet. Sciomyzidae have the same general appearance as the compost-breeding Sapromyzidae and Helomyzidae, but have progressed a stage further. They were long regarded as organic scavengers and carrion feeders, like the Phoridae, and like them fond of feeding and pupating in dead snails. It was known that they sometimes attacked living snails, including aquatic ones, and in recent years Sciomyzidae have attracted a good deal of attention as potential killers of watersnails.

The work of Professor C. O. Berg and his collaborators has exposed an interesting picture of Sciomyzidae as a family in process of adapting themselves from a saprophagous diet, feeding on any decaying animal remains, first to specializing in dead snails, like many Phoridae; then to killing a snail for their own use; and finally to preying actively on a succession of living snails. Thus Neff and Berg [214] say: 'Larvae belonging to the genera *Sepedon* and *Dictya* characterize the aquatic, predatory Tetanocerinae, which kill their prey quickly, operate with efficiency in water as well as on land, and commonly kill more than a dozen snails during the growth of each larva. At the other extreme, terrestrial, parasitoid larvae of some Sciomyzinae may feed on a snail for seven or eight days before killing it, and confine all of their feeding to one snail. Other species have intermediate and less fixed habits. The larvae of *Atrichomelina pubera*, for instance, resort to predatory, parasitoid, or even saprophagous feeding, depending on the food available and on the intensity of the intraspecific competition.'

When we speak of 'aquatic' Sciomyzid larvae we use the word loosely for those that kill aquatic prey. The larvae still need to breathe air through their posterior spiracles, and if the snail sinks and drags them down they must soon abandon it. It has been shown that in certain circumstances the snail and larva between them can retain enough air to support them both, aided by the resistance to wetting of the area surrounding the posterior spiracles of the larva.

Experiments in the effective control of the snail vectors of schisto-somiasis have been inconclusive so far. Some of the more powerful operculate snails – that is, those that can close the opening of the shell with a rigid disc called the operculum – react to attack by doing this with such force that the wretched larva may be cut in two. *Sepedon macropus* has been introduced into Hawaii in an attempt to control the snails that carry the liver-fluke *Fasciola gigantea*, a most important pest of beef and dairy cattle.

Milichiidae and Carnidae are small, often tiny, blackish flies which breed in dung, or scavenge in burrows and nests. Adult flies feed from flowers, but are so tiny that they are seldom noticed unless they are numerous. They also like oozing organic liquids, and some of them allow themselves to be transported by predatory insects such as robber-flies, or bloodsucking bugs, or by spiders, for the sake of the spoils when prey is killed and blood flows. They have been called jackals of the insect world. *Desmometopa sordida* is said to have been found licking the pollen from the leg of a worker bee, and certain African species of *Milichia* behave like the mosquito *Harpagomyia* that we have already met, and persuade *Cremastogaster* ants to regurgitate for them.

Other, more domestic species are better known to the householder. *Madiza glabra* is a tiny fly with a shining black body, which some-times appears indoors in great numbers for no obvious reason. It seems to result from a mass hatching out of some unsuspected pocket of animal matter, rather than to be a mere trapping of wandering flies like the hibernating swarms we shall mention later. No simple explanation has yet been found for this, nor for the Borborid *Leptocera caenosa* that equally mysteriously appears indoors. Various species of *Meoneura* that come into houses are even more tiny, little more than 1 mm long, and yet so perfectly constructed on their Lilli-putian scale that their complicated male genital organs provide excellent characters by which to distinguish the different species.

In the family Carnidae to which *Meoneura* belongs we suddenly come upon one of those freaks of evolution that should deter us from accepting too easy an explanation of the process. Not content with a larval life in the debris of a nest, the adults of *Carnus hemapterus* settle down in the plumage of a nestling bird, and commit themselves to it by breaking off their own wings at the base. This is a trick that we shall meet again in the deer ked, *Lipoptena cervi*, but that is a moderately big fly, and is exposed to the danger of losing contact with an active host; *Carnus hemapterus* is a tiny fly less than 2 mm long, and is completely lost among the downy feathers of the nestling bird. It would seem, in fact, that the loss of wings might be less to avoid flight, than to make it easier for the fly to crawl between the bases of the feathers.

Both sexes are fully winged at the time of emergence from the puparium, and they can fly well in their search for a nestling. After the wings have been lost, the flies become physogastric, that is the membranous areas of the abdomen distend, leaving the sclerotized plates isolated from each other. As might be expected, this effect is more pronounced in the females than in the males, and compares with what takes place in certain Phoridae and Borboridae in comparable circumstances. It is commonly said that the adult flies suck the blood of the bird, but Bequaert, who studied this fly in some detail, though not in life, concluded that this is probably not so. The proboscis is stiffened in the stem of the labium, but the tip has soft hairs, and does not seem capable of piercing skin. Bequaert thought that it is probably used to scrape greasy debris from the skin.

A similar nutritional pressure has produced a different response in other families. Thyreophoridae are little known, but appear to be widespread, and to flourish on the cadavers of larger mammals, such as horses, asses, and dogs. Only about a dozen species of this family are known, and so few scattered over the world suggests a group that is shrinking towards extinction. The males have the scutellum of the thorax grotesquely drawn out to cover most of the abdomen. Apart from these peculiarities, Thyreophoridae are a sideline from the compost-feeding Lauxaniidae, etc, and so apparently is another uniquely isolated nest-fly, *Neottiophilum praeustum*.

The nests of birds are attractive to a variety of insects, because they provide at the same time shelter, concealment, a supply of highly nutritious guano, and the opportunity to attack the birds themselves. Nestling birds are particularly vulnerable, because they are able neither to escape nor adequately to defend themselves. *Carnus hemapterus*, which we have just met, has taken perhaps the obvious line of adaptation, becoming a scavenger in the nest as a larva, and transferring itself to the birds themselves after emerging as an adult insect. This is also what the Hippoboscidae have done, except that they have given up larval feeding altogether.

The nest-fly, *Neottiophilum praeustum*, has taken quite a different line. The larva of this fly emerges from hiding in the nest material, feeds upon the blood of a wretched nestling, and crawls back into hiding to digest its meal. It feeds intermittently until it is fully grown, and having then taken its fill of protein food, it passes the adult life as a harmless, flower-feeding fly, hardly to be distinguished from the dung-breeding Dryomyzidae.

Those ornithophilous flies that spend their adult life on the bird have the advantage that they are distributed without any effort. A blood-sucking larva that only visits its victim has to solve the problem of dispersal. The larvae of *Neottiophilum* do not move from one nest to another, and the dispersal of the species – indeed its continued survival – depend upon the diligence of the egg-laying female in visiting a series of birds' nests, instead of laying all its eggs in one. These nest-bound larvae, as one might expect, are to be found chiefly in the nests of the common, small birds, blackbirds, thrushes, finches, and so on, which are easily visited by a hedgerow-haunting fly. *Carnus hemapterus*, with its free adult transport, has a wider range, and also occurs on birds of prey, to which it has transferred when its first host was killed.

Last of this series of families come the predatory Chamaemyiidae, inconspicuous, greyish flies, that are little known as adults. Their larvae prey on plant-lice and scale-insects, and move about the colony feeding, and in their turn sometimes being eaten by various Syrphid larvae. The adult flies are said to suck honey-dew from the surface of leaves and stems, and it is possible that the larva, too, takes a mixed diet. Some Chamaemyiid larvae live in the galls of other flies, Cecidomyiids, and some acalyptrates – eg in the reed *Phragmites communis* – and it is probable that the Chamaemyiid larva is a predator or 'parasitoid' upon the original occupant of the gall.

An interesting off-shoot of this family is the bee-louse, *Braula caeca*, a tiny, brown, mite-like insect, with a round, hairy body, strong legs, small eyes, and neither wings nor halteres. There is only the one species, which is found clinging to the bodies of the honey-bees. Though its thoracic structure shows it to be a wingless fly, the appearance of *Braula* is entirely adaptive, suiting its way of life, but giving no help in relating it to other, more normal flies. For over a century it was placed in the group Pupipara, with the wingless bat-flies, Nycteribiidae, until it was found to lay eggs and have a normal maggot-like larva. The most comprehensive recent study of

Braula was made by Imms, who decided that its nearest relatives were the predatory Chamaemyiidae, and cleared up a number of other doubtful points about its life-history.

The eggs and larvae are to be found on the inner surface of honey-cells, and Imms describes an ingenious way of exposing them. The caps of a series of cells are cut off in a slice, and then placed just outside a hive, when the bees will clean off all the honey and leave the wax tunnels in which the larvae are found. The young larvae of *Braula* tunnel into the wax cap of the cell, and may pass from one cell to another. The mass of tubes in a badly infested comb can be seen from the outside, before the cap is cut off. The larva swallows wax grains and some pollen, and apparently digests the wax with the aid of intestinal micro-organisms in the same way as the wax moth, *Galleria mellonella*. One of the main reasons why Imms thought that *Braula* was related to the Chamaemyiidae is that the mouth-hooks are similar, and suggest that perhaps *Braula* once had a carnivorous ancestor. There are three larval instars, in the typical way of the Cyclorrhopha, and the pupa is formed within the last larval skin, though this is not hardened into the customary barrel-shape. The larva has already weakened a part of the overlying cell-cap, and when the adult insect inflates its ptilinum, this forces an opening through which the insect can crawl.

The adult *Braula* is said to cling mostly to the junction between thorax and abdomen of the bee, but to move towards the head when it wants to feed, and there to take saliva from the mouthparts of the bee. This reminds us of the many insects that take food from ants, and indeed none of the evolutionary devices of *Braula* is unique. The interest of this fly is in its lonely evolution, which raises again the familiar question: how did it get started in this direction, and why have not others done the same?

This is a convenient place to mention another solitary, peculiar almost wingless fly, by way of comparison with *Braula*. It was discovered in the mid-nineteen-thirties, at a place called Ukazzi Hill in Kenya, and this is still the only known locality. On top of the hill is a large boulder, or rocky outcrop, which is split from top to bottom by a narrow crack about a yard wide. Bats roost in this crevice, and with the bats live the flies. Both are normally inaccessible, but when a heavy tropical storm breaks over the hill a mass of bats' dung is washed down, together with hundreds of eggs, larvae, pupae, and adults of the fly.

Mr H. B. Sharpe first sent two males of the fly to the late Major E. E. Austen, with whom I shared a room at the time, and very excited he was about it, too. He called the fly *Mormotomyia hirsuta*, 'the hairy terrible fly'. Nearly fifteen years later Dr V. G. L. van Someren was able to go there under the right conditions, and collected a good series of all stages, and the late Dr F. I. van Emden subjected them to careful systematic analysis.

The adult fly is quite big, nearly an inch long, and looks like a hairy, six-legged spider. It has a big head, with small compound eyes; long, hairy legs; and wings reduced to mere rudiments. The early stages show that *Mormotomyia* is a normal cyclorrhaphous fly, and van Emden demonstrated that the adult has a ptilinum in the head. From a number of small pointers he concluded that the family Mormotomyiidae is intermediate in evolution between the acalyptrate families, and the dung-flies or Coryluridae (Scatophagidae), which begin the series of calyptrates.

The larvae feed in the dung of the bats, and the adults appear to suck sweat and other body-secretions from the bats themselves. Biologically, therefore, there is an obvious parallel with *Carnus hemapterus*, but instead of having wings and then breaking them off, *Mormotomyia* is able to dispense with them entirely. In fact this fly is admirably adapted to its little world, and evidently flourishes there. But what interesting speculations it raises. Does it occur anywhere else, or has it really gone through all its long evolution in that one crack. One cannot believe that. It might be possible to find other colonies of the fly if one could follow bats to see if they had alternative roosting-places, but in fact no one yet knows what species of bat is involved. The bat, too, might be new to science. Dr van Someren tried to shoot a bat by firing a gun into the crack, like shooting an owl in a chimney, but without success.

Such a crowded mass of eggs, larvae, and pupae as was found here would surely attract parasites and predators. It seems a melancholy probability that Ukazzi Hill is one of the last, if not the very last colony of this weird fly in existence. In Samoa, Buxton found unidentified larvae in bats' dung, but as he reported no adults, it seems unlikely that they belonged to a *Mormotomyia*.

Chapter 15

Vegetable- and Fruit-Flies

THERE IS NO REAL division between these families and those of the preceding chapter. Lines of evolution in acalyptrate flies are little understood, and undoubtedly such larval habits as that of feeding in the tissues of plants have evolved independently a number of times. The families that I have left for this chapter are some of the most clearly defined of acalyptrates, both structurally and in their natural history.

We may begin with the versatile and successful family Ephydridae, sometimes called shore-flies, because many of them haunt the edges of ponds, streams, marshes, and so on. In a number of ways they parallel the Dolichopodidae: they even look like them, and it is very easy to mistake one momentarily and waste a long time trying to identify it in the wrong family. Examined in more detail, particularly about the head, Ephydridae can be recognized as more successful relatives of the Dryomyzids and Sapromyzids (plate 21).

It is said that nearly 1,000 species of Ephydridae are known, and in view of the versatility of the family it is certain that many more will be found, especially in tropical countries. Starting as usual in decaying materials, there are one or two Ephydrids that are highly successful there, provided that the organic content is high. Crisp and Lloyd [53] did not find any in the 'woodland mud', but larvae of *Scatella* appeared when the mud became fouler, and the authors record great numbers of adult flies at times from sewage filters. They quote an account by Terry of as many as 750 adult flies per square foot of filter surface per week, producing clusters of flies that could be seen from a distance encrusting walls and posts.

It is an easy step from algal feeding to sewage, and thence to carrion. Bohart and Gressitt [24] in their macabre pilgrimage to liberated Guam Island, found *Hecamede persimilis* and *Chlorichaeta tuberculosa* breeding not only in excrement, but also underneath the corpses that were still unburied. Larvae of *Discomyza maculipennis* were no mere surface feeders, but penetrated deep into muscular tissue. Perhaps the most unsavoury habit is that of the almost cosmopolitan *Teichomyza fusca*, which appears wherever there are cesspits, woodwork and other materials soaked in urine, or cadavers at the early stage of putrescence when the collapsing tissues ooze a nutritious liquid.

These larvae, too, are remarkably gregarious, and may assemble in a compact mass that can block drains and septic tanks. I have seen such a mass taken from medieval woodwork that was found in a London excavation, thereby giving a strong hint as to the purpose for which the woodwork had been used. Pupae of *Teichomyza* have tracheal gills, a device which is developed into obvious 'horns' in the pupa of *Brachydeutera*. Larvae of *Teichomyza* are sometimes accidentally swallowed with foul water, bringing about myiasis, as we shall see later. These carrion-feeders are approaching near to parasitism, and this stage is reached in *Discomyza incurva* which is recorded as a parasite of land-snails of the genus *Helix*.

Ephydrid larvae generally have the two posterior spiracles at the tips of a forked process, and readily adapt themselves to liquid habitats. Those we have mentioned obviously profit from this ability to exploit organic liquids, but a great many Ephydrids have gone into properly aquatic habitats. Larvae of Ephydridae have developed prolegs with which they crawl about under water, and when they pupate they cleverly modify the last two of these to grip a submerged stem, like a perching bird. Some feed on algae, but there is a strong tendency to turn to carnivorous habits, which is intensified in water that is brackish and then markedly salt. It is the ability to make this change of diet that makes it possible to colonize saline waters, since the only source of food is the corpses of other small animals that have fallen in and died.

Equally essential is the physiological ability to withstand the osmotic pressure of salt water, without losing too much water through the skin. Ephydrids are pre-eminently successful at this, and colonize salt-marshes and salt-pans all over the world. Thorpe [288] calculated that Mono Lake in California is so salt that its osmotic pressure must be 50 atm: 'The only aquatic insect definitely known from Mono Lake appears to be *Ephydra hians*.'

The star performer in this line is the Ephydrid *Psilopa petrolei*, also studied in detail by Thorpe [287]. The larva lives in the pools of crude petroleum that ooze from the ground in the Californian oil fields, and is remarkable in two ways. One is in being able to avoid being overwhelmed by the oil. We remember that all aquatic larvae, including the near relatives of this one, struggle against being wetted by water, and use the repellent action of an oily secretion to help them in this. We kill larvae of mosquitoes by suddenly replacing the water with oil. How, then, did *Psilopa petrolei* ever make the decisive step from one to the other ? Thorpe suggests that it may have evolved directly from a terrestrial, soil-living ancestor, and not via one that was aquatic. The secondary problem of swallowing oil is less difficult. The intestine is full of it, and the carnivorous larva lives, as it were, on a diet of sardines in oil. The oil is believed not to be digested, but is kept away from the lining of the intestine by the peritrophic membrane, an inner tube of material pervious to dissolved nutrients, which is continuously renewed, and which is secreted by many insects.

The other remarkable feature of *Psilopa petrolei* is its geographical isolation. We shall hear at several places in this book of single species of flies that are now known from only one locality in the world. Usually we explain this as the last outpost, and imagine that the species was once widespread. It is difficult to imagine this of the petroleum fly, since there can surely never have been a much greater abundance of suitable localities than there are now. It would seem that this is an instance of local evolution, an extreme example of parochialism in insects. Thorpe reminds us that the tar pit at Rancho la Brea has existed since the days of the sabre-toothed tiger, and so the fly has had similar conditions for a very long time. Like us, he was impressed by the versatility of the Ephydridae, and says: '. . . if life was possible at all in an oil pool, an Ephydrid would be the first to discover the fact!' He also established that the larvae could not live on filtered oil unless also supplied with animal food such as small insect larvae.

A dramatic ending to this story was supplied by Thorpe's observation that the adult flies could walk on the surface of the oil just as long as only the tarsi of their feet touched it. If any other part of the body made contact the fly was at once trapped, and was devoured by its own offspring!

It seems that the aquatic larvae are closely related to those that mine the stems of water-plants. *Notiphila*, nearly related to *Psilopa*, mines in *Potamogeton*, reeds and water-lilies. Plant-feeding Ephydridae are not confined to water-plants: though many species of

Hydrellia are aquatic, some are terrestrial, mining not only *Potamogeton*, water-cress and duckweed, but grass and other meadow-plants as well. *Psilopa leucostoma*, placed in the same genus as the far-off petroleum fly, has been bred from mines in *Chenopodium*.

Berg has given a most interesting account of the association between larvae of *Hydrellia* and other Ephydridae and *Potamogeton*. He states that besides getting food and protection from the plant, there is circumstantial evidence that the larvae 'normally obtain oxygen from the intercellular gas spaces of these plants . . . *Notiphila loewi* larvae regularly depend on these plants for sufficient oxygen to maintain life, and specimens of *Hydromyza confluens* less frequently do the same'.

The larvae of *Hydrellia* are metapneustic, having open spiracles only on the posterior segment. These spiracles open near the tips of two sharp, hollow spines, which are usually inserted into the tissue of the plant, and are presumed to draw oxygen from it. The pupae are very similar, and the emergent adult flies are able to get out of the water without being trapped, and afterwards to walk on the surface film without getting wet. The larval mines are linear, usually parallel to the midrib if in a leaf. Puparia are formed within the larval mine, and the respiratory spines are inserted into vascular structures such as the midrib itself.

The eggs of *Notiphila* are laid on water-plants, and the larvae have the hind spiracles opening on to spines with which they penetrate aquatic plants, but, according to Berg, none of this genus is a leaf-miner. Instead, they live in the soil at the bottom of lakes, ponds, and streams. It is not clear what the larvae feed on. We might see in this genus a clue to the origin of the plant-feeding Ephydridae from forms living in wet humus of high organic content and poor in oxygen. 'Embedded in this material, and thus isolated from circulating water, these insects have no apparent source of oxygen except from the roots of plants' [Berg]. We might therefore see the trick of tapping plants for air as the first step, followed later by making use of them for food; and perhaps later still by a movement to terrestrial plants.

Clearly, then, Ephydridae are nothing if not versatile. The adults vary in size from the robust *Ephydra riparia*, about as big as a house-fly, down to *Chlorichaeta tuberculosa*, less than 2 mm long. The latter makes a nuisance of itself by clogging the eyes and nostrils of sweating persons in hot countries, just like the 'eye-flies' that we shall meet presently in the Chloropidae. Adult Ephydridae are found in damp localities on land, and on water-plants, feeding on dung or on liquid food materials, or attacking other insects. *Rhynchopsilopa apicalis* in Nigeria is even said to take food from the gullible ant *Cremastogaster*, like several other flies we have already met. The carnivorous Ephydrids are familiar insects of the countryside, and are often armed with enlarged, spiny, nutcracker-like legs. *Ochthera mantis* is so called because it snatches its prey as a praying mantis does, and it may even alight on the surface of water to do so.

Evidently we are seeing in the Ephydridae a family of flies in the full flower of its evolution, and as such they offer attractive material for study, not only to the dipterist, but also to the student of insect physiology and behaviour.

The movement towards plant-feeding that we have seen in the Ephydridae is evident in all the families that remain for this chapter, and in some it has become the dominant source of larval food. Only recently the transition has been demonstrated in the stalk-eyed flies of the family Diopsidae, whose biology was for long a mystery.

The stalked eyes that are so conspicuous a feature would seem to give these flies a wonderful judgment of distance by stereoscopic vision, if the fields of view of the two eyes overlapped sufficiently. The inclination of the eyes, however, suggests that only an insignificant overlap can exist, and moreover the length of the stalks is variable, even in one species, a fact that has led to much confusion in classification. In spite of this fact, the way in which the adult flies sit about on damp vegetation is so reminiscent of the predatory flies of the families Dolichopodidae and Ephydridae that it has been difficult to avoid assuming that adult Diopsidae must live by catching some sort of prey.

A recent paper by Descamps [63] has given us the first precise answer to these speculations. First, the eyestalks. These are not visible in the pupa, and come into being by what Descamps calls '*le déploiement des tiges oculaires*'. While the newly emerged adult fly is distending and hardening its body, the eyestalks unroll with a twisting motion. This process is obviously very much affected by the conditions under which emergence takes place, and it is not surprising that, by small variations in the tempo of emergence, there are produced considerable variations in the length of the eyestalk, even among individuals of the same species.

The more valuable part of Descamps' paper explains in detail how the larvae of the Diopsidae bridge the gap between living in an earthy vegetable compost, and mining into the stems of plants. Larvae of *Diopsis tenuipes* invading a plant that was already diseased would eat the rotting tissue and leave the rest, but larvae placed on a healthy plant would mine into it. He considers that these larvae are in process of evolving from compost-feeders to true plant-feeders. Other species are equally ambivalent. *Diasemopsis silvatica* breeds in moist earth, vegetable debris, and healthy plants, but *Diasemopsis fasciata* in plants only where there are holes already made by caterpillars. Larvae of *Diasemopsis thoracica* mine in the stems of rice, living there for about a month and then pupating in a cavity in the stem. There is never more than one larva per stem, although no evidence of cannibalism could be found. Only young plants were affected, and once they had formed the panicle they became immune to attack.

Other plants attacked by larvae of Diopsidae were sorghum, *Jardinea congoiensis*, *Panicum repens*, *Cymbopogon giganteum*, and millet. Some species had two generations per year, others four or five.

Descamps had nothing to throw light on the feeding habits of adult Diopsidae. He found no evidence of carnivorous habits, and suggests that they may mop up a varied range of semi-liquid foods from leaves: honeydew, the droppings of ants and so on. Diopsidae are very common in many tropical and sub-tropical countries, but seem to do little except sit about. I watched them a great deal in West Africa, without seeing them take any kind of food. In North America, *Sphyracephala brevicornis* overwinters successfully as an adult even as far north as Massachusetts, and its larvae are apparently semi-aquatic.

A few isolated members of other families of flies have the eyes separated far enough to be called stalk-eyed: *Achias* of the family Platystomidae, the Ortalid *Gorgonopsis*, and a few more. In true Diopsidae the antennae as well as the eyes are widely separated, whereas in other families the antennae remain in their normal position, and only the outer margins of the head are drawn apart.

The important family Chloropidae takes up where the Ephydridae leaves off, as it were, because it has a few larvae that are parasitic or predatory, but it is as a family of plant-feeders that it comes into prominence. These are all smaller flies 2 or 3 mm long, often with the green eyes from which the family gets its name, and broadly divided into the Chloropinae, which are mostly striped in yellow and black, and the Oscinellinae which are typically black. Such tiny flies escape notice as individuals, so that although over 1,000 species are scattered all over the world, and clearly very firmly established, the only Chloropidae that are generally known are those that occur in bands, mistakenly called 'swarms'.

By far the most annoying of these are the 'eye-flies' of the tropics, belonging to the genera *Hippelates* and *Siphunculina*. The adult flies are particularly eager to suck up liquid secretions, and apparently do not get enough for their requirements from honey-dew and the moisture of plants. They mass around any animals, settling on the eyes, the nose, and lips, and on the skin where it is moist with perspiration. They obviously must spread conjunctivitis and skin diseases in those areas where people are crowded together and have no effective means of driving the flies away.

Less virulent, but more annoying in their way, are the masses of *Thaumatomyia notata* that make their way into houses in the autumn in temperate countries. The behaviour of these flies is part of the study of swarming that I hope to make in a later chapter. At present we are interested in their life-history. The eggs are laid in the soil around growing plants, including grasses, and the larvae feed on the root-aphis, *Pemphigius bursarius*. There are two generations during the summer, the second of which normally spends the winter as pupae in the soil, from which adults emerge in the spring.

Apparently a certain number of pupae always hatch prematurely, especially in a warm autumn, and from them emerges a 'doomed generation' of adult flies. Driven away from the ground by falling temperatures, they take to the air, and find their way into houses in myriads. It is recorded that 27 litres of the flies were collected in 1927 from one house in Brno in Czechoslovakia! Like others of this family, they are susceptible to both heat and cold, and most of them perish during the winter. The species is continued by the pupae that remain in the soil until spring.

The main interest in this family Chloropidae, however, is in those which have plant-feeding larvae. The frit fly of oats, *Oscinella frit*, and the gout fly of barley, *Chlorops pumilionis* (= *taeniopus*) are only the best known of a series of species that damage cereal crops by feeding as larvae in the young plants. A typical species of this group will overwinter as larvae in the stems of grasses, or of autumn-sown grain, and give rise to adult flies in the spring. After mating, these flies lay eggs at the base of young spring cereals, whence the larva climbs up into the plant and descends into the sheath of leaves, eating its way down the main stem. It pupates in the withering plant, and a second generation of adults emerges.

In many species, especially in higher latitudes, the second generation of adults arrives in time only to lay its eggs, and hatches larvae that can pass the winter snugly feeding, as did their grandparents, in the stems of plants. Some, however, including the frit fly over much of its range, manage to fit in a third generation that feeds on the ripening grain. This generation is less damaging than its predecessor, since it can do only fractional damage to the total yield of the crop.

These two types of life-history, the carnivorous and the plant-feeding, are only two examples of a wide range of larval habitats in this one family. Going back, once again, to compost-feeding, larvae of the genus *Elachiptera* are found in leaf-mould, rotting vegetation, decaying wood and so on, but are best known as feeding on the damaged tissues of plants that have already been attacked by other Chloropidae such as the frit fly. Some other species of *Oscinella* do the same thing, so that in this one genus we have some species that attack healthy plants and some that require them to be already damaged.

The gout fly gets its name because of the swellings that it produces in the stems of the plants in which the larvae feed. Barley is the crop principally attacked, but in some countries wheat is also badly affected. The eggs are laid over a long period in the spring, and rarely more than one per plant, not on the soil, but attached to one of the upper leaves. The eggs are easily killed by drought, moisture, or fungus, so that the damage is not as heavy as it might be. At first the larva behaves as an ordinary plant-feeder, eating its way down the main stem until it reaches the first node. Here the larva stops, and settles down, making itself a cavity, and feeding on the sap; in response the plant enlarges the node into a swelling. This is effectively a plant-gall, a sort of tumour produced in the plant-cells by the irritation of the larval feeding. As the rest of the plant grows, the main stem remains small and deformed, and may break off.

The larva pupates in the stem, a little higher up, and a second generation of adults is ready to lay eggs in September–October, spending the winter as a larva in winter cereals, especially rye, and in couch grass, *Agropyrum repens*. This larva burrows deep down the stem, below ground level, and so is protected against frosts. When winter cereals resume their growth in the spring they develop the characteristic swellings at nodes.

We met gall-forming flies in the Nematocera, where Cecidomyiidae make a speciality of them. The acalyptrate flies have rediscovered the art. We have already noted *Halidayella* of the Lauxaniidae, making galls in clovers and in *Viola*. In Chloropidae, *Lipara lucens* makes a large, elongate gall in the reed *Arundo*, and we shall meet more gall-makers in the family Trypetidae.

The relationship between a gall-making insect and its host-plant is a curious one, not quite like any other biological association. It is mutual, in the sense that both parties gain to some extent. The plant obviously would be better off without the insect at all, but the formation of a gall is an act of containment. If the insect can be kept to a gall, without any of the migrations up or down the stem, and without the meanderings of a mine, the plant may escape with fairly minor damage. The larva profits from the shelter, protection against enemies – with important exception of those parasites and inquilines that are clever enough to get themselves into the gall – and a continuous supply of food without effort.

Séguy comments of galls that they: '... *intéressent toujours les tissues jeunes, en pleine croissance, jamais les organes ayant achevé leur développement*'. In fact the insect has become a true parasite, feeding upon plant tissues without killing it, and indeed may acquire the characteristics of a larva that is parasitic in animals. Thus the larva of *Chlorops speciosa* has almost lost its hind spiracles, and apparently breathes through the cuticle, though it still has large mouth-hooks and takes food into the intestine in the normal way. The larva of *Chlorops strigula* has gone further, and has reduced the mouth-hooks as well, down to vanishing-point. The young larva seems to have stirred up the plant-cells to such purpose that they produce enough food for the older larva to feed as well as breathe through its skin.

All stages between this and normal plant-feeding can be seen in one species or another of the genus *Chlorops*. In the same family we have met the carnivorous larvae of *Thaumatomyia notata*, and those of the genera *Oscinosoma* and *Siphonella* are parasitic in the egg-masses of spiders, as well as attacking certain beetles, *Apion* and *Baridius*. So yet another versatile family of flies runs the whole gamut from compost-feeding to parasitism.

Much of what has been said about Chloropidae applies to *Opomyza* and *Geomyza*, tiny flies, often with spotted wings, which are placed in a separate family Opomyzidae. Not as widely known as the frit fly and gout fly, Opomyzidae are principally damaging to crops grown on a large scale, as in eastern Europe and Russia, but occasionally they give trouble to small-scale farmers in Western Europe.

Once or twice already we have mentioned leaf-mining as a larval habit, and in the present group we have the family Agromyzidae as the leading exponents of it. Leaf-miners normally feed upon the parenchyma of leaves, leaving the epidermis of both surfaces intact (figure 32), but a few epidermal miners occur in the tropics. Professor Martin Hering has long been the principal student of leaf-mining insects, and his book that is listed in the bibliography gives a

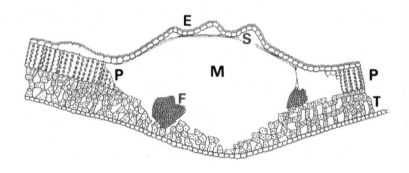

Figure 32. Section through a leaf-mine: M, general cavity of mine; S, silk lining produced by larva; F, frass of larva; E, epidermis of leaf; P, palisade parenchyma; T, spongy parenchyma. After Hering.

most comprehensive analysis of this biological niche. The essential of a mine is that both surfaces must remain intact, so that the larva is provided not only with food, but with shelter as well. He emphasises that only larval insects make mines, and no perfect insect has been known to do so. The mine need not necessarily be in a leaf; it may be made in a fruit, a stem, or a root, provided that it is a channel in the parenchyma, and not in deeper tissues. Leaves are the most eminently suitable material for mining, and the colour of the mine, in contrast with the rest of the leaf, indicates how much tissue has been consumed, and whether the mine is entirely in the palisade parenchyma, or extends also in the spongy parenchyma below it. 'Leaf mines are the most primitive form of leaf galls; and conversely, leaf galls are highly specialized mines.'

True leaf-miners occur in Sciaridae, Cecidomyiidae, and Chironomidae among Nematocera; Syrphidae of the Aschiza; Agromyzidae, Trypetidae, Lauxaniidae, Psilidae, Drosophilidae, and Ephydridae of the Acalypterae; and Muscidae and Scatophagidae of the Calypterae. Certain Lauxaniidae mine in dead leaves, but though they technically make a mine, they obtain quite different food-materials from those that feed on living plant tissue, and they are, indeed, only a special case of the primitive, compost-feeding type of larva.

Broadly speaking, mines can be divided into two types: blotch mines, where the larva feeds over a continuous area from one centre, thus producing an irregular blotch of paler colour; and tunnels or serpentine mines, where the larva moves forward all the time, producing a pattern of convolutions.

The story of Agromyzidae has much in common with that of Cecidomyiidae. Adult flies are small to minute, brown or black, and seldom seen except when they have been bred from plant-material. The family is studied by collecting leaves upon which mines can be seen, like the familiar leaves of holly that show the intricate mines of *Phytomyza ilicis*. It is easier to classify the mines and the host-plants than the flies themselves, and so necessarily the classification of Agromyzidae has lagged behind that of many Diptera. More particularly so since Agromyzidae, like Cecidomyiidae, have only a limited structural range, and are lacking in useful systematic characters. Besides Hering in Germany, Frick in America, and Spencer in England are rapidly extending our knowledge of Agromyzidae from all over the world.

Larvae of Agromyzidae are easily recognized by the fact that the prothoracic spiracles are placed close together, not in the usual widely separated lateral positions. Many pupate in the mine, and obviously it is worth collecting only those leaves that show mines intact, with a larva or a pupa visible by transmitted light. A mine with a dark puparium, but also with an exit hole, is one from which the fly has emerged.

The plant-feeding families so far discussed have laid their eggs on the surface of the soil, or of leaves and stems. An obvious step forward is the development of a rigid ovipositor with which the eggs can be placed in some hidden crevice, perhaps even inserted into plant-tissues, thus giving them more protection, and ensuring a higher rate of survival. This would seem a device particularly appropriate to plant-living larvae, but it is not confined to these, as we have already seen in certain Asilidae.

The conspicuous fruit-flies of the family Trypetidae make good use of such an ovipositor – like a miniature fountain-pen in shape – to insert their eggs below the rind of young fruit, or into the flower of composite flowers. Trypetidae have turned over almost completely to feeding on living plants, usually feeding as larvae in the flowers, fruits, or seed-pods. A few bore in the stems, or form galls at the tips of the twigs. As with Agromyzidae, the detailed association between plant and insect is an intricate and fascinating study, of which we have space here only for the merest outline.

Typical of the Ceratitini, which live in fleshy, succulent fruits, is the Mediterranean fruit-fly, *Ceratitis capitata*, now a widespread pest in all fruit-growing countries that have a Mediterranean type of climate. The egg is laid in a crack of the rind of the ripening fruit, and the larvae burrow in and feed on the flesh near the middle, eventually causing the fruit to fall. The larvae then escape and pupate in the soil.

In spite of the efforts of growers, larvae of the Mediterranean fruit-fly are often found in oranges and other citrous fruits imported into temperate countries. Sometimes there is a large area of damaged fruit, but often there is no reason why most of the fruit should not be eaten, if the larvae are cut out. Munro [211] makes the point that these fruit-feeding larvae do not usually damage the seeds, and so in a wild tree they would have little effect. Their evil reputation is commercial rather than biological.

Larvae of Trypetidae infest the flowers of Compositae, and a few make galls at the tips of twigs of bushy plants. These 'terminal galls' as made, for instance, by the genus *Spathulina*, have thin walls and a small hole. The puparium, when formed, is shining black, and lies loosely in the cavity, and the emergent adult comes out through the existing hole. One of the best-known Trypetidae is the European *Urophora jaceana*, which makes galls in the flowerheads of the knapweed, *Centaurea jacea*, described in detail by Varley [295], and which has been introduced into Canada and the USA with the knapweed in recent years; other *Urophora* make galls in thistles. These are unusual in that they may contain more than one larva, a condition that arises from the merging of several adjacent galls. A few Trypetidae damage garden vegetables in Europe: e.g. the larvae of *Philo-*

phylla heraclei mine in the leaves of lettuce and celery, and not only disfigure these, but may retard or kill the plant.

A large group of Dacinae specialize in feeding in cucurbitaceous plants, melons, squashes, gourds, etc. Their larvae face an additional hazard because these fruits are much harder than the tree-fruits favoured by the Ceratitini, and have a harder rind. In one of his papers Dr Munro quotes an instance in which over fifty adult flies were found dead inside a melon because the larvae had found the tissue too hard to bore through when they were ready to pupate. Consequently they pupated in the cavity they had already excavated, and the emerging adult flies were trapped. It is noteworthy that they were unable to mate and reproduce there, as a group of Phoridae would have.

Only a few Trypetidae stray from this restricted range of larval habitats. One of these few is *Rioxa termitoxena*, which breeds in the burrows of the Australian tree-living termites *Mastotermes darwiniensis* and *Calotermes irregularis*. The termites enter the tree-trunk below the level of the soil, and their tunnel gradually comes out on to the surface of the trunk, when it is roofed over with an earthy cement to make a covered gallery running up the bark of the tree. The larvae of the fly live and feed in an evil-smelling liquid that fills much of the tunnel, and oozes through the bark where the tunnel is near the surface. The adult flies are to be seen at this point of the trunk, and it is an interesting possibility that the attraction of the adult flies may have been the first step that led the larvae to take up this substance as food.

Adult Trypetidae, like those of Drosophilidae we shall see presently, are strongly attracted by fermenting substances, which give off vapours of alcohols and esters. I was particularly struck by this when I was collecting in the African rain-forest. Having been puzzled for a time at a rich, wine-like smell, I traced it to a small palm that had been cut and left to drip to provide a liquor for making palm-wine. Here I was caught up in a cloud of adult Drosophilids and Trypetids, flying and sitting all round, and apparently revelling – if that is the word – in the intoxicating smell. Munro comments on the strength of this attraction, considering that the larvae of Trypetidae do not live in decaying materials. The example of *Rioxa termitoxena* shows a trend in this direction, and an interesting comparison with the Drosophilidae, whose larvae do live in these fermenting liquids.

The ovipositor that is developed to a greater or lesser extent in

Trypetidae is not used very actively to pierce tissues, but merely serves to insert the eggs into the flower-head, or through a crack in the rind of a fruit. It is a feature of this group of families, along with conspicuously banded or spotted wings. The pattern of the wings is most intricate, and is displayed when the flies walk slowly about, raising and lowering the wings. As we have seen in, say, Sepsidae, this can have a part in courtship behaviour, but it is difficult to believe that such highly complicated patterns as many of these flies display can be significant to the flies themselves. Look at the wing of the Mediterranean fruit-fly shown in plate 20, and remember that there are vivid differences of colour as well as of tone. Systematists find it convenient to make use of minute differences in the size and spacing of individual wing-spots, if these are constant in each species, but it seems improbable that the flies themselves can be conscious of anything more than the most general pattern. Dr Paterson, of Johannesburg, recently described to me how *Afrocneros mundus* uses a small ball of foam as a mating lure, in the manner of the Empids: it is produced from the mouth of the male and offered to the female, who eats it.

The related families that share the general appearance and behaviour of Trypetidae are Otitidae (better known, perhaps as Ortalidae), Lonchaeidae, and a number of small 'splinter groups': Pallopteridae, Ulidiidae, and so on. Except to the specialist, they are difficult to separate from Trypetidae, but biologically their interest for us is that, as far as is known, their larvae still live in decaying vegetable matter, or in dung. That is, they take us a stage further back before the sophisticated breeding habits of the Trypetidae had been evolved.

Among these are the odd stalk-eyed flies – *Achias* and so on – that we have already mentioned as not belonging to the stalk-eyed family Diopsidae. In these genera, the separation of the eyes seems to be even less useful than in Diopsidae, and confirms the impression that these stalked eyes are no more than an outburst of evolutionary extravagance. This is carried to a further absurdity in the genus *Phytalmia*, the males of which sport grotesque 'horns' growing out from the cheeks.

The last group of acalyptrate flies centres round the genus *Drosophila*. *D. melanogaster* is probably the best-known insect in the world, since such a tremendous volume of experimental work on genetics and heredity has been carried out upon it. To us, however, it is just one of a large family, the one that has 'made good', as it were. The names 'fruit-flies', 'vinegar flies', 'wine flies', 'pomace flies' all show how this family is linked with fermentation and fruity odours. Sometimes when you are sitting over a glass of wine you will see a little tiny fly hovering clumsily and tantalizingly before your eyes. Don't dismiss it as one of the elusive *muscae volitantes*, or a djinn out of the bottle; it is probably *Drosophila melanogaster*, or the less welcome *D. repleta*.

Species of *Drosophila* go a stage further than Trypetidae, in that not only are they attracted as adults by the smell of fermentation products, but their larvae actually feed in the medium in which the fermentation is occurring. *D. melanogaster* is used as a laboratory animal because it is small, prolific, and easily reared on mixtures of bran, fruit pulp, and fermenting substances of like kind. Colonies are easily maintained because larvae, pupae, and adults will live happily together in one jar, and no elaborate organization is needed to provide for differing requirements at successive stages of the life-history.

This fact has also made various species of *Drosophila* as much at home in the town as in the country, indoors or out, and these flies are often a domestic nuisance. *D. repleta* made a dramatic appearance in London restaurants and kitchens during the war, and the fact that it has bred in hospital bed-pans makes it one fly that even a dipterist does not greatly welcome. *D. busckii* and *D. funebris* quickly find any milk bottle that is exposed with dregs of milk still in it, and in a few days in summer the larvae have pupated and stuck themselves firmly to the glass with a casein-type glue that often defies the hot caustic solutions of the washing-plant.

The various types of habitat favoured by wild *Drosophila* in Scotland are analysed by Basden [13], who defines the natural habitat thus: '. . . the denser, shadier, and more mature deciduous woods are usually more productive of specimens and species than the more open or younger ones'. He associates this with abundance of (a) leaf-mould and other mould-generating media, and (b) sap-exudation and honey-dew. Open, grass-covered ground yielded fewer Drosophilidae.

Basden divides his catches into fruit-haunting species; those which are found where fermenting sap oozes from a tree; those which are bred from mines in plants; and those known to him only from specimens taken on windows. This last group reminds us that the fruit-flies are not the whole of the family Drosophilidae, and that this family has a bundle of 'brushwood' of its own, genera such as *Camilla*, *Diastata*, *Aulacigaster*, and so on. Among the systematists the 'lumpers' put all these into the one family, while the 'splitters' make separate families for some of them; biologically this consensus of genera repeats the pattern now familiar to us.

Larvae of *Diastata* and *Camilla* probably represent the basic, compost-feeding type. *Parascaptomyza* feeds in decaying vegetable matter, and may invade the mines that other insects have made, but does not make mines of its own. From this habit perhaps originated the true leaf-mining of the genus *Scaptomyza*. Some larvae, as we have seen, take the fermenting sap oozing from trees, and from these may have evolved *Drosophila* proper, with its *penchant* for fermenting materials of all kinds. Today the sap-feeders include genera like *Aulacigaster*, and also true *Drosophila* of the *subobscura* group. Perhaps from sap-feeders also arose the carnivorous larvae, which prey mainly on plant-bugs. *Gitona* eats aphids and coccids, *Acletoxenus* takes *Aleurodids*, and the very remarkable *Cryptochaetum* is an internal parasite of scale-insects.

The biology of *Cryptochaetum* is a study in itself, related in detail by Thorpe [288A]. The egg is pushed through the skin of the scale-insect, and the larva lives in its blood. The minute first-stage larvae can get enough to eat and to breathe through its skin, but more mature larvae gradually develop the digestive system as they progress from plasma-feeding to eating the fat-body and other tissues. Most remarkable of all, the larva develops long, whip-like tracheal gills, as its demands for oxygen grow greater and greater. Finally, the larvae, of which a dozen or more may be found in one host, pupate in the dead body of their victim. *Cryptochaetum iceryae* featured in one of the classic examples of biological control, when this fly was introduced from Australia to California to combat the 'cottony cushion scale', *Icerya purchasi*, on the citrus trees there.

As a final example of Drosophilidae, and of acalyptrate flies, *Cacoxenus indagator* lives as a larva in the nests of various carpenter bees and mason wasps. It is referred to as a 'parasite', but is more correctly a 'social parasite', a general scavenger among the nutritious crumbs that are dropped by its hosts.

Chapter 16

The House-Fly and its Relatives

THE REMAINING FLIES make up a group known as the calyptrates, and are mostly bristly flies, short and plump of body, with broad, efficient wings, buzzing their way actively about. They are equivalent, not to the whole of the acalyptrates, but to one of its evolutionary branches. In contrast to the many and confusing families of acalyptrates, the calyptrates fall neatly into two groups, the Muscoidea, or flies like the house-fly, and the Calliphoroidea, or flies related to the blow-flies.

These two groups clearly had a single origin, and must have come from some compost-feeding ancestor, perhaps somewhat like the present-day Helomyzidae. They have been evolving for a long time, and have gone further than any acalyptrates, except possibly the Ephydridae. In number of species they are just about equal to the acalyptrates, but these species are grouped into only about six families.

Starting from a compost-feeding larva, of which many still persist, Muscoidea have enriched their diet on the one hand by attacking living plants, and on the other by seeking dung, urine, guano, carrion, by preying on other insect larvae, and even by sucking blood from man and other warm-blooded animals. All this we have seen in some acalyptrates but Muscoidea go about it with a greater efficiency, and Calliphoroidea more grimly still. Since we left the Brachycera, adult food has been of lesser importance, except in a few families such as Ephydridae and some Chloropidae; in Muscoidea the adults again begin to feed more urgently, and so these flies are notorious pests for the way in which they fly round humans and animals, seeking perspiration and other bodily secretions. A few have even turned to sucking blood, and since they have long lost the primitive piercing organs they have had to develop new ones. They still have sponge-like pseudotracheae like those of the hover-flies (figure 28), with which they can mop up blood that is flowing from wounds, or from the punctures made by other biting insects. Some Muscoidea help the blood to flow by scratching at the skin with hardened pieces of the end of the labial channel ('prestomal teeth'), or developed from the ring of a pseudotrachea ('pseudotracheal teeth'). Others have stiffened the whole stem of the labium until it becomes a stiletto, as in the stable flies and tsetse-flies that we shall see presently.

Garcia and Radovsky [322] summarize this transition from mopping up blood to actively piercing and sucking.

The fly that is typical of the Muscoidea is the best known and most notorious of all flies, *Musca domestica* of Linnaeus, the house-fly, or just 'the fly' to millions of people. This fly has gone everywhere with man as far back as memory goes, and although the urbanization and public hygiene of the present century have made it less familiar in many countries, its day is by no means over.

Larvae of *M. domestica* live in compost-like mixtures of decaying vegetable material, preferably enriched with dung or animal matter – household garbage in a primitive culture is ideal. It happens that the larvae prefer the dung of pigs, humans, and horses to that of cows, which is preferred by the closely related *M. autumnalis*. On this simple fact may depend the whole history of *M. domestica* as the house-fly, since it has tended to keep *domestica* about man's house and yard throughout the summer, when *autumnalis* is scattered in the pastures.

M. domestica is not a biting fly, though some of its near relatives in the tropics have developed the scratching habit that we have referred to above. When people are bitten by what appear to be house-flies, the culprit is the stable fly, *Stomoxys calcitrans*, sometimes called the biting house-fly. The true house-fly merely mops up its food, and is very catholic in its tastes, taking any liquid food, or anything that it can liquefy by regurgitating upon it. This, together with its house-haunting habits, causes the house-fly to be suspected of carrying many human diseases, especially those of the respiratory and intestinal tracts, but we shall discuss these in a latter chapter.

The house-fly is now cosmopolitan, having travelled everywhere that man has settled. Its larvae are rather limited in the range of temperature that they can tolerate, and are immobilized or killed by either heat or cold. They like most household waste materials, and if the temperature indoors is kept up to about 60°F they will breed continuously throughout the year. At this range they take about three weeks from egg to adult, but in tropical heat may develop in less than one week. Though as a carrion-feeder the house-fly is less well known than the blow-flies, its larvae will infest wounds, or wound-dressings, live in birds' nests, dead and living snails, and even in the egg-pods of locusts.

The house-fly flourishes in the tropics as well as in temperate countries, the tropical flies being recognizable as different, mainly in colour. Besides the shortened development, Reid [241] mentions the interesting fact that in severe tropical heat house-flies come out at night to rest on vegetation, whereas we in temperate countries see our flies coming indoors at night to shelter from the cold. The reason why the house-fly is so well known an insect in cooler countries is probably that it is exceptionally efficient in adapting its behaviour to the more exacting conditions there.

In temperate and cold countries, house-flies disappear in the winter, make a slow and late start each spring, and reach their maximum numbers in late summer and early autumn. 'Where do flies go in the winter-time?' This was the theme of a comic song, but the joke is against the entomologist, who still cannot give an unqualified reply. Recent work by Sømme [269] in Norway has established that overwintering as larvae and pupae may be impossible in countries that have a cold, continental type of winter. From a combination of observation and experiment he is convinced that a small number of adults survive each winter in cow-houses. Even if these houses have no artificial heating, the body-heat of the beasts themselves may give local temperatures of 15–18°C [59–64°F], and Sømme considers that there is a slow, continuous breeding through the winter months. He established that the number of house-flies in a given cow-house during the late summer was directly related to the number of living adult flies that he could find in the same cow-house during the winter.

Kobayashi in Korea also regarded the adult as the important overwintering stage, partly by continuous breeding in suitable places, and partly by survival of hibernating females. Recent Russian workers, on the other hand, have found larvae and pupae all through the winter, even in exposed places where the adult flies were not active. The only conclusion one can come to is that *Musca domestica* is a highly versatile insect, and suits its behaviour to the local conditions as it finds them.

The fecundity of house-flies is proverbial, and many authors of books have amused themselves by estimating how many descendants one pair might have in a single summer, if they all survived. One published claim, often quoted, is that there would be enough flies to cover the earth to a depth of 47 ft! Incredulous, I recalculated this, and decided that a layer of such a thickness would cover only an area the size of Germany: but that is still a lot of flies.

Heavy losses of larvae through unsuitable temperatures, and by the depredations of barnyard fowls, reduce the numbers of larvae very severely in practice, but even so the build-up of house-flies in a favourable area is impressive during the last two months of summer. Fortunately the autumnal drop in temperature cuts the population down again to the very small number of adults that can find a cosy refuge for the winter.

People bothered by house-flies in the summer often ask how far they can fly from their breeding-place. Like most such questions this cannot be answered precisely. It seems that most flies stay within a mile or so of their breeding place, and many within half a mile, but radio-active flies have been traced over 20 miles. This latter figure seems to be increased by wind-drift, like the seaweed flies that travelled far inland in England in 1954, carried by a southerly wind. The fly need not be entirely helpless in the wind, like aphids or small Chloropidae: it may fly quite purposefully over short distances, but unless it comes back each time to some sort of marker it will gradually drift downwind.

M. autumnalis is another species that closely resembles the house-fly, though systematically it is placed in a different group of the genus. It is much less widely distributed, and until recent years it was confined to Europe and Asia, and unknown in North America. Over the last fifty years it has become well known to entomologists as a nuisance to cattle in the summer, and as a nuisance to householders in the autumn, when it shelters in large numbers in attics and roof-spaces. In fact, its behaviour is apparently complementary to that of the house-fly, since it comes in when the house-fly is dying down, and leaves again in a great swarm before the house-fly has re-appeared for the summer.

It seems as if *autumnalis* is tolerant of a wider range of temperature than *domestica*, since it will sit out in hot sun on the cattle in summer, when the house-fly is seeking shade; and not only survives in the attic in winter, but revives its activity on every mild day. I shall have more to say about this activity when discussing swarms.

M. autumnalis made its first appearance in North America in 1952, having been introduced from Europe. Specimens that had been previously reported from North America proved on examination to be *domestica*. In a few years it has become such a pest of stock that it has been given the name of the 'face fly', and articles about it appear in all the American journals of economic entomology. Cow-dung is its principal breeding medium.

The family Muscidae is a very large one, and does not fall readily into groups that can be recognized without careful study. One cannot do justice to its complexity and intense entomological interest without going into a mass of detail: Séguy's masterly volume on Muscidae in the *Genera Insectorum* series is of a folio size, and several inches thick! The present brief summary must, therefore, give only an inadequate picture.

To summarize the habits, one might say that the larvae of a majority of genera fall into the category of saprophagous, dung-breeding species, of which the house-fly is a fair representative. There is, however, in the group a constant urge to enrich this diet with a greater content of animal protein. Keilin showed that the larvae of a number of species become carnivorous as they grow older, and end up by preying on the smaller larvae in their food material. In particular this was recorded of *Polietes*, *Hydrotaea*, *Myospila*, and *Muscina*, but it seems likely to be a widespread habit. *Fucellia* is a grey fly common on the beach about heaps of stranded seaweed, and is more obvious that the true seaweed fly, *Coelopa*, because the adults fly more readily. The larvae are predacious, and in North America have been recorded as preying on the eggs of a marine fish, *Leuresthes tenuis*, which lays them in the sand above high water mark. The larva of *Passeromyia* sucks the blood of nestling birds.

Limnophora, as its name implies, is another shore-living Muscid, and its carnivorous larvae are aquatic or semi-aquatic. Maritime species live between tidemarks, under pebbles, and feed on other organisms there. They have been said to penetrate into barnacles when these are submerged and open, and to devour the soft tissues.

We have already mentioned several flies that exploit the highly nutritive organic materials to be found in birds' nests, apart from those that actively feed on the birds themselves. Among Muscidae the most conspicuous bird-nesters are *Anthomyia procellaris* and *A. pluvialis*, the adults of which are strikingly patterned in grey with black spots. Though their larvae feed in nests it is not certain that

205

LIVERPOOL JOHN MOORES UNIVERSITY
LEARNING SERVICES

they are entirely confined to them, and they have at various times been reported from other materials: fungi, rotting fruit, wounds, and even as causing myiasis of the intestine.

Highly nitrogenous materials are also exploited by flies of the genus *Fannia*, of which there are a number of poorly known species, and two notorious ones: *F. scalaris*, the latrine fly of temperate countries – tropical latrines have a different fly, *Chrysomyia megacephala* – and *Fannia canicularis*, the lesser house-fly. The latter are the familiar little flies that circle about in the middle of the room,

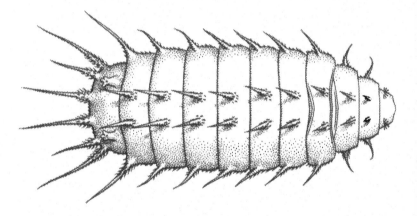

Figure 33. Larva of the Muscid fly *Fannia canicularis*, with fringed processes suited to life in a wet medium.

playing 'kiss-in-the-ring'. This circling flight is one of the variations of the swarming/nuptial flight behaviour to which I shall turn later. It seems to need an overhead marker, and usually takes place beneath the hanging light fitting. Outdoors they can be seen flying in the same way under a tree.

The larvae of *Fannia* are extraordinary among Muscid larvae in having abandoned the maggot-like shape. They have become flattened, and fringed with rows of branching processes along the upper and lateral surfaces of the body. Apart from helping in the identification of species, these processes act as a flotation device by increasing the surface area of the body. Larvae of *Fannia* like highly nitrogenous food, especially urine-soaked materials, and *F. scalaris* gets its name of the latrine fly from its fondness for cess-pits and *'fosses d'aisances'*. The larvae still need to breathe atmospheric air, and the body-processes are believed to help in keeping them from going under.

F. canicularis gets its name because it is particularly associated with the Dog Days of summer. In the early months, when *Musca domestica* is recovering slowly from its winter recess, the lesser house-fly has a clear lead, and is the fly most common in houses until July, when true house-flies begin to appear. Hence the popular but erroneous belief that the little flies of the spring, if not killed, will become the big ones of the summer. Domestic *canicularis* may have come from a rubbish heap of a rather drier kind than that favoured by the house-fly, but they flourish particularly on bird-dung, and are common in chicken-houses and deep litter. Among the more *outré* breeding places indoors are, urine-soaked cot-blankets, and containers used for storing babies' nappies until there are enough to justify starting the washing-machine. A comprehensive review of this group of Muscidae, with many valuable comments on evolution, distribution, and habits, is given by Chillcott [40A].

Turning momentarily to a less repulsive subject, a number of Muscid larvae have become plant-feeders. Though not numerous in species, they are highly efficient. Several damage commercial crops, and have the distinction of being the subject of a Ministry of Agriculture Advisory Leaflet: the wheat bulb-fly, *Leptohylemyia coarctata*; the onion fly, *Delia antiqua*; the mangold [beet: spinach] fly, *Pegomyia hyoscyami*; and the infamous cabbage root fly, *Erioischia brassicae*. Like most insects that are of any practical importance, they suffer from frequent changes of name, so do not be surprised if you find them called something else in an agricultural report.

Adult flies of this family commonly take to some degree of predation. We have seen that they have prestomal teeth which may be developed until they may be used to draw blood from vertebrate animals, or to crush and suck small insects. The sub-family Lispinae specialize in catching other small insects, and are said even to dart down on to the surface of water and 'fish' for mosquito larvae. The Stomoxydinae go further, and have developed the proboscis into a stiff, hard, piercing organ. *Stomoxys calcitrans* is the common stable fly of temperate countries, often called the biting house-fly because it may be mistaken for a true house-fly until it inconsiderately bites through one's sock.

Larvae of *Stomoxys* prefer a dung that is generously mingled with straw, and farmyard manure, either horse- or cow-dung, is their favourite material. Like house-flies, they will also breed in a variety of other materials, including birds' nests. The adult flies lurk in stables, cowsheds, and other buildings, including the farmhouse itself, where their droppings, stained with undigested blood, make a mass of dark brown specks on the curtains and windows. Out in the fields, adult *Stomoxys* cluster in the shade of the bellies of grazing

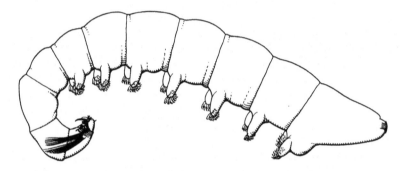

Figure 34. Larva of the Muscid fly *Phaonia keilini*, which preys on mosquito larvae.

animals, and bite them unmercifully. *S. nigra* is suspected of transmitting diseases of livestock. *Lyperosia* and *Haematobia* are smaller flies of similar habit, *Lyperosia* being known as the horn fly, because of the way in which it clusters round the horns of cattle.

The African *Stomoxys ochrosoma* gives a remarkable example of the evolutionary ingenuity of flies. The larva of this species lives in the nests of the ferocious driver-ants. Many dipterous larvae live in ants' nests, as we have seen, and the egg-laying females have to run the gauntlet of the ants each time they approach, unless they are small and tolerated, like the Phoridae. *S. ochrosoma* hovers over the column of ants as they carry their burdens back to the nest, and finds one ant that is carrying nothing. The fly drops its egg in front of this ant, which, in a blind, instinctive way, picks it up and carries it in procession to the nest [290].

Again in Africa, the tsetse-flies are among the most completely studied of insects, making up in our knowledge of their biology and natural history for what we know only experimentally about *Droso-phila*. There is only one genus, *Glossina*, now confined to the African continent south of the Sahara and southern Arabia, and containing fewer than twenty-five species. All of them suck blood as adults, and all of them are viviparous, retaining the larva within the body of the mother, and nourishing it on internal secretions until it is fully grown and ready to pupate. The large larva is then dropped, and burrows quickly into the soil, where it pupates.

These flies have thus eliminated a free larval stage, and sacrificed the advantages of a wide range of larval food-materials. They have become completely dependent on the feeding of the adult: the oppo-site of the Chironomid midges, and of certain flies in the next group, where all feeding is carried out by the larvae.

Adult tsetses feed exclusively on the blood of vertebrates, mam-mals, birds, and reptiles, and they feed from a living host. Occa-sionally they may probe recently killed animals, but apparently they do not take blood from open wounds, as the horse-flies do. Nor do they take plant food, in spite of reports to the contrary. None is restricted to one particular host animal, but most of them show some preference, possibly dictated as much by the size and movements of the host as by its zoological classification.

Tsetse are known as day-biters, but some at least bite a little at night. Their principal hosts are ungulate mammals, and the flies fall into three well-marked groups according to their habits and habitat.

The *morsitans-* group are the game tsetse of more or less open country, woodland, thorn bush, savannah. These occur over wide tracts of land, but are entirely dependent upon the game animals, and disappear if the game is killed or driven away. Hence the agoniz-ing decision in some areas, whether to destroy the game for the benefit of the cattle-raisers, or to preserve the game to please every-body else. The *palpalis-* group live in denser, wetter, heavily forested areas, either in the equatorial rain-forest proper, or in local patches like the gallery forests that follow the banks of streams in arid coun-try. While they will bite ungulates, their mainstay is reptiles, par-ticularly crocodiles and monitor lizards, and so they are not extermi-nated by destruction of the ungulates. The *fusca-* group are larger and more primitive flies, which feed on small animals in the forest, and so are of less practical importance than the others.

The reason why so much money has been spent on the study of tsetse-flies is, of course, the blood parasites that they transmit.

Trypanosomes are flagellate Protozoa, single-celled animals with a whip-like flagellum, which live in the blood of many animals, including man. Some, like *Trypanosoma evansi* which causes surra in horses and camels, are mechanically transmitted by any biting fly; others have become adapted to passing part of their life-cycle in the body of tsetse-flies, which thus are an integral part of the cycle of the disease. The *morsitans-* group are responsible for *nagana*, a fatal disease of cattle and horses, and both they and the *palpalis-* group are responsible for outbreaks of human sleeping sickness.

Compare again, the habits of the two groups. The *palpalis-* group lurks in forests, where the human population is small, and scattered in small units. Sleeping sickness is therefore endemic at a low level, and occasionally breaks out in local epidemics. Such a disease is susceptible to shock treatment with modern drugs. But the *morsitans-* group live over wide areas, and have the wild game as a reservoir of trypanosomes, though the game animals themselves have developed immunity to the symptoms of disease. Hence the tsetses of the *morsitans-* group have the power to exclude domestic cattle and horses entirely from vast areas of the African savannah country.

Buxton [37] goes so far as to blame *Glossina morsitans* and its relatives for the fact that Africa has remained more static than any other continent, except the Amazon basin, and says: 'We believe it to be legitimate to say that civilization could not have developed under this immense handicap.' This endorses the views of earlier travellers, which I quote in Chapter 19.

Since there is no larval feeding stage to be considered, the distribution of the various species of *Glossina* is entirely dictated by the behaviour of the adult flies. Early explorers noted that there are 'fly-belts', and areas free from fly. In fact, many species require two types of vegetation, a more open one in which to seek their victims, and a denser one in which to drop their larvae. For this purpose the female fly sits in the shade, sometimes under an overhanging rock, in a shallow cave, or even in a tree-hole. The larvae drops to the ground, and immediately burrows in the soil, where it disappears in about a minute.

The pupal stage is passed in the soil, where it is subject to the local conditions of heat and humidity. It appears to prefer a high atmospheric humidity – hence the shady site – but it is drowned by flooding. The effect of intense heat from sun beating on the ground varies in different species, and is one factor which determines the type of oviposition site. Detailed studies of these requirements have made it possible to kill off tsetse just by clearing enough of the vegetation to alter the character of the site, without necessarily clearing it completely.

Female tsetse mate as soon as they are hardened, or even before, and the larva takes about a fortnight to develop. This cycle is repeated immediately, and continues to the end of the fly's life, during which it is uncommon to find an unfertilized female. There is no fixed number of larvae per female, but even so the rate of reproduction is painfully slow compared with that of *Musca* or *Stomoxys*. Survival depends on keeping down wastage far below that of most flies, which, as we have seen in the house-fly, are extravagantly prolific. Given settled conditions this is an elegant method of reproduction, but in a changing world the tsetse is obviously vulnerable.

A feature of particular interest in the natural history of tsetses is the ease with which different species mate with each other and produce hybrids. This is particularly so between *morsitans* and *swynnertoni*, and Vanderplank has published a series of papers elucidating the combinations that are possible. When hybrids mate with each other, or back to one of the parents, eggs develop to varying stages, sometimes to a larva, sometimes even to a pupa. Apparently no hybrids of the *morsitans*- group have developed further than the pupa of the second generation, but some *palpalis*- group have produced an adult of F_2 hybrid generation, a remarkable feat. This readiness to hybridize makes tsetse particularly suitable for experiments in control by irradiating and releasing males, as mentioned in a later chapter.

Returning from the veldt to the farmyard, one of the most familiar flies of temperate countries is the yellow dung fly, *Scatophaga stercoraria*. These are the furry yellow flies that cluster over horse- or cowdung, scatter temporarily at one's approach, and move back almost at once. The ones with bright yellow hairs are males; the females are fewer and more drab. These flies are well named, because the dung is everything to them. Their eggs are laid upon it, and have two ridges that hold a plastron of air, which is intimately concerned with ensuring a supply of moisture and air to the egg, as described by Hinton [136]. The larvae feed in the dung, and pupate in the soil beneath. The adult flies not only assemble on the dung for mating purposes, but also prey upon the many other flies that come there to lay eggs or to feed.

Adult *stercoraria* of both sexes have a sucking proboscis, essentially similar to that of the house-fly. Among Muscidae we have seen various modifications in flies that have taken to animal food as adults. The prestomal teeth, which some species of *Musca*, and others of the family, use to rasp the skin of their victims, are highly developed in *Scatophaga*, and form a most efficient organ of predation. So effective is it, indeed, that *Scatophaga* can tackle house-flies and blue-bottles, bulky, clumsy flies bigger than itself, which must be extremely difficult to seize and hold. The prestomal teeth are used to cut a hole in the membrane of the neck, severing the nerve cord, and crippling the victim, and at the same time releasing the body fluids which are sucked out of the head and thorax. The abdomen is then tackled separately.

Scatophaga stercoraria is one of those grimly efficient flies that arouse one's admiration. It exploits a common, widespread medium, which ensures a regular supply of food for larvae and adults. Breeding goes on smoothly all through the summer half of the year – April to October in the Northern Hemisphere – with four or five generations. Like the house-fly, its winter behaviour is elastic, and varies with the severity of the weather. Adult flies can be seen on warm days, and possibly some larvae and pupae also overwinter beneath cowpats.

The yellow dung fly is only one of a considerable family of flies, called the Scatophagidae, or Cordyluridae, which for structural reasons is regarded as being intermediate between Muscidae and the acalyptrates. Even among students of flies, few could instantly name any others of this family, yet there are about 500 known species. *Scatophaga* is world-wide, but most of the others are found principally in the North Temperate Zone. Not all are predaceous as adults, and not all larvae feed on dung. Some live in other decaying materials, some in plants, either in galls and tunnels made by other insects, or actively damaging the plants themselves. Like the larvae of Muscidae they may become carnivorous and cannibalistic. Séguy says of them: '*si la nourriture vient à manquer elles deviennent zoophages et pour finir elles se dévorent entre elles*'. Just like Little Billie, '. . . us must eat we'.

Chapter 17

Blow-Flies, Bots and Warbles

THE LAST GREAT GROUP of flies that remain to be considered might collectively be known as 'flesh-flies', if that term were not already in use for a limited section of them. Their common characteristic is that their larvae feed on animal protein in one form or another, living or dead, without the vegetarian members that occur in the Muscidae. Because of their widely varied structure and habits they have been classified in a number of different ways, but we may conveniently consider them as composed of four families. Calliphoridae are the common blow-flies, the bluebottle and greenbottle and many others, basically carrion-feeders, with a few experiments in bloodsucking, parasitism, and viviparity. Oestridae are a highly specialized off-shoot, living as bots and warbles in the bodies of vertebrate animals. Sarcophagidae are the ones usually called 'flesh-flies', and have a wide range of larval habitats, from dung and carrion to parasitism, while Tachinidae have given themselves over wholly to living as parasites of insects and other arthropods.

The English words 'bot', 'warble' and 'blow' are of ancient origin, and the dictionaries can give little explanation of them except as 'old natural history'; '-bottle', as in bluebottle, is sometimes explained as a diminutive of bot, but the connection is unlikely unless there is some faint implication that eating 'blown' meat might give one stomach bots like those of horses. The term 'blown' refers to meat that has had eggs of blow-flies deposited on it, but in common use 'fly-blown' has been broadened to refer to any neglected room, window or display, which is dirty and covered with the droppings of flies, house-flies, and stable flies in particular.

The common blow-flies of Europe and North America are the bluebottles, *Calliphora erythrocephala* (or *vicina*) and *C. vomitoria*, and the greenbottle, *Lucilia* (or *Phaenicia*) *sericata*. *Calliphora* comes indoors more readily than *Lucilia*, and makes so much noise that the

213

room seems full of them. It is less dangerous to health than the house-fly, because it does not feed so greedily from human food-stuffs. It will sip sugary substances, but the main object of its rest-less movement is to find a place to lay its eggs. Throughout this group, except for the Tachinidae perhaps, there is an evident urge to speed up reproduction. Great masses of eggs develop and threaten to hatch, and the wretched fly obviously feels 'Time's wingèd chariot hurrying near', as she searches frantically for a suitable car-cass on which to drop them. Her movements are largely guided by smell, but sight obviously plays some part, especially in the attrac-tion to light.

At various times I have watched bluebottles in my house, and tried to find some simple explanation of their movements, but with-out much success. I noticed that they will fly ceaselessly to and fro between a pair of lamps in the room, and it looked as if the point-source effect of the more distant lamp had a more powerful attrac-tion than the general illumination of the nearer. Then I found that they flew between the same two lamps in the daytime! A bright light certainly attracts them: the easiest way to get rid of one is to darken the room and open a door to a lighted hall. Yet they do not spend all their time trying to get out of the window, as do many flies, and instead, they will leave the window and fly into dark corners of the room, or settle down behind a chair. The only thing one can say about a bluebottle is that it is a prey to conflicting desires, and poten-tially neurotic.

Calliphora larvae feed naturally in animal tissues that are recently dead. In suitable weather the arrival of egg-laying females is prompt, and the larvae begin to feed even before natural decay is evident in the tissues. Hatching of the eggs and survival of the young larvae are very much dependent on humidity. The necessary relative humi-dity has been placed as high as 90%, and Hinton has said that the eggs of *Calliphora* might almost be classed as aquatic. The meta-bolism of the egg, and especially its respiration, are closely bound up with the functioning of a plastron, or thin layer of air acting as a lung, and moisture is an essential part of this mechanism.

Hobson [139] described the way in which the larvae attack animal tissue. A newly dead body develops *rigor mortis*, a condition in which the tissues are acid, and then not digestible by the larva. At first the larvae feed on the fluid serum between the muscle fibres, meanwhile discharging their excreta, which contain ammonia as well as protein-digesting enzymes. It is the accumulation of these, and the increasing alkalinity, that dissolve the intramuscular tissue, and the sarcolemma of the muscle-fibres, and it is only when the fibres themselves have been exposed that the larvae begin to lacerate them with their mouth-hooks. Thus the massing of maggots, so common a feature of 'blown' meat, helps the process by concentrating the alkali from their combined droppings.

Larvae of blow-flies will attack cooked meats, and those that have been frozen and thawed out, but they lose interest in a natural corpse after the liquefaction has proceeded to the point at which the material is more suited to specially adapted larvae like those of *Fannia*, with their fringed processes, that we have previously mentioned.

Besides natural tissues, blow-flies will also feed on animal products such as cheese. One of the most thriving communities of *Calliphora* larvae that have been sent to me were living in a cheese-and-tomato sandwich that had been kept too long in a café. They ate the cheese completely and grew fat on it, but ignored the tomato.

When they are fully fed, the larvae leave the food material, and seek a dry, sheltered place in which to pupate. In warehouses, shops, and factories there are always a few dead rats, mice, or birds upon which a batch of larvae may develop, and when they are ready to pupate they will crawl quite long distances along the floor, to hide in the most incongruous places. They creep into narrow spaces, and so they like packaged goods in store. I have had them from packets of tobacco and tea, wearing-apparel such as scarves and gloves, and in bandages and other surgical dressings. Sometimes young people move into a new house and are dismayed to find pupae of *Calliphora* in the cupboards and drawers, imagining the wood to be infested. Fortunately in these cases the infestation is harmless, and temporary.

The greenbottle, *Lucilia*, has habits that are generally similar, but it is apparently attracted by a different set of odours, and so is not so prone to come indoors, and blow human foodstuffs. *Lucilia* breeds more freely in offals, and is very common round dustbins, *abattoirs*, and places where meat and fish are handled. Of particular attraction to *L. sericata* and *L. cuprina* is moist wool, and once these flies have established themselves in the fleece of a sheep, round the anus and hindquarters, the ammoniated excreta attract more and more egg-laying females. Soon the larvae begin to attack the flesh of the animal, and another case of sheep strike has begun.

L. sericata is a notorious pest of living sheep in countries as far apart as Australia and Scotland, though in Australia *L. cuprina* is even more damaging. Mackerras and Freney [191] thus describe the attack: 'There are normally two stages in the development of a primary strike, the first stage being from hatching up to the time the larvae attack the skin. . . . During this stage they must feed, if at all, on materials already present. The second stage is from the commencement of an actual skin lesion up to the full development of the maggots. During this stage there is a more or less copious serous exudation, which has been shown to be an adequate food for the full development of the maggots. Faeces-staining, presence of exudate due to a prior lesion, and products of wool hydrolysis have been shown to be adequate to carry the larva through the first stage. Wool hydrolysis is probably not an important factor on the living sheep, judging by the examination of wool samples, but we have isolated organisms which, when growing on a nutrient medium, are capable of disintegrating wool fibre.'

There is a vast literature on this subject, which is of the greatest practical importance in sheep-rearing areas, but we have no space to pursue it here. *The Review of Applied Entomology*, series B, is the best means of following the latest developments.

A particular habit of the larvae of *Lucilia*, which has interested me personally for many years, is that of suddenly emerging from the soil of gardens after heavy rain in late summer. In England this happens every year, and almost always just outside the kitchen door. The larvae are nearly always those of *Lucilia sericata* (only occasionally those of *Calliphora*) and they are certainly in the soil in great numbers. They come out because a sudden rain-storm causes the soil to become waterlogged, and so the larvae come up for air. They are seen by householders, either wriggling across the concrete strip beside the house, or sometimes actually coming into the house, beneath the door (figures 38, 39).

This much has been established. What has not been decided is why they are there, and what they are feeding upon. The orthodox explanation would be that they have bred in the dustbin and have moved into the soil to pupate, but the circumstantial evidence is all against this. The householders concerned always look first to the dustbin and see that it is free from larvae. Those in the soil are not particularly near the dustbin, and there is an obvious disparity between the numbers in the soil and the few in the dustbin that might have been overlooked.

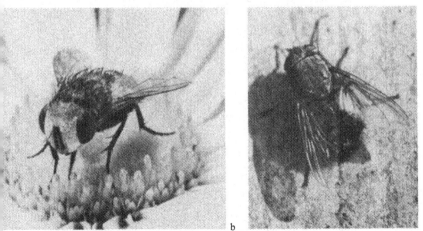

21. (a) A typical acalyptrate fly, *Stenodryomyza formosa*, from Japan; (b and c) typical calyptrate flies, the South African Tachinid *Billaea*, and the Japanese bluebottle *Calliphora lata*.

22. Head of the yellow dung-fly, *Scatophaga stercoraria*, a fierce
predator among small insects.

23. (a) A tsetse-fly in the act of biting; (b) the blow-fly *Aldrichia grahami* on the entrails of a fish; (c) the greenbottle *Lucilia illustris* laying eggs on the head of a sardine.

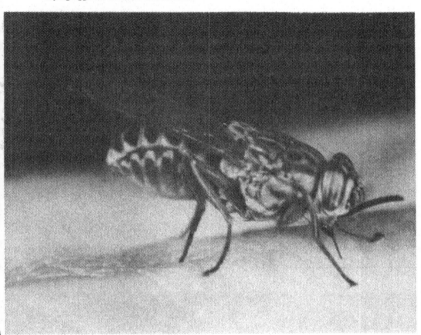

24. Eggs of carrion-feeding blow-flies: (a) laid among the hairs of a dead mole; (b) removed from the eye of a human corpse.

My own view is that the circumstances strongly suggest that there is a seepage of organic matter from a drain into the soil, and that this is attracting and nourishing the larvae. So far I have been unable to get any direct proof of this, since one hesitates to advise people to dig up their drains if it can be avoided! It is striking, however, that such masses of larvae seem always to come from soil in the vicinity of drains, and that the only cases I have known in which they occurred at the front of the house involved soil beneath which drains were buried.

Calliphora and *Lucilia* are world-wide, but are reinforced in the tropics by *Chrysomyia* and *Hemipyrellia* respectively. The distinction between 'bluebottle' and 'greenbottle' is not so obvious in these tropical forms, which belie the beauty of their colouring by the viciousness of their habits. Many *Chrysomyia* behave like *Calliphora* while others have the most varied habits, infesting wounds in cattle, sheep, and man. *Chrysomyia bessiana* is known as the Old World screw-worm, and may be considered along with *Callitroga americana* and *C. macellaria* the screw-worms proper of the New World. These are truly fearsome flies which lay eggs on any wound, sore, or even contusion, and the voracious larvae devour the living tissue to an extent that may cause disfigurement or even death. Any part of the body may be attacked.

In these flies we see the transition from scavenging and carrion-feeding to parasitism. *Callitroga macellaria* is a carrion-feeder and a secondary invader of injured animals, but *C. americana* and *Chrysomyia bezziana* are obligatory parasites, which cannot feed except in living tissues. They are extremely prolific, and it is said that a large open wound may have from 800 to 1,000 eggs deposited in it.

After this gruesome picture, those larvae that merely hang on and suck blood like so many leeches seem comparatively harmless. In the more temperate regions *Protocalliphora sordida* (or *azurea*) lives as a larva in the nests of a large number of small hedgerow birds, where it feeds by sucking the blood of nestlings. Several hundred have been found in one nest, and nestlings may be killed by the loss of blood, though this does not often happen. It is interesting to note the parallel with the acalyptrate *Neottiophilum praeustum*, the two species having reached the same stage of life quite independently.

The congo floor maggot, *Auchmeromyia luteola*, has evolved the same habit in relation to man living under primitive conditions in Africa south of the Sahara. The adult fly lives round about African villages, keeping to the shade and laying its eggs in cracks in the mud floor. The larva sucks blood from people who sleep on the bare floor or on matting, but cannot climb even a short distance on to a bed. This species is said to be restricted to a diet of human blood, while related flies attack pigs and wart-hogs in the same way. Here we have what is apparently the only fly that is strictly evolved along with man, and an interesting commentary on the current theory that Africa was the original home of the human species.

A stage further, perhaps, in evolution, produced the larvae that pierce the skin and take up their position beneath it, with hind spiracles opening to the exterior, and mouth-hooks at the other end happily feeding on muscular tissue and blood. Several well-known tropical Calliphoridae have taken this step. In Africa the Tumbu fly, *Cordylobia anthropophaga*, has as its natural hosts rats and dogs, and *Stasisia rodhaini*, sometimes called the larva of Lund, infests antelopes and the giant rats, *Cricetomys*. Either of these maggots may infest man, and the Tumbu fly commonly does so. The growing larva forms a furuncle, or boil-like swelling, much as does the warble-fly of cattle, but the infestation is purely local, and there is none of the movement through the tissues that is characteristic of the warble-flies. Indeed, so far as a single larva is concerned the parasitism is local, temporary, and of small importance, though painful. The first entrance of the larva may not be noticed, but as it grows the pain becomes intense. The larva can be expelled by painting the boil with liquid paraffin, which clogs the hind spiracles, and compels it to come out in search of air. The wound heals well provided that it avoids secondary infection, but small animals with multiple infestations may suffer severely.

Even elephant-hide is not proof against such attacks, and in the original account of *Elephantoloemus indicus*, Austen quoted a report that 'some parts of the [elephant's] body were so closely pitted as to impart to the skin a honeycomb appearance, showing that thousands of the parasites had burrowed out'.

Another tropical Calliphorid at this stage of evolution is *Dermatobia hominis*, the human bot-fly of South America. A unique feature of its life history is that it does not seek out its own victim, but attaches its eggs to a mosquito, *Stomoxys* or other biting fly, which then carries these until they reach a suitable victim. Sometimes *Dermatobia* lays its eggs on a leaf, where the larvae have to wait until a vertebrate animal brushes against them. Although this sounds an improbable story, there are plenty of specimens in the museums to be seen and photographed. In Central and South America this larva is an important pest of cattle, and from Dutch Guiana Bruyning told a grisly story of a man who had a larva wandering under the skin of his head, where it 'caused bone conduction to the auditory centre by scratching against the skull'.

We are led to wonder how such a device could ever have evolved. Admittedly, once the eggs are safely attached to a biting fly there is a good chance that they will reach a suitable host, but how did the fly ever choose suitable carriers? Perhaps even now this is a matter of chance, and that any flying insect of approximately the right size and movement is accepted. If *Dermatobia* lays a number of eggs even approaching that of *Calliphora* it can afford a great many failures.

The early larva of *Dermatobia* is curiously flask-shaped with the narrow end towards the open air. A specimen was taken from the knee of a young man in an English hospital who had recently returned from Honduras.

Dermatobia belongs to a New World group of Calliphoridae transitional in habit towards the related family Oestridae, which includes the well-known bot-flies and warble-flies, and it is convenient to digress at this point to look at these flies, returning to the rest of the Calliphoridae later. The Cuterebrinae, to which *Dermatobia* belongs, are considered by Zumpt [316] to be a line of Calliphoridae that arose in the Palaeocene period along with the rodents, of which they became specific parasites. From their basic stock arose the family Oestridae, the bots and warbles, characterized by the reduction of the adult mouth-parts, and loss of the power of feeding after they become adult. This is a significant step in their biology, which marks the success of the parasitic larva in providing all the food that the fly is going to need for pupation, adult activities, and provisioning the eggs of the next generation. From the point of view of the collector of flies this means also that, though larvae may be plentiful, adult flies are elusive, because they are short-lived, furtive, and always in close attendance on the host animal.

Zumpt recognizes four groups of Oestridae: Oestrinae and Cephenomyinae, which live as larvae in the nasal cavities of various grazing mammals; Hypodermatinae, the warble-flies, with their larvae beneath the skin of the back of various even-toed ungulate mammals and of rodents; and Gasterophilidae, mostly in the stomachs of horses, rhinos, and elephants. Excepting the warbles, all the others are known as bot-flies.

Cephenomyia, the nasal bots of deer, are widespread in the Northern Hemisphere, from Scotland to New Mexico. It was in New Mexico that Townsend made his notorious observation of a fly that he said was the male of *C. pratti*, and which he said covered 400 yd in one second. By simple arithmetic this is 818 mph, and was quoted as such for many years in sporting diaries and other works of reference. Long before supersonic speeds were possible to man this fabulous fly was discussed, but all the argument centred upon whether a large, hairy, bluebottle-like fly could be propelled through the air at supersonic speed, and if it could, whether the necessary energy could be generated in such a small body. No one appears to have said bluntly that the original observation was made without instruments, and was just a wild guess.

The adults of the Oestrinae have a most curious appearance, rather bare, and very much wrinkled and mottled, as if they were suffering from leprosy. Patton thought that this served to conceal them among the wool of sheep, but they also attack many other animals with much shorter pelts, where the cryptic concealment is not so effective. It is possible that their colour may serve as a 'dazzle pattern' when the egg-laying females approach their victims in bright sunlight. It is not clear why they should look like this, but their appearance is very characteristic. The best known of them is the sheep nostril fly, *Oestrus ovis*, essentially a fly of the shepherding countries of the Old World, of the countries so familiar in the Bible scene. Now it has been spread wherever domestic sheep have been taken.

Larvae of *Oestrus ovis* also attack goats and dogs, and sometimes man, settling in the nose and throat and causing intense pain. Young larvae are also found in the eyes of men who associate with sheep, and it is thought that the female fly projects its eggs into the eye by flying sharply towards the victim's face. Larvae in the eye do damage to the cornea, but do not normally develop beyond the first stage. Evidence is conflicting whether the larvae from the eye may move down into the nasal passages, or whether actively feeding larvae in man come from eggs deposited directly into the mouth and nose.

There are a number of related genera and species of nose-bots, in various antelopes, wart-hogs, pigs, hippopotami, horses, and zebra.

Cephalopina titillator is the head-bot of camels, and *Pharyngobolus africanus* lives in the gullet of the African elephant, so the group has invaded almost all the large mammals of the African continent. In other continents they are not so well known, probably because their study, like that of the gall-making insects, is a slow business. It is difficult either to catch or to breed the adults, and until this has been done the larvae are difficult to classify. Dr Zumpt, indeed, rightly urges that names should not be given to newly discovered species until adults as well as larvae are known. That Oestridae have flourished as long as mammals have existed is suggested by the fact that the kangaroos have a species of their own, put into a separate genus as *Tracheomyia macropi*, living not in the head sinuses, but in the mucus in the windpipe.

These larvae feed primarily on the mucus present in the head sinuses and throat, but they have powerful mouth-hooks, sharp and strongly curved, and they use these freely to scrape the mucus membrane and so to stimulate it to secrete more vigorously. Thus they cause inflammation and severe pain, and are usually expelled by coughing, sneezing, or vomiting, to pupate in the ground. Patton describes how camels kneeling round a feeding trough may suddenly be overcome by a fit of sneezing, and expel larvae that can be seen wriggling in the food. The larvae seem to become active and provoke expulsion mainly towards evening. Infested animals usually recover quickly after the larvae have gone, but sometimes the pain and perhaps internal damage may bring about convulsions or even death.

Warble-flies make up the sub-family Hypodermatinae, and are well-known agricultural pests in northern countries. Unlike many flies, they do not survive if they are accidentally transported to other parts of the world: for example, there are no *Hypoderma* in Africa south of the Sahara. The eggs are laid on the outside of the body of a grazing animal, usually on its legs. The young larvae penetrate the skin, but, unlike such larvae as the Tumbu fly, they do not remain in one place throughout their life. On the contrary, they immediately begin to wander through the body of the host animal. The eggs of the cattle warble *Hypoderma bovis* are laid in the hottest part of the summer, and those of *H. lineatum* a little earlier. The larvae disappear until autumn, when they are to be found in the connective tissue of the wall of the oesophagus or gullet. 'They work their way towards the diaphragmatic end of the oesophagus, pass under the pleura, across the diaphragm, until they gain the cartilages of the ribs. They follow the vessels between the intercostal muscles, and travel towards the dorsal integument, often passing through the spinal canal between the periosteum and the dura mater. In February or March the larva reaches the spot for its final development, entering the hide from inside, and beginning to form a "warble"' (Keilin [159]). Here they lie as do the larvae of the Calliphoridae, breathing air through the hind spiracles, which are applied to a hole in the skin of the host. When they are fully fed they leave the warble and fall to the ground, where they pupate.

Besides causing pain to the animal, and promoting restlessness, the warbles leave a hide pitted with holes, or marred with healed sores. They occur in the middle of a hide, just where flaws are of the greatest practical nuisance.

Adult warble-flies are big, hairy, brightly patterned flies, that look like bumblebees. Those which fly round cattle, particularly *Hypoderma bovis* and *H. lineatum* are not often seen, though they must be quite common, judging from the number of warbles that are found in hides, and the official Regulations that exist to combat them. There is evidence, largely brought together by my late colleague Major E. E. Austen, that the humming of warble-flies causes cattle to 'gad': that is, to rush about wildly, holding their tails high in the air. This disturbance of grazing cattle is usually blamed upon 'gad-flies', which are presumed to sting or bite, linking 'gad-fly' with 'goad'; but *Hypoderma* does not bite, indeed does not feed at all. The 'gad-flies' of common speech are Tabanidae, horse-flies which together with the biting muscids, *Stomoxys*, *Lyperosia*, and *Haematobia*, do make life a misery for cattle in summer. Yet cattle accept the biting flies with a dull apathy, and can be seen covered with them and giving only a lazy swish of the tail.

Why, then, does the approach of a single *Hypoderma* send cattle into such a paroxysm? It is one of the strange mysteries of behaviour, since the beast can have no way of knowing why this particular fly is a menace to it. A number of witnesses have testified that they have seen cattle provoked into gadding by an imitation of the hum of a *Hypoderma*, or have even done so themselves. There may be a physical basis in the fact that *Hypoderma* has one of the highest rates of wing-beat frequency, upwards of 350 per second, and possibly even higher harmonics may be present, like the famous Galton dog-whistle.

Reindeer have a warble of their own, *Oedemagena tarandi*, a magnificent fly. Since the adult flies are much commoner than are adult *Hypoderma* it looks as if *Oedemagena* does not meet with much opposition, though my colleague Mr M. E. Bacchus tells me that a tethered reindeer became very disturbed by these flies. The damaged hides, of course, are a loss to the owner of the animals.

Plate 26 shows a piece of skin of the Tibetan gazelle, *Procapra picticaudata*, beneath which are larvae of warbles, though I have not yet come upon any recorded from this mammal.

Finally, there are the stomach bots of horses and a few other mammals. Post-mortem examination of a working horse of the country, one that has spent its life in the fields, and not one that

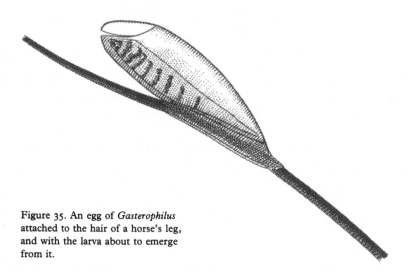

Figure 35. An egg of *Gasterophilus* attached to the hair of a horse's leg, and with the larva about to emerge from it.

draws a brewer's dray in London, will generally show a number of barrel-shaped objects attached to the lining of the stomach. These are larvae of *Gasterophilus intestinalis*, the horse bot-fly.

The adult fly is furry yellow, and bee-like, the female equipped with a long ovipositor. She disposes of her eggs by attaching them singly to the hairs of a horse, usually to the legs, sometimes to other parts of the body that can be reached by the horse's mouth. During egg-laying she hovers with great precision, and the sound is said to disturb horses in much the same way that *Hypoderma* affects cattle. There is method in the choice of a site for the eggs. The larvae form within them, but do not emerge until the horse licks them, when the young larvae come out and bore into the epithelium of the tongue, move along it, and pass down the throat into the stomach. Patton found that he could make larvae emerge from ripe eggs by stroking them with a moistened finger.

This seems a beautifully perfected mechanism of infestation, until we learn that the other common species, *Gasterophilus nasalis*, lays its eggs beneath the chin of the horse, where they cannot be licked off! The larvae of this species have to emerge without any special stimulus, and to make their own way to the mouth.

The mouth-hooks of the larvae of *Gasterophilus* are sickle-shaped and very sharp, and look like offensive weapons. Actually they are anchors which attach the larva to the wall of the stomach, and resist the powerful forces of peristalsis, and the pressure of the bulky food material as it sweeps past. The larva is thus not a true parasite, but only a 'kleptoparasite', which lives by stealing a small part of the food of the host. It is still an air-breathing larva, but there is little fresh air in a horse's stomach. The hind spiracles of the larva are protected by being sunk into a spiracular pouch, the lips of which open when air is available, and close for protection when the bubble has passed. Breathing is thus irregularly intermittent, but the best use of what air available is ensured by the presence of red cells containing haemoglobin, which are amply supplied with small tracheal branches.

Thus two groups of flies as far apart in the scale of evolution as *Gasterophilus* and the non-biting midges of the family Chironomidae have arrived at the combination of adults that do not feed, and larvae that use haemoglobin to enable them to breathe in a deficiency of oxygen.

The correct placing of *Gasterophilus* in the classification of flies has been much disputed. Fairly recently it was regarded as having no close relationship with *Hypoderma* and *Oestrus*, the similarity of habit being attributed to convergent evolution, and it was placed as a family of the acalyptrate flies. Zumpt has recently returned to what appears to be the commonsense view. He makes Gasterophilidae a family close to the Oestridae, and includes in it *Gasterophilus*, the stomach bot of horses, donkeys, mules, and zebras; *Gyrostigma*, the huge and picturesque stomach bot of rhinoceroses (plate 26c); *Cobboldia*, the stomach bot of elephants; and, for structural reasons rather than biological ones, two other parasites of the elephant, *Neocuterebra* which lives in the sole of the foot, and *Ruttenia* in the skin of the back. Adult *Cobboldia* are moderately sized flies, beautifully coloured in black and white, which lay their eggs in neat stacks at the bases of the tusks.

Although these stomach bots are not really parasites, they do cause some distress to their hosts. There must be some indigestion and malaise, and there is some injury to the stomach wall when the mouth-hooks pierce, and are eventually torn away. One rhinoceros may have two or three hundred of them, and perhaps this explains why such a powerful vegetarian animal, without any obvious need either to defend itself or to attack others, should be so chronically bad-tempered.

Even elephants are not too grand to notice these larvae, though their bots are smaller both as larvae and as adults than those of rhinos. Patton wrote of the effect of *Cobboldia* larvae on working elephants in India: '. . . heavily infested elephants suffer from digestive troubles and are often unable to work, and the younger animals become anaemic and often die'. One calf elephant had its stomach 'so full of larvae that there was not room for a pin to pass'.

Now let us return to the Calliphoridae, with their diversity of habits, and trace them in another direction, that of parasitism of cold-blooded animals, including other insects. A development in this direction is always imminent among larvae that feed in carrion or in animal secretions. An example is *Lucilia bufonivora*, another species of greenbottle that specializes in feeding in the heads of toads, frogs, salamanders and even newts. The larvae live for a number of days and finally kill or severely maim their hosts. This species is notorious for this habit, but I have also had specimens of *L. porphyrina* from India where they occasionally attack toads. James has recorded a fatal infestation of a toad, with larvae of *Bufolucilia elongata* in the flank of the body. Perhaps toads are particularly vulnerable to attack because they produce so much mucus on the head and body.

One Calliphorid or another has discovered almost every kind of animal that is vulnerable to attack, and all stages of relationship exist, from carrion-feeding, by way of feeding on bodily secretions, to the pus of wounds, and finally to parasitism. The interest lies in the degree of parasitism with which some Calliphorids confine themselves to a single host species.

Thus *Melinda cognata* was studied by Keilin, who found that it confined its activity to the common European snail *Helicella* (*Helicomanes*) *virgata*, ignoring five other common snails, some of the same genus, which were to be found in the same places. The eggs are laid in the mantle cavity of the snail, and the young larva bores into the kidney and feeds upon it. At first this has no apparent effect on the snail, but it becomes progressively less active, as it is slowly devoured from inside. There are thought to be three generations a year, and the fly spends the winter as a pupa.

Pollenia rudis is well known as an adult fly, because this is the cluster fly which accumulates in attics and roof-spaces in the autumn, along with *Musca autumnalis* and *Dasyphora cyanella* as we shall see in a later chapter. It was Keilin, again, who showed that the larvae live as parasites in earthworms of the genus *Allolobophora*. He describes this life-history in great detail in several of his papers, and later workers have confirmed the fact, though differing in their interpretation of some of the details. Since there is only one larva per earthworm, it seems incredible that such vast numbers of flies can breed in such a laborious way, until we remember Darwin's account of the great numbers of earthworms in the soil, perhaps 50,000 per acre.

Stomorina lunata is a strange Calliphorid, the larvae of which feed in the egg-cases of locusts. The fact has been known for some time, and sporadic appearances of the fly in England, such as those reported in 1947 by Fonseca and by Colyer, are attributed to the same conditions of wind and weather that bring us our occasional migratory locust. Dr D. J. Greathead very kindly allowed me to read an unpublished thesis of his in which he described his own observations of this fly in Somalia.

The ovipositing female sits on vegetation close to groups of ovipositing locusts, and periodically comes down and examines depressions in the soil to see if there are egg-pods of locusts in them. When she finds one, a female fly lays her own egg upon it, and covers it with soil. The flies move on when the locusts move. The larva simply preys on the eggs, and since it breaks more than it eats, putrefaction sets in and attracts other flies, notably the Drosophilid *Cyrtonotum cuthbertsoni*. In Greathead's view the fly attacks only four species of locust: *Locusta migratoria*, the migratory locust; *Schistocerca gregaria*, the desert locust; *Nomadacris septemfasciata*; and *Locustana pardalina*.

Finally, Calliphoridae, like so many other families of flies, have not overlooked the ants and termites. The tropical yellow flies of the genus *Bengalia* not only feed as larvae in nests of termites, but as adults hover over marching columns of ants and snatch from them eggs, pupae, or food material, whichever the ant happens to be carrying.

I have left myself little space in this chapter to deal with two large groups of flies, the Sarcophagidae and the Tachinidae. The Sarcophagidae are mostly grey, checkered flies known as flesh-flies, a translation of their scientific name that is appropriate to their habits. The big *Sarcophaga* are common enough in the countryside and in gardens, though they seldom come indoors. As a group they carry to extremes the habit of that section of blow-flies that feeds actively on carrion, and is so eager to drop its eggs that it can scarcely control itself. *Sarcophaga* in fact is effectively viviparous and normally the eggs hatch in the uterus of the female before they can be laid. As an example of eagerness, I have known a female *Sarcophaga* to fly to a coat, the sleeve of which had been soiled with blood from a shot pheasant, and immediately lay a batch of young larvae upon it.

Larvae of *Sarcophaga* are huge maggots, bigger than those of *Calliphora*, and are remarkable by having the slits of the hind spiracles nearly vertical, and deeply sunken in a posterior cavity. Dealers in 'gentles', the larvae of *Calliphora* bred for use as bait by fishermen, look longingly at the juicy *Sarcophaga* maggots, but so far have not found a way to breed them commercially.

This is a large and difficult group that occurs all over the world, and a great deal of research is still needed before it is properly understood. Yet for our present purpose all that we can say is that the larvae are voracious; they will take anything of animal origin that they can get, living or dead; and that the adults are of uniform appearance, with infinite variations in detail. One group of special interest are the Miltogramminae, which lay their eggs on the food that bees and wasps provide for their young. Hatching inside the closed cell, the fly larva grows more quickly than the other, and eats all its food; an entomological cuckoo in the nest. *Termitometopia skaifei*, described by Dr Zumpt, breeds in the nests of the termite, *Amitermes atlanticus*, in South Africa: 'the maggots are fed and cared for by the termites, which seem to get from them a fatty exudation of which they are very fond'.

I am afraid that I shall have to dismiss in the same summary way the large and important family Tachinidae, which spend their larval lives as internal parasites in other insects, spiders, woodlice or centipedes. A whole book could be written about these flies, giving details of their ways, and those of their hosts, classifying them according to their selection of host, the way in which they attack it, and the ingenious devices developed in various larvae to overcome their special problems. Respiration in particular is a great problem, rather like that which faces aquatic insect larvae, and sometimes resulting in somewhat similar developments.

Several excellent classifications of the habits of Tachinidae have been made, notably those by Lundbeck in his *Diptera Danica*, by Mesnil in his *Essai sur les Tachinaires*, by van Emden in the *Handbooks for the Identification of British Insects*, vol. X (4), and in various papers by W. R. Thompson. Herting's *Biologie der Westpaläarktischen Raupenfliegen* [126] contains a valuable section on *Die Tachinenlarve als Endoparasit* and a comprehensive bibliography.

Tachinidae play a role in biological control of other insects no less important than that of the hymenopterous parasites, the ichneumons, braconids, chalcids, and proctotrupids. Their practical effect upon the balance of nature is fundamental. The adults are among the most bristly of bristly flies, and provide a fascinating subject for the systematists, just as their larvae do for the ecologist and applied entomologist. I can only regret that there is no space to begin to discuss them in this book.

At the end of this Chapter may be a convenient place to mention a group of South American flies which have developed maternal incubation of the larva in a way that is less complete and less efficient than that of the tsetses. One large larva at a time is retained in the abdomen of the female parent until the second instar. Part of the larva can be seen protruding from the tip of the abdomen, which is simplified to enable a large larva to pass out of it. There is no evidence that any food is provided by the mother.

These flies, which are abundant in the rain-forests of tropical America, seem to combine features of Muscidae and of the Calliphorid blow-flies of the following Chapter. They are usually placed in a subfamily Mesembrinellinae of the Calliphoridae, but it seems likely on biological as well as on structural grounds that they may be a small evolutionary twig arising from the common origin of the two great calyptrate stocks.

Chapter 18

Parasites of Mammals and Birds

WE STARTED THIS survey of the flies with the crane-flies as an example of a primitive family. Their larvae have a well-developed head and chewing mouthparts, and live on a variety of vegetable food materials, in a wide range of different situations. The adult flies are clumsily built, slow-flying, pleasant, but ineffective.

The final group of flies, at the other end of the series, is very different. The larvae have almost no independent life, each being fed individually in its mother's womb, and becoming a pupa as soon as it is liberated. The adult flies do all the feeding, and subsist on a pure diet of blood, living as external parasites of mammals and birds. They are very considerably modified in structure to suit this way of life, and seldom or never leave their host.

Since animals that give birth to living young are said to be viviparous, these flies that appear to give birth to pupae are known as Pupipara. Although this term is inaccurate, and the flies concerned are apparently not even a natural group, Pupipara is a convenient term by which to refer to them collectively.

There are three families of Pupipara: Hippoboscidae feed upon many different mammals and birds; Streblidae and Nycteribiidae are confined to bats. It appears most probable that these three families came from at least two different ancestors, Hippoboscidae perhaps from flies living in birds' nests, and Streblidae and Nycteribiidae from flies that bred in the roosts of cave-dwelling bats. Hippoboscidae have much in common with tsetse-flies which, as we have already seen, also drop fully fed larvae, and have adults that feed only on blood. The obvious difference is that tsetse-flies have retained their independence, and only visit their victims when they want to feed, whereas Hippoboscidae spend their adult lives on their host. The two families have enough in common to suggest that in the remote past they might have arisen from a common ancestor among the muscoid flies. In particular, their piercing proboscis is formed from the stem of the labium, and the labella, which are spongy lobes in house-flies, are hard, small, and insignificant in Hippoboscidae; whereas in Streblidae and Nycteribiidae it is the labella themselves that are drawn out into a piercing organ.

There are about a hundred species of Hippoboscidae scattered throughout the world, and almost everything of interest about them concerns the adult flies. They are always flattened dorso-ventrally, with a tough, leathery appearance, and strong legs equipped with long, curved claws. They are called 'louse-flies' when they occur on birds, and 'keds' on mammals, the latter especially the wingless species. There is a growing fashion for calling them 'flat-flies', an ugly name that should be discouraged.

The adaptation of these flies to a parasitic life has evidently been going on for some time, and is very complete, affecting every detail of their external appearance. The flattened shape and strong, spider-like legs enable them to move quickly about on the skin of the host, pulling themselves along by grasping the hairs or feathers, with a crab-like motion.

The thorax is a solid box with its component sclerites more firmly united than in most flies, but the abdomen is soft, and mainly membranous. The sclerites of the abdomen are reduced in number as well as in size, and show a progressive reduction from the primitive to the advanced members of the family, and more reduction in females than in males. The eyes are also smaller than they are in flies that live a free life in the open air, but not as greatly reduced as they are in the other two families of Pupipara.

Most Hippoboscidae have wings, and many fly well. Broadly speaking, those that fly actively are found on a variety of hosts, and few, if any, are physiologically confined to one host in the sense of being unable to tolerate any other blood. As with fleas, and perhaps other parasites, it is really the host's habitat to which the parasite is adapted, and it will accept other hosts if they occur in similar surroundings. Thus Hill has recently defined the habitats of the three species of *Ornithomyia* that occur in the British Isles: *avicularia* favours the larger woodland birds; *fringillina* the small birds of the hedgerows; and *lagopodis*, a fly first recognized on the red grouse in Scotland, is associated with moorland birds in general. The distributions of the two last in the British Isles broadly agree with the geographer's division into Lowland Britain and Highland Britain respectively.

Thus the number of hosts available to any particular Hippoboscid is determined by the type of habitat. Such lonely creatures as the frigate birds of tropical seas carry with them their own parasite, *Olfersia spinifera*, which they pass between themselves when they nest. On the rare occasions when these magnificent birds are carried away by storms and turn up, for example, in Scotland, they usually have one or more parasitic flies with them. *Crataerina* and *Stenepteryx* are Hippoboscids with narrow and pointed wings, or very short ones, which in either case cannot be used for flight; this sacrifice of the powers of flight is linked with the fact that these flies live exclusively on swallows, martins, and swifts, birds that spend most of their time on the wing, and which return year after year to the same nesting colonies. It would therefore seem reasonable to say that loss of flight is an advantage, reducing the risk of separation from such a high-speed, streamlined host. The process of reduction has gone even further in more tropical members of this group. *Myiophthiria* in the Old World, and *Brachypteromyia* in the New have ludicrous little pads, with only a caricature of wing-venation, like the wings of fairies in a pantomime. Unfortunately for theory, *Ornithomyia biloba* and perhaps one or two others also live on swallows, but have the fully winged mobility of their genus. Once again we wonder why one insect has slowly evolved what seems to be an elaborate adaptation to a particular way of life, while another insect does the same job without adaptation.

Crataerina pallida is surprisingly common in the nests of the European swift, and may weaken or even kill nestlings by its voracious appetite for blood. Pupae overwinter in the deserted nest, and the new generation of adult flies emerges after the birds have returned in the following summer. *Crataerina* is thus obviously adapted to birds that return to the same nest, or at least to the same colony. If the nest is not reoccupied the hungry flies crawl – being unable to fly – in search of food, and they often appear in the bedrooms of houses that have swifts nesting under the eaves. These tick-like flies are repulsive to look at, and their claws make one's flesh creep if they walk over one's skin. They can bite through a human skin, and do so on occasion, but they are not aggressive. Their habit of squeezing into narrow spaces causes them sometimes to creep between the bed-sheets, where they are definitely unwelcome.

All the Hippoboscidae of birds are likely to be taken on birds of prey, to which they have transferred when their original host was killed, so that hawks, owls, and falcons appear to be infested with a great variety of parasites. One of these is the pigeon fly, *Pseudolynchia canariensis*, the natural hosts of which are wild pigeons of tropical and sub-tropical countries. This parasite has spread with the domestic pigeon into most of the warmer countries, though it does not penetrate far into colder ones, and since domestic pigeons are easy meat for the local birds of prey, the hosts of the pigeon fly include a series of hawks and owls.

An amusing sidelight on the ways of birds is that *Olfersia fumipennis*, a specific parasite of the osprey or fish-eagle, is also regularly taken from the American bald eagle, now alas itself becoming rare. The bald eagle is a lazy fisherman, and often gets his meal by robbing an osprey of the fish that it is carrying back to its nest, and swooping down on it in the air. Apparently the brief contact is enough for the fly to move from one host to the other.

The Hippoboscidae that live on mammals are the lesser section of the family, and apparently a later evolutionary experiment. Like the bird flies, they have their active, fully winged members, and their flightless ones. The genus *Hippobosca* itself consists of relatively large, handsome flies, with a striking pattern of brown and yellow, and with fully developed wings. They seem at first sight to have some obscure connection with domestic animals: *Hippobosca equina* with horses and mules, *H. variegata* (*maculata*) with cattle and horses, *H. camelina* with camels, and *H. longipennis*, surprisingly, with dogs. The dog fly occurs from the Mediterranean to China, was known in Europe in classical times, and is mentioned in Chinese literature. It is likely that the link between the hosts of *Hippobosca* is not directly their association with man, but that they are inhabitants of steppes and semi-deserts, areas in which man first became a nomadic herdsman, and domesticated a number of animals for his own advantage. Other species of *Hippobosca* live on various antelopes and gazelles and one species, *H. struthionis*, has transferred itself to the ostrich.

This last example gives convincing support to the idea that it is the habitat that is significant to the fly, and not the specific nature of the host itself. Apparently to *Hippobosca* the ostrich is not a bird, a member of another zoological class of animals, but is simply a gazelle with feathers!

The active mammals of the steppes have active, fully winged

25. Part of a human thigh, at the stage where larvae of *Piophila nigriceps* feed on it.

26. Flies living as larvae in the bodies of living mammals:
(a) *Gasterophilus* in the stomach of a zebra; (b) warbles beneath the
skin of a Tibetan gazelle, *Procapra picticaudata*; (c) *Gyrostigma*
from the stomach of a black rhinoceros.

a

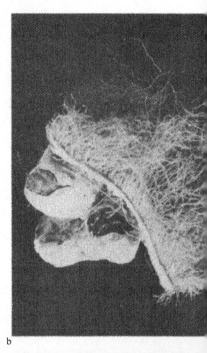
b

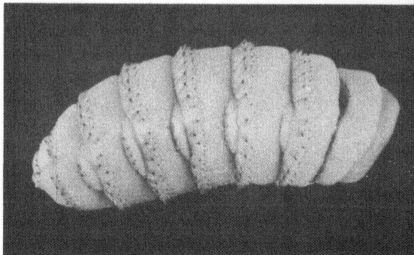
c

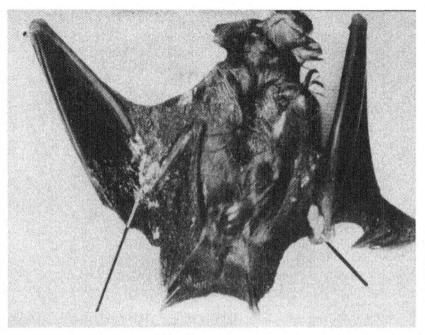

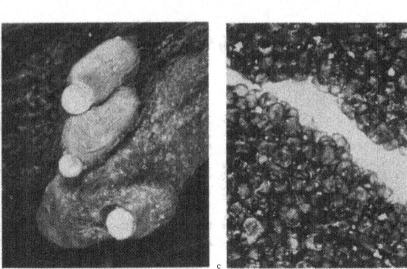

27. Bat-parasites of the family Streblidae: (a) adult females of
Ascodipteron jonesi embedded in the skin of a bat's wings,
indicated by black lines; (b) close-up of the flies *in situ*; (c) an
incrustation of puparia of *Trichobius phyllostoma* taken from
the walls of Tamara Cave, Central Trinidad, by Dr Theresa Clay.

28. *Musca sorbens* feeding on discharges from ulcerated skin:
(a) on the legs of a child; (b) on a horse.

Hippoboscids. The less active deer of woodland and open forest have their own group of flies, which have developed the curious habit of breaking off the wings, a habit that we have seen previously in the bird-feeding acalyptrate *Carnus hemapterus*. Adults of both sexes are fully winged for mating and finding a suitable host, but then they break off the wings near the base, and commit themselves to the fortunes of their host. They are able fliers – I have seen a winged *Lipoptena cervi* from a garden in London more than a mile from Richmond Park, where deer are abundant – and it is not easy to imagine why they should have evolved the mechanism that is needed in order to shed the wings in this way. It is usually said that they benefit by being able to move freely about their host, unimpeded by the wings, but it is not obvious why *Lipoptena*, and the African *Echestypus*, should need this advantage, when *Hippobosca* is able to keep its wings.

There is something to be said for the complete loss of wings, as in the sheep ked, *Melophagus ovinus*. Here the parasite has to live in a dense, tangled fleece, creeping about as if on the floor of a primeval forest. *Melophagus* has gone to the extreme of adaptation to this life, not only becoming wingless, but losing the halteres too, and reducing the eyes further than any other Hippoboscid.

These wingless, flattened flies look like ticks, and are sometimes confused with the true sheep tick, *Ixodes ricinus*. Their proper name is the sheep ked, and in a similar way *Lipoptena cervi* is known as the deer ked, especially after it has shed its wings.

The evolution of the sheep ked is obscure. It is believed that at the present day there is no truly wild population of the fly, and that where it does occasionally turn up on the native sheep of any country it has always spread to these from strayed domestic sheep. There are two other species of *Melophagus*, *M. rupicaprinus* on the chamois of Central Europe (but not those of other areas), and *M. antilopes* which is known only from two specimens collected long ago from the Zeren, *Procapra gutturosa* in Mongolia. Almost certainly there are others still lurking on rare antelopes and goats in the various mountainous areas of Asia. The sheep ked itself has been found on chamois in the Caucasus, and possibly came originally from a goat, not a sheep. As these local mammals dwindle it becomes all the more urgent that their parasites should be recorded, before these too become extinct.

Hippoboscidae infest the kangaroos and wallabies of Australia, and the lemurs of Madagascar, mammals that are at opposite ends of their evolutionary line. The stage of evolution reached by the respective Hippoboscidae cannot be matched with that of their hosts, and once again it looks as if the flies spread into a habitat rather than on to a host, and made do with the mammals that happened to be there, regardless of their zoological classification. Bequaert suggests that the wallaby-flies have changed less from the ancestral bird-feeding

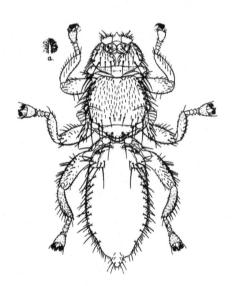

Figure 36. A bat-parasite, of the family Streblidae, for comparison with figure 37. The wings of this fly have been removed to show the compact appearance.

Hippoboscidae than some other mammal-flies, and so must be of more recent origin.

With the last two families we almost leave the world of real flies, and enter that of fantasy. Streblidae are slightly the more normal of the two. Most of them can be recognized as flies, though the females of *Ascodipteron* make up for this by looking quite unlike insects at all (plate 27). Instead of being flattened like Hippoboscidae, most Streblidae of the Old World have a cylindrical body, the thorax being nearly globular. Many of the Streblidae of the New World are more flattened, and *Nycterophila coxata* could be mistaken for a flea. The tips of the tarsi are flattened, with powerful claws, and the whole body is covered with a neat array of bristles. Mr Boris Jobling has made a lifetime's study of this family, and I am privileged to be allowed to borrow one or two of his beautiful drawings, which show the arrangement of bristles perfectly.

Most Old World Streblids have wings, and fly well on occasion, and their wing-venation is not widely different from that of other flies. They solve the problem of what to do with their wings by folding them in pleats along the back, in a way that is very convenient for them as they move about the fur of their host bats, but difficult for the systematist who wants to examine the wing in detail. They are found only on bats, and then only in warm countries. Jobling [150] showed that their distribution falls neatly within the winter isotherm of $+10°C$ in the Northern and Southern Hemispheres, this being the critical temperature at which the bats begin to hibernate (figure 40). During hibernation the body-temperature of the bats falls to within a fraction of a degree of that of their surroundings, and this is too much for the Streblidae: in the few instances where they have occurred outside the limits shown in the map there have been special circumstances, an exceptional year, or the protection of a cave against low temperatures.

Jobling also notes that the Streblidae of the Oriental and Australasian regions are isolated from those of Africa and the Mediterranean. He attributes this to the low winter temperatures of the Iranian Plateau, where the bats are forced into hibernation, an invisible barrier that is none the less effective.

There are almost a hundred species known, in about twenty genera, and there is a sharp difference between those of the Old World and the New. While the number of species in the two hemispheres is about the same, the Old World has only four genera, while the New World has sixteen. Moreover there are differences between these two groups which suggest that they are descended from two distinct ancestral lines which diverged a long time ago. Since the American Streblidae have evolved so many genera of such diversified shape, they must have progressed more rapidly at first, whereas the Streblidae of the Old World are still at the species stage.

The relationships of Streblidae with their hosts, like those of Hippoboscidae with theirs, seem to depend more on the habits of the bats than upon their zoological relationships. Streblids are most abundant on bats that roost in caves, congregating in masses, often of more than one species of bat, and sharing their parasites as well as their shelter. When a species of Streblid is known only from one species of bat it is generally a bat that roosts in isolation, or with only members of its own species. Fruit-eating, forest-dwelling bats live more isolated lives than cave-bats, and tend to have their specific parasites, though sometimes they share these with cave-bats which make excursions into the forest and may even roost there sometimes.

The most remarkable Streblid, and one of the weirdest of all flies, is the genus *Ascodipteron* (plate 27). The female fly not only remains on its host, but burrows into it with enlarged mouthparts until the parasite almost disappears from sight. Wings and legs are shed, and the sclerotized body is reduced to an almost unrecognizable remnant; on the other hand, the first abdominal segment swells and becomes a membranous bag, which envelops the rest of the fly, and converts it into a flask-shaped object from which the scientific name is derived. The male flies remain normal in appearance, and fly actively.

Ascodipteron drops its larva to the ground where it pupates, in the same way as the larvae of Hippoboscidae and of tsetse-flies. All other female Streblidae attach their mature larvae to a wall or other surface, and this becomes encrusted with puparia (plate 27) from which the adult flies emerge in due course. Since they are already in the roosting place of the bats they have no difficulty in finding a fresh host.

Jobling considers that Streblidae probably arose from some ancestral muscoid fly that bred in the dung of bats in caves, but I am not sure that I agree. It is difficult to see why they should abandon a larval feeding medium that is at once organically rich and unlimited in amount. *Mormotomyia* has not done so, as we have seen, but lives all its life in the dung, not bothering about the bats themselves. I should have thought it more likely that the ancestral flies first developed the blood-sucking habit, like tsetse, and dropped their larvae to the ground in the same way, and as *Ascodipteron* still does. As the family flourished best on bats that lived a communal life in caves, the flies gradually came to deposit their larvae in the cave too, and so transferred their whole life into the roosting-place of their hosts.

A similar origin perhaps may be postulated for the Nycteribiidae, also restricted to bats. The fact that these two families share the same hosts only serves to emphasize the striking differences between Streblidae and Nycteribiidae. Nycteribiidae never have wings, and have evidently been without them for a very long time, because the thorax has lost its usual box-like structure as the flight-muscles dwindled. Streblidae still look like flies (except for *Ascodipteron*), but Nycteribiidae look like six-legged spiders. The upper surface of the thorax is little more than a framework of hard chitin, joined together with large areas of soft membrane, and the head is a grotesque structure apparently sitting on top of the thorax (figure 37). Indeed, any one seeing a Nycteribiid for the first time is likely to mistake the under surface for the upper, and fail to find the head at all! The eyes are greatly reduced, and may be absent altogether; when they are present they are quite unlike those of other adult flies, being either a single, round facet, or two little lenses on a black mount.

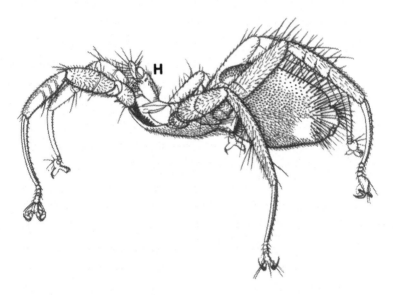

Figure 37. A wingless bat-parasite of the family Nycteribiidae. Note the long legs, the small head (H), placed dorsally, and the hooked claws.

The female flies leave their hosts and attach their larvae to surrounding surfaces, the rock walls of caves, or trees where the bats roost, or to the timbers of roof-spaces in buildings. Some have been reported to attach their larvae to the hairs of the host bat, but this is not the general rule. Apart from this brief excursion, Nycteribiidae never leave their host except to move over to another bat with which it is in contact.

Rodhain and Bequaert kept living fruit-eating bats in wire cages and studied the behaviour of the Nycteribiids that fed on them.

Cyclopodia greefi, placed upon the bat *Cynonycteris straminea*, fed almost continuously, and deposition of a larva, followed by a fresh mating, took place at intervals of six days or less. This, I think is one of the surprising facts about these surprising flies: living such a protected, even luxurious, life one would have expected them perhaps to breed in a slow, leisurely way. This undignified haste is more suited to flies like the bluebottles which breed in a rapidly changing medium.

Streblidae, as we saw, are more diverse in the New World than in the Old. Nycteribiidae on the contrary are much more diverse in the Old World than in the New, where, apart from two primitive species of *Hershkovitzia*, all the species belong to one genus, *Basilia*, and live on vespertilionine bats. In the Old World Nycteribiidae are widespread, and appear to be immune from the temperature limits that confine the Streblidae. Indeed some species have been recorded from Borneo to the Belgian Congo, but recently more careful work, like that of Theodor, has indicated that the various zoo-geographical regions have their characteristic Nycteribiidae, just like other families of flies. Even so, some species remain widespread.

As in the other two families of Pupipara, host-selection is not very precise, and seems to be more of a habitat-preference than a definite choice of a breeding host. As with Streblidae, Nycteribiidae that live on small bats congregating in caves are found to occur on a wider range of different species of bat than those living on relatively solitary hosts, such as fruit bats. The genus *Cyclopodia* favours the great flying foxes (*Pteropus*) while *Eucampsipoda hyrtlei* and some other *Cyclopodia* infest the allied fruit bats of the genus *Rousettus*. The Nycteribiidae do not appear to harm the bats and are not known to carry any disease.

A world catalogue of Nycteribiidae by Professor Theodor is about to go to press at the same time as this book.

Part Three

Chapter 19

Flies and Man

> Marcus: *Alas, my lord, I have but killed a fly.*
> Titus: *But how if that fly had a father and a mother?*
> *How would he hang his slender gilded wings*
> *And buzz lamenting doings in the air!*
> *Poor harmless fly,*
> *That with his pretty buzzing melody*
> *Came here to make us merry! And thou*
> *Hast killed him.*
> 'Titus Andronicus', Act III, Scene 2

> *Childe Harolde basked him in the noontide sun*
> *Disporting there like any other fly;*
> *Nor deem'd before his little day was done*
> *One blast might chill him into misery.*
> BYRON, 'Childe Harolde's Pilgrimage', Canto I, stanza IV

THE NATURAL HISTORY of flies is interesting to those who have sufficient detachment and intellectual curiosity, but it is the impact of flies upon Man's affairs that makes them important. When Pope said: 'The proper study of mankind is Man' he was expressing the same viewpoint as those people who ask us: 'What good are flies?', meaning what good are they to Man.

There are no flies whose lives are entirely bound up with the human person. This is rather surprising, because a number are specific to one kind of animal, either externally, like the sheep ked, or internally, like the bot-flies of the horse or of the rhinoceros. The closest dependence, perhaps, is that of the Congo floor maggot, sucking the blood of sleeping people in the African forest. The fact that related larvae attack various pigs and pig-like animals puts it on a level with the human flea.

All the other so-called 'synanthropic' flies, those which flourish particularly in association with Man, are attracted and encouraged by the food and shelter that he unintentionally provides for them. Like the nests of birds, and the burrows of many other mammals, the habitations of Man have their own specialized fauna. Man, moreover, is an outstanding host, because he not only continues to increase in number more and more rapidly, but he also alters the physical environment at a speed incomparably greater than the slow progress of geological time. Hence the future of all animals is dominated by Man, and none more so than the flies. Those flies that can adapt themselves to Man's ways will flourish, and those that remain aloof and rely on the continuance of their own favourite conditions, or their chosen host animal, will ultimately decline.

Seen from Man's own viewpoint, the flies that matter to him are those that attack him, those that merely annoy him, and those that eat his food. Any of these may also infect him with the organisms of disease. We think first of 'biting flies', remembering, however, that almost no adult flies really bite, in the sense of chewing and masticating their food. They stab, pierce the skin, suck blood, and are properly known as 'bloodsucking flies'. Larvae, and a few adult flies may scrape and lacerate tissues, feeding upon the blood or other fluid materials that are thereby released.

Considering how long Man has been plagued by flies it is startling to realize how very recently we have learned anything precise about the relationships between bloodsucking flies and disease. Table 1 gives the dates of a number of significant discoveries, all of them within the last hundred years, and many within living memory. Herms makes the interesting point that, in general, biting insects that carry a malignant disease cause less distress by the immediate effects of their bite than do harmless 'wild' species. This is not a sinister diabolical device, but is a sign of mutual adaptation between the host animal – as a species, of course, not as an individual – and the attacking fly, and the parasitic organism that causes the disease. Thus *Anopheles maculipennis*, which can give a man malaria, causes him less pain when biting than does a harmless salt-marsh species.

Mosquitoes, of course, head the list of bloodsucking flies. Malaria is the most important insect-borne disease, because it is almost worldwide in distribution, and, paradoxically, because it kills only a relatively small number of its victims. Human populations can quickly replace huge losses, as the wars and other slaughters of the twentieth century have shown, but the burden of chronic illness is much more crippling. Malaria is one of the great chronic diseases, with various forms of recurrent attack, and an insidious debilitating effect upon its sufferers.

In wars throughout the ages more men have been put out of action by malaria than by their enemies. Just one example at random was the disastrous Walcheren Expedition of 1809, during the Napoleonic wars, when so many men died that they had to be buried secretly at night to avoid spreading panic among the survivors, most of whom were in no shape to carry out their duties. Soon a fifth of the force was immobilized, and in some units not a man was fit for duty; an original force of 40,000 men was soon reduced to 24,000. Many survivors of the First World War in the Middle East had chronic malaria as a souvenir, and with this in mind, in the 1939–45 war the spray gun was as important as more conventional weapons. In the South Pacific theatre alone 4,407 persons were directly employed on anti-malarial activities (Cushing [57]). Ancient civilizations, the Roman among them, are believed by some to have been hastened to their decline by the insidious toll of malaria.

Human malaria is a cyclical disease carried only by anopheline mosquitoes, and only by some species of these. Something like seventy species are capable of developing and passing on the malarial parasite, but other conditions must be fulfilled before the disease can flourish. There must be enough mosquitoes to make it statistically probable that the same mosquito will bite an infected and then a healthy person after the correct interval for the parasite to have developed. Effective control does not demand that all the mosquitoes be destroyed, but only that contacts between flies and people should fall below a certain critical number. The converse is also true. Primitive, scattered communities may suffer little so long as they do not actually live in the marshy areas where the flies are very abundant, and which primitive people learn to avoid, and often protect by taboo. When the people begin to gather into large communities, and scoff at primitive fears, then malaria may begin to gain a footing.

Anopheline mosquitoes belong to the hemisynanthropic group of Gregor and Povolny, and flourish only through the secondary effects of Man's own evolution. Although Indian writers of the fifth century speculated that the bites of mosquitoes might spread malaria, this was not established as a fact until the end of the nineteenth century. Since then malaria has been reduced in most parts of the world, partly by improved drainage and public hygiene, and mainly by a

243

highly organized attack on the breeding and resting places of anopheline mosquitoes. At the present time two opposite views are held: the optimists look forward to the ultimate extermination of malaria; the pessimists think that the mosquitoes will eventually win, aided by the rapid decline of European influence in other continents.

Human malaria is carried exclusively by anophelines, but bird malaria is carried by culicine mosquitoes, and some monkey malaria has now been traced to biting midges, as we shall see presently.

Close behind malaria comes yellow fever, which is now known to exist in two forms. 'Ordinary', or urban yellow fever is a highly malignant virus of tropical America and Africa. Everyone has read the story of how 'yellow jack' delayed the building of the Panama Canal, and nearly stopped it altogether. Then after Reed and others had studied the disease in Cuba, and traced it to the culicine mosquito *Aedes aegypti*, and Gorgas had been able to eliminate it in the island, he applied similar techniques in the Isthmus, with spectacular results.

The character of urban yellow fever is set by the extremely domesticated habits of the mosquito *A. aegypti*, which breeds in artificial receptacles round human habitations, and seldom flies far away. It bites in daylight, and rarely attacks domestic animals. Since 1928 it has been known that many monkeys are susceptible. The original experimental transmissions were carried out on Indian rhesus monkeys, which proved to be easily infected, whereas African and South American monkeys are usually immune. It appears now that this is not because they do not take the disease, but because in those areas they have a naturally acquired immunity, in the way that many humans have immunity to poliomyelitis. There is thus a reservoir of the virus in the monkey population, carried from one to another by certain tree-top living mosquitoes that seldom come to the ground. On occasion a man becomes infected through accidental contact with this environment, and the virus can then be spread to other humans by the local mosquitoes of a different species. Then occurs an outbreak of jungle yellow fever. The extensive work with tree-top platforms carried out in the forests of Uganda by Haddow, Lumsden, and their colleagues has brilliantly demonstrated this mechanism.

Yellow fever does not occur in either of its forms east of the African continent, although *A. aegypti* and other 'suitable' mosquitoes are widespread there. The reason must be that neither among humans nor among monkeys has there yet built up a reservoir of virus sufficiently concentrated to start off the cycle of transmission. With modern air travel it has become a danger that such a reservoir might accidentally occur, and once the disease took a hold it might be impossible to eliminate. According to W. W. Macdonald *A. aegypti* was principally found in coastal towns in Malaya fifty years ago, but it has now reached a high density in many inland towns.

To embark on a catalogue of all the diseases carried by mosquitoes would take up too much of our space in this book, and a textbook of medical entomology should be consulted. We may, however, mention dengue, or break-bone fever, because after malaria this was one of the most serious of pestilences among troops operating in tropical and sub-tropical theatres of the Second World War. It is carried by *A. aegypti* and a few other mosquitoes, and unlike the virus of yellow fever it can also be transmitted mechanically. According to figures quoted by Cushing [57], for the period 1940–45 for all theatres of war, the US Army had 546,230 cases of malaria and 107,893 cases of dengue. While seldom fatal, dengue is very effective in crippling an army as a fighting unit.

One other notable infection carried by mosquitoes is the nematode worm *Filaria* (or *Wuchereria*) *bancrofti*. Filariasis is the general term for all infections due to filarial worms, but is often used in a restricted sense for this particular tropical disease. The worms block the lymphatic glands, and as they accumulate over a period of years a gradual swelling of the tissues leads in extreme cases to those grotesque examples of 'elephantiasis' that one sees in tropical countries, with a limb or an organ almost as big as all the rest of the body. The connection of mosquitoes with filariasis was one of the first discoveries made about these flies, as Table 1 shows.

Much elephantiasis in the Indo-Malayan region is caused by the worm *Brugia malayi*, carried by *Mansonia* mosquitoes, the larvae of which draw air from the roots of water-plants. Unfortunately killing of the adult *Mansonia* mosquitoes by spraying may bring about an increase in numbers of *Culex fatigans*, another mosquito which carries *Wuchereria bancrofti*, so the final result is just as bad.

Other diseases that flare up among men herded together in sub-tropical countries, whether as soldiers or as civilian labourers, are those carried by the tiny sand-flies, *Phlebotomus*. As we have seen, these flies live in arid, poorly developed areas, and spend most of their time hiding in crevices away from the heat and drought. In the Mediterranean, sand-fly fever is endemic, and smoulders away with little publicity, but during the Sicilian campaign it put several thousand combat troops out of action. The sand-flies also spread leishmaniasis (kala-azar and Oriental sore), a bacillary disease of the skin and of the intestinal tract, as well as verruga peruana in South America. These sand-fly-borne diseases come into Mackerras's second group, which have an animal reservoir, and when they reach man flare up as a violent eruption.

In contrast onchocerciasis, transmitted by black-flies of the genus *Simulium*, is slow to develop. The human parasite *Onchocerca volvulus* lives in nodules beneath the skin of infected persons, and is transmitted cyclically by the fly, after an interval of ten to fourteen days. Out of nearly sixty African species of *Simulium* only five are known to bite man, and only two, *S. damnosum* and *S. neavei* seem to carry the disease to man. In Central America some three species are involved.

Like the other filarial worm that slowly builds up to elephantiasis, *Onchocerca* has only minor effects for a long time, but eventually the optic nerve may be affected, and total blindness often results. Indeed this is more than a theoretical possibility: in the countries south of the Sahara, and the lake region of Uganda and Kenya, 'river blindness' is so prevalent that in recent years voluntary organizations have conducted special campaigns to supplement the official medical services. Onchocerciasis is also a serious problem in Central America, where living conditions differ from those in Africa, and where study of the disease and its vectors is actively developing at the present time.

Choyce (1962), however, suggests that there is little real evidence, derived from actual cases, that *Onchocerca* ever causes blindness either by destroying the optic nerve, or by damage to the eye itself.

Simulium is a very painful biter, said to be the most painful of all. Remembering that I quoted earlier from Herms that a well-adapted bloodsucking insect ought not to alarm its victim by a painful bite, it looks as if *Simulium* is not yet fully adapted to taking blood from man. And that is true of all its victims. The celebrated story of the Golubatz fly at the Iron Gates of the Danube, related in Chapter 7, is only one of many instances of the wholesale weakening and killing of domestic animals by a mere bite of the fly. Birds, too, are attacked, and *Simulium* is said to be able to cut the egg-production of hens by more than half. The fly carries cyclically a *Leucocytozoon* which is lethal to turkeys and ducks. In Australia it is suspected, but without confirmation, that *Simulium* may carry *Onchocerca gibbonsi* which in Queensland cattle has caused losses of up to £500,000 per annum.

The claim to be man's most painful biter is certainly challenged by the tiny biting midges, the 'no-see-ums' of the genus *Culicoides*. These widespread flies are even less efficiently adapted to their victims, to whom they are mainly a painful nuisance. In Africa, *Culicoides grahami* and *C. austeni* transmit the filarial worm *Dipetalonema perstans* which is to be found in the blood of every person in some parts of West Africa, apparently without any ill effects. *Culicoides* is now known to carry certain malaria-like organisms in animals other than man: a *Haemoproteus* in Canadian birds (Fallis and Bennet, 1961) and *Hepatocystis kochi*, a form of malaria in monkeys (Garnham and others, 1961). The latter, at least, is not a specific relationship, and the parasite may be carried by other bloodsucking flies, including Hippoboscids.

All the biters we have met so far have been Nematocera, and have been 'primitive biters': ie they are using equipment handed down from ancestral flies, and following an archaic habit. This is true, too, of the bloodsucking Brachycera, horse-flies and a few snipe-flies. These bigger insects are more gross feeders than the others, except perhaps the Simuliidae, and are mainly a menace because of their numbers, and of the sheer quantities of blood that they extract. I have given some examples of this in Chapter 8. Some are suspected of spreading disease mechanically, especially the trypanosome of surra in horses and camels in the Middle and Far East. We will come back to that presently. The only truly cyclical transmission of disease by a Tabanid is the filarial worm Loa loa, carried in the West African forest by the 'red flies', *Chrysops silacea* and *C. dimidiata*. This is a large worm, up to an inch long, which lives in the connective tissue underneath the skin, and like the other filariae, is insidious in its effects. When a worm crosses the cornea of the eye there are alarming, but transient symptoms, and sufferers from time to time experience 'Calabar swellings' at the wrists and ankles, or wherever there is constriction. Another *Chrysops*, the North American *C. discalis*, is a regular vector of the bacillus of tularaemia, but this is a highly contagious disease of rodents, easily caught by those who trap and skin small mammals, and the *Chrysops* is only one of many means of spread. Tularaemia also occurs in northern Europe and Asia, where no horse-flies have yet been implicated.

For the remaining direct enemies of man and his domestic animals we have to take a big step forward in evolution, to the advanced muscoid flies or calyptrates. The ancestors of these flies had long lost the primitive equipment for piercing, and, as we have seen, had cultivated the sucking sponge of the type shown in figure 29. Some of these, from Dolichopodidae onwards, have encouraged the flow of liquid by scratching with small teeth, called prestomal or pseudo-tracheal, according to their derivation. Within the living family Muscidae we can see all stages by which this equipment has been developed until it enables certain flies to pierce and suck blood with the best. We can see, too, how the habit may have arisen, as we watch the 'spongers' – well named, indeed – gather round the wound caused by a biter, even crowding him out, mopping up the blood with their spongy labella, and those with teeth scratching to keep up the flow.

The biters of this group are the stable fly, *Stomoxys*, its close relatives the horn-flies, *Lyperosia* and *Haematobia* (*Siphona*), and finally the tsetse-flies of the genus *Glossina*. All of these have evolved a thin, hard proboscis, with its teeth at the tip, not suited for any other diet except blood. Moreover, having made a fresh start, they are not bound by the original limitation of biting to the female sex, and both males and females take blood exclusively.

The stable flies and horn-flies are the more general feeders, and occur anywhere, usually in great numbers. The horn-flies – so-called because they rest on the horns of cattle between feeds – feature regularly in the magazines of economic entomology, and *Stomoxys* has long been regarded with suspicion as a possible mechanical carrier of many diseases of man and animals, particularly of poliomyelitis. If this last were true, this fly would at once become one of the most vital to man, since polio is one of the great medical problems of the day. Yet, in spite of some experimental successes, the evidence from epidemics strongly suggests that no bloodsucking fly is responsible.

The tsetse-flies of Africa are a specially fascinating group. For one thing they are now confined to Africa south of the Sahara, plus a tiny area of the Aden Peninsula, though this gives them a range of over 4 million square miles. There is slight evidence that they once existed in North America, but the living species, and the parasitic trypanosomes that they carry, have evolved in concert with the game animals of Africa.

The first European travellers in Africa tell of the 'fly', which killed

29. (a) Blow-flies, *Chrysomyia pinguis*, sunning themselves on leaves; (b) the common fate of many muscoid flies is to be killed by an attack of the fungus *Empusa muscae*.

30. (a) A piece of 'kungu-cake', made up entirely from the bodies of small midges, *Chaoborus edulis*, from an East African lake; (b) a fly, *Muscina stabulans*, entirely covered with mites, which are not parasites, but use the fly for transport (phoresis); (c) puparia of the Phorid fly *Paraspiniphora bergenstammi* in an empty milk bottle.

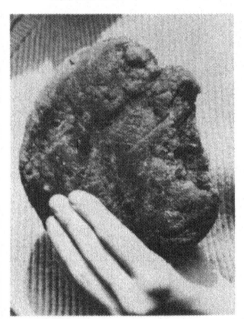

b

31. Damage to an onion by larvae of the onion fly, *Hylemyia
antiqua*, family Muscidae.

32. A narcissus bulb damaged by larvae of the Syrphid fly
Merodon equestris.

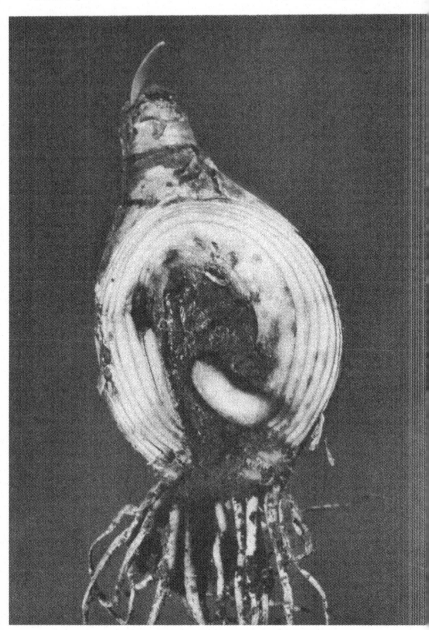

off their beasts of burden, and made movement slow and difficult, if not impossible. James Bruce, in his *Travels to discover the source of the Nile* (1790) has much to say about the 'Tsaltsalya', and gives an enlarged drawing of it that is recognizably either *Stomoxys* or *Glossina*, the tsetse-fly. Bearing in mind the northerly area in which Bruce moved it seems more likely that his insect is a confusion of *Stomoxys*, *Tabanus*, perhaps *Philoliche magrettii*, a horse-fly with an obvious proboscis, and perhaps even Hippoboscids. It is noteworthy that he was fully convinced that the 'fly' 'prohibited absolutely those inhabitants of the fat earth . . . from enjoying the help or labour of beasts of burden'.

A century later, Sir Harry Johnston, of okapi fame, wrote of the tsetse: 'It is difficult to overestimate the importance of the part played by this noxious little insect in preventing the opening up of Central Africa. This was first experienced by the earlier Portuguese expeditions of five hundred or six hundred mounted men, which would set out from Sena on the Lower Zambezi in the 16th and 17th centuries to secure the gold mines on the north and south. We read in Portuguese records how their horses soon succumbed to the attacks of a fly. But for the tsetse-fly the whole history of South-Central Africa would be different. It would have been rapidly traversed by mounted men, not nearly so much ill-health would have pursued explorers and pioneers forced to move on foot, and the whole question of transport would be rendered infinitely more easy, as coaches and waggons could run, and huge numbers of pack animals – horses, mules and oxen – might convey goods which at present are carried on porters' heads.' Johnston also suggested that 'undoubtedly the tsetse has checked the southward range of Muhammadan raiders from the north . . . [with] their sturdy little ponies'.

Sir Richard Burton (1860) produced a plaintive variant on the theme of 'what good are flies?': 'It is difficult to conceive the purpose for which this plague was placed in a land so eminently fitted for breeding cattle and for agriculture, which without animals cannot be greatly extended, except as an exercise for human ingenuity to remove'!

We have said something in Chapter 16 about the various tsetse and the trypanosomes they carry. Human trypanosomiasis, or sleeping sickness, was not part of the early explorers' complaint, which was directed entirely against *nagana* in cattle and horses. Sleeping sickness was first studied in a great epidemic to the north of Lake

Victoria, which lasted from 1902–10, until the area had to be evacuated of its remaining people. *Trypanosoma gambiense*, the organism responsible, is carried by *Glossina palpalis*. In 1908 a second form of sleeping sickness appeared in Northern Rhodesia, when a mineral prospector returned from an extended tour. The organism responsible, *Trypanosoma rhodesiense*, is indistinguishable from *T. brucei*, the agent of nagana, except that it will attack man, and *brucei* will not. Since then this form of sleeping sickness has appeared further and further north, till it has now reached the original area of the Gambian form. In the region of overlap it appears that *Glossina palpalis* may transmit both forms of sleeping sickness.

In a recent paper Ormerod [223] writes: 'Epidemic Rhodesian sleeping sickness appears to be a new disease, which has evolved during the present century. Previous historical reviews [of sleeping sickness] assume that *T. rhodesiense*, the causative organism, has remained the same throughout its history. There is evidence to suggest that this is by no means the case, and that the disease has become more active as it has spread from its original focus.' The view taken is that *rhodesiense* has evolved from *brucei*, the game trypanosome which has a permanent reservoir in the game animals. So we can see a progression, from *Trypanosoma evansi*, the surra organism of the East, mechanically spread; through *brucei*, cyclically spread to animals by tsetse, and possibly mechanically by *Tabanus* and *Stomoxys*; to *rhodesiense* spread cyclically by tsetse to man. Mechanical transmission in man is less probable because he does not normally accept biting flies as passively as grazing animals do.

Man's relations with biting flies get the more spectacular press, but his relations with certain non-biting groups are no less intimate. Man and his domestic animals are attacked by the larvae of flies as well as by the adults. Only one larva sucks his blood externally, the Congo floor maggot previously mentioned. Many other larvae invade the body to a varying degree, the logical progression being from feeding on bodily secretions such as those of the eyes, nostrils, throat, and head sinuses, to the pus of wounds. Then the sharp mouth-hooks are used to stimulate a more copious flow of secretion, or to enlarge existing wounds. Active destruction of tissues is the next stage, and finally come larvae that live an entirely parasitic life, buried in the musculature, feeding upon tissue, and breathing by applying the posterior spiracles to a small hole in the skin. A second channel of invasion is by way of the mouth, gullet, and intestine.

The invasion of a vertebrate body by larvae of flies is called *myiasis*, and we can see that this term covers a wide range of differing degrees of malignancy, from an accidental swallowing and temporary passage through the intestine, or a casual scavenging on waste-products of the body, down to a completely dependent parasitism. James [148], in his most stimulating monograph of myiasis-producing flies, offers two ways of classifying them, either according to the part of the body they attack, or by the degree of interdependence between larva and host. The latter, Patton's grouping, is the more helpful to us here, and recognizes three classes of myiasis: *specific* – ie larvae that can live only in the living tissues of a particular host; *semispecific* – larvae that mostly live in decaying animal, or even vegetable matter, but will sometimes attack living tissues; and *accidental* myiasis by a miscellany of flies whose larvae find their way into the body, particularly by being swallowed.

There is no space here to attempt a catalogue of the flies that invade the human body, which are admirably presented in James's monograph; the paper by Jacobs [147] gives some early reports of this occurrence. I shall content myself with generalities and a few examples. The first two of the three categories are made up entirely of calyptrate muscoid flies, such as we discussed in Chapters 16 and 17, with the one exception of the bird-feeding larvae of the acalyptrate *Neottiophilum praeustum*. We have already visualized the evolution of the flesh-feeding habit in larvae as part of the search for more concentrated protein.

The most highly evolved, no doubt, are the Oestridae, the larvae of which, as we have seen, live entirely within their host, either in head cavities, like *Oestrus* and *Cephenomyia*, in the intestine, like *Gasterophilus* and *Gyrostigma*, or in a variety of internal tissues, ending up beneath the skin, as in *Hypoderma*. Their high degree of specific adaptation has been reached only in relation to animals other than man, and we have already mentioned how much damage they do to domestic animals. Invasion of the human body by Oestrids is accidental.

The larvae of certain Calliphoridae and Sarcophagidae are more menacing. Once again, the closer the specific adaptation, the less the direct injury, like the relatively painless bite of the highly adapted malaria-carrying mosquitoes. The African Tumbu fly, *Cordylobia anthropophaga*, as its name implies, regularly 'eats man', the larvae living in a boil-like swelling or tumour under the skin. *Dermatobia hominis* does the same in the New World. The swelling is painful, but temporary, and complications only occur normally if the larva should die and putrefy *in situ*.

The really destructive maggots are those that infest wounds and sores. They have not reached the biological equilibrium, the 'live and let live' of *Cordylobia* and *Dermatobia*. They are really voracious feeders in carrion, or in any animal secretion, slime or blood-clot. In most warm countries it is inviting trouble to sleep in the open if one has any scratch, wound, or discharge from nose or ear. The discharge attracts egg-laying females, and the early symptoms of maggot infestation are easily overlooked or misunderstood. After a period of medical treatment for ordinary cold or catarrh, it is suddenly discovered that the discharge is stimulated by larvae, which by then are already at work destroying living tissues.

Cases have been reported in which as many as 250 large larvae were driven out by douching, and even then it was difficult to be sure that all had gone, since they hide in awkward corners. Infestations of the head are particularly dangerous, and may result in collapse of the nose, destruction of various septa of the head, deafness, blindness, delirium, and sometimes death.

There is much variation in the extent to which this behaviour is 'obligatory' to the larvae concerned. Some, like the screw-worm, make this their normal mode of life; others, like the sheep blow-flies, can live entirely on waste matter, but sometimes break out as destructive sheep-maggots. Since there is a continuous transition between carrion-feeding and attacking living tissues, there have been disputes where particular species will or will not attack healthy tissue. At one time, after the First World War, larvae of certain blow-flies were deliberately introduced into wounds in the belief that they would eat only the dead matter, and so promote healing growth. This method had obvious dangers of introducing accidental infections, as well as the chance that the theory might, after all, be wrong. Modern sulpha-drugs and other synthetic organic compounds have made this piece of applied entomology obsolete.

Larvae that are swallowed, usually by accident in food, are likely to set up intestinal disturbances, mainly transitory, but sometimes persisting. A fair number of reported cases of intestinal myiasis are mistaken, larvae being seen in stools, but not having in fact passed through the intestine. We have remarked how quickly flies come to a suitable egg-laying medium, and how prone are *Calliphora*, and especially *Sarcophaga* to drop living larvae in place of eggs.

There is no doubt that some larvae are easily killed by human digestive juices. For example larvae of the Mediterranean fruit-fly, *Ceratitis capitata* (plate 20) are not uncommonly found in oranges that are being eaten, and they must fairly often be accidentally swallowed. I have never had a report of living *capitata* larvae having been passed by any adult human being, though they do survive the weaker digestive juices of very young infants. When juice is squeezed from fresh oranges, it is turbid, and small larvae in it are easily overlooked.

There are, however, many dipterous larvae that can survive adult digestion. Some cases have been reported involving blow-flies, *Calliphora*, *Lucilia*, or *Sarcophaga*, and these strong maggots, equipped with powerful mouth-hooks, can set up acute indigestion and nausea, and even cause haemorrhage of the intestinal wall. They do not readily tolerate lack of oxygen, and they become violent when they are short of air, provoking retching and vomiting, and so partially emptying the stomach. Such larvae are not often swallowed by mistake, because they are conspicuous in the uncooked meat in which they are most likely to occur.

One of the oddest cases that came to my notice was one in which a housewife complained that when she pricked sausages that were cooking in the frying-pan, maggots popped out of the holes! It was eventually discovered that a small part of the sausage-making machine was inaccessible when the machine was dismantled for cleaning, and the residue of sausage-meat in a crevice had become 'blown'. The larvae feeding here had been neatly packed into a fresh sausage when the machine was next used.

If larvae are able to survive in the intestine there is no reason why they should not continue their normal development there. The regular inhabitants, if we may call them that, such as the bot-fly of horses (*Gasterophilus*), eventually release their hold and pass out before pupating in the ground. Some cases are on record where chance inhabitants of the intestine have continued to develop to the adult stage, and a few have even continued to breed there. Needless to say these were *Phoridae* (Chapter 12), which can live almost anywhere.

This being so, it is surprising that there are people who actually swallow larvae voluntarily. At times, cheeses in which larvae of *Piophila casei* were growing have been considered a delicacy for gourmets, and have been eaten, larvae and all. A colleague of mine who tried this to see what it was like did not recommend the experience.

We have not yet touched upon the most obvious encounter between flies and men, the one that most inspired the quotations at the head of this chapter: flies as a pest of man, buzzing round his head, and crawling over his person and his food. In the next chapter I hope to discuss swarms of flies, and what brings them together in numbers: for the present we shall consider briefly which flies pester us, and why.

House-flies come into our homes because they like our way of life, and they are the most accommodating of guests. Haines [106] studied the flies that bred in twenty different kinds of waste material that are normally produced around a farm and the farm-house in Georgia, and found interesting differences between them when they were grouped into five categories. The house-fly was the only species that would breed in all the five types. Haines also disputes a common idea that house-flies have become less numerous as horses have become fewer. In rural communities the house-fly is still the most common fly at all seasons of the year. That the house-fly has become less common in towns and 'conurbations' must be explained by better disposal of waste, since all Haines' materials are the sort of thing that modern urban district councils dispose of, or at least dump outside city limits.

The next most common indoor fly is *Fannia canicularis*, the lesser house-fly, the males of which are so maddening when they play 'kiss-in-the-ring' endlessly in the middle of the room. As we have already seen, *Fannia* breeds in nitrogen-rich materials, especially urine and bird-guano, and is abundant when there are chickens. Having a more even distribution through the summer than *Musca domestica*, *Fannia canicularis* is the more obvious of the two until the true house-fly gets into its stride in July.

The other indoor flies are a mixed lot. On farms the stable fly *Stomoxys calcitrans* crawls on the windows, soiling the curtains with tiny brown spots. The occasional *Calliphora* and *Lucilia* come in to buzz about in a tiresome way. There is nearly always an odd *Drosophila* or two about, and one will mysteriously appear as soon as you begin to eat fruit or to drink wine or cider. When a cloud of them arrives it is usually a sign that you have forgotten a fruit, or a few vegetables, and these have begun to ferment and generate moulds and yeasts. A *Drosophila* with a brown thorax covered with black spots is likely to be either *repleta* or *hydei*, and has probably come from some kitchen scraps. One such infestation in a restaurant was traced to the used bottles of tomato ketchup, which were left with that infuriating last inch of sauce in them that is impossible to shake out.

The flies that congregate in attics, in roof-spaces and on the ceilings of bedrooms are not really dwellers indoors, but are outdoor flies taking shelter for the winter. We shall see in the next chapter how they assemble. One of these is a very close relative of the housefly, *Musca autumnalis*, for which over here we have invented the name 'autumn-fly'. In the United States, where it has only recently arrived, it is called the 'face fly' because it is such a tiresome pest of cattle, crawling over their faces. The 'sweat-flies' of man, which make life a misery for the country labourer and the country lover are other muscid flies of the genus *Hydrotaea*, settling on the skin to suck up the perspiration. A similar habit in the 'eye-gnats' of the family Chloropidae, genera *Hippelates* and *Siphunculina* is believed to help in the spread of conjunctivitis.

The fact that these flies do not pierce the skin does not preclude their carrying disease, and in fact they were the earliest kinds of flies that were suspected of doing so. In the biblical account of the plague of flies, the fourth of the plagues of Egypt: 'The land was corrupted by reason of the swarms of flies.' In modern times, books, pamphlets, and posters show enlarged drawings of a fly's foot, and ask us to imagine how easily it can carry bacteria on to our food.

Yao, Yuan, and Hine studied 384,193 flies, of which 98·4% were *Musca domestica*, and found that they carried an average of 3,683,000 bacteria per fly in a slummy district. The further statement that in the cleanest district there were only 1,941,000 bacteria per fly merely adds artistic verisimilitude. The view of these authors that the average number of bacteria paralleled the specific death-rate from gastro-intestinal diseases is shared by Peffly [228], who states that infant mortality in one village in a sub-tropical country dropped in one year from 227 to 115 per thousand live births when the flies were controlled by chlordane; when the flies became resistant and the fly population rose, the number of deaths rose too.

It seems obvious that all flies that frequent dung and other animal wastes, and are also attracted to the bodies of living animals must pick up and transport many organisms of disease. When contagious disease is rife the flies must certainly help to spread it more quickly. 'Swat that fly' is always sound advice. Yet as naturalists we ask ourselves to what extent the flies are an integral part of the cycle of the disease, even as mechanical agents: whether in fact we can really say that the flies 'cause' the disease. We have to give a qualified reply.

The whole question of the relation between non-biting flies and disease was admirably summarized by Lindsay and Scudder [180], who arrived at the measured judgment that flies '. . . under certain optimum conditions can be as effective in spreading infection as are fingers, dirty eating utensils, and contaminated food'. The real cause of an outbreak of a contagious disease, or of one that gains entry to the body through the mouth, nose, or throat, is bad hygiene. A large population of flies is possible only if waste materials are there for them to breed in, and such conditions also make it easier for the disease to spread by any mechanical means. It does not follow that if the flies were all killed the disease would vanish, like malaria or sleeping-sickness.

Moreover house-flies have proved that they can quickly become resistant to any one insecticide. Lindquist and Knipling [179] say: 'House-flies have developed such a high degree of resistance to DDT and related materials that satisfactory control is impossible in most areas', and they suggest that organic phosphorus compounds may suffer the same decline. When we say that flies 'develop' a resistance we do not mean, of course, that individual flies do so. No poison kills every individual, and the few that survive include those that have a natural resistance, by some accident of physiology. If these survivors are so few or so scattered that they cannot keep the race going, then a complete kill will result. But if the survivors do manage to breed, their offspring will be artificially selected for resistance to that particular insecticide, and a resistant population will slowly grow, undetected at first, but presently bursting out with renewed menace. This is the sad fate of all the new 'wonder-killers' after a few years.

When the house-fly develops resistance in this way it is demonstrating the evolutionary vitality that has enabled it in the past to develop along with man, and to learn to exploit such a variety of different breeding materials. Lindquist and Knipling note that 'As yet no reports of resistance of horn-flies, horse-flies, deer-flies, stable flies, or sheep-keds . . . have appeared.' These are less plastic species, each much more conservative in its ways than the house-fly, and thus more vulnerable to our new weapons.

A matter of controversy for many years has been the possible link between the house-fly and poliomyelitis, one of the problem diseases of the present day. Lindsay and Scudder discussed this matter at length, and gave a convincingly negative answer. Certain experiments seemed suggestive. Polio virus could be found in human faeces, and in sewage; it could be found on flies caught wild, and communicated to flies bred in the laboratory; and fly-contaminated food when fed to chimpanzees gave rise to polio virus in their stools, though not to any clinical symptoms of the disease.

All this made it seem very likely that the flies might spread the disease. Yet by chance an outbreak of polio occurred in Hidalgo County, Texas, at the very time when experiments in fly-control were in progress, and showed conclusively that there was no link between the two. The rates of paralytic polio cases were nearly the same in places with and without fly-control; whereas the attack rate per thousand was twelve times greater among individuals who had been in contact with known polio cases than in those who had no apparent contact. This experience seems to show that polio only spreads between persons who are close together. If house-flies have any part to play they are only the most minor of supplementary agents, and their distribution has no appreciable effect on the progress of the epidemic.

The above remarks, of course, apply to *paralytic* polio. It is a curious aspect of this dreaded disease that most adults seem to have acquired immunity to it, hence the name 'infantile paralysis'. It has recently been suggested that this immunity might come from eating fly-contaminated food, and thereby receiving in effect an oral vaccine. This would help to explain why paralytic cases tend to be more numerous among people with a higher standard of hygiene, who would protect their children against flies.

Turning now, finally, to flies as destroyers of man's food, or of the food of his domestic animals, the relationship is even more a matter of chance. For example, the Piophilidae feed on rather dry animal tissues, carcasses that have reached the fatty acid, or even the mummified stage. A leg of ham or a side of bacon is a great find for these flies, and the larvae will eat away inside until they have made a cavity that ruins the meat. Female blow-flies, as we have seen, are impatient to get rid of their eggs, and will smell out meat in the larder and 'blow' it in no time. They also find any dead mouse, rat, or bird in a house or other building, and oviposit in that. When the larvae have finished feeding they leave the carcass and crawl quite a long way in search of a dry shelter in which to pupate. If they crawl into a packet of some dry goods in the larder the shocked housewife throws it out.

As we have heard already, many dipterous larvae attack living plants, sometimes killing them, and if they happen to be cultivated plants we regard those flies as pests, ignoring the many relatives that live only in wild plants. It is thus difficult to bring such species together briefly in a general account, but we can mention certain families that are most often concerned.

Among Nematocera the gall-midges, Cecidomyiidae, are the principal offenders. Most of them live as larvae in plant-galls, and naturally many galls are made on food-plants or ornamental trees. I referred earlier to Barnes' summary in a series of books, *The Gall Midges of Economic Importance*. A tiresome one in the garden is the pear-midge, which causes very young pears to shrivel and blacken in a distressingly unsightly way each spring; the actual effect seems to be negligible, since pears are so wasteful in any event.

More annoying to the gardener are the leather-jackets, the soil-living larvae of Tipulidae, which can do a lot of damage to a lawn, and to other small plants. The Syrphidae that destroy bulbs, *Eumerus*, and *Merodon* (plate 32) cause the loss of many a favourite bloom. Both leather-jackets and bulb-flies cause losses to commercial growers, but can be kept in check by a suitable spraying routine.

Larvae of many Trypetidae live in growing fruits, and can cause heavy losses to the commercial grower, mainly in the tropics and sub-tropics; they seem to find the apples and pears of cooler climates little to their taste. The various countries with a Mediterranean type of climate, which have developed the growing of citrus fruit, stoned fruits like peaches and apricots, and cucurbitaceous plants like melons, squashes and gourds, are vulnerable to Trypetid larvae. Air transport, and the tremendous increase in tourist travel in recent years, have given the flies a chance to spread more rapidly than ever before. No wonder that Australia, South Africa, and California have strict quarantine laws, and act ruthlessly when a new arrival is detected.

A few special crops have their own fly attacking them. The Ephydrid *Hydrellia nasturtii* mines the stems of watercress, as does the Drosophilid *Scaptomyza*, and the Psilid *Psila rosae*, the carrot fly, will also attack celery, parsley, and parsnips.

The leaf-mining Agromyzidae disfigure some ornamental shrubs and trees, but seldom defoliate them to the point of danger; others attack commercial crops such as tea, coffee, and leguminous fodder plants. The family Chloropidae and a few close relatives damage cereals as well as living in wild grasses, examples being the frit fly of oats and the gout fly of barley, previously described. The ever-versatile Muscidae, having on the one hand given us the stable fly and the tsetse, on the other give us the cabbage root fly, *Erioischia brassicae*, the onion fly, *Hylemyia antiqua*, the wheat bulb fly, *Hylemyia coarctata*, and the mangold fly, *Pegomyia hyoscyami* var. *betae*. I have mentioned some of these in earlier chapters, and full details of their various methods of attack, and the damage they do can be had in the economic leaflets published in most countries.

An article by A. H. Strickland [276] gives a useful summary of losses in agriculture through the activities of flies.

We must not forget the beneficial activities of some flies, though these are as much an accidental impact upon Man's affairs as are the 'malevolent' activities already discussed.

Many flies help to fertilize flowers, notably hover-flies and some Bibionids, both shown in plate 9. Sergent and Rougebief [330] claimed that the visits of adult *Drosophila* to grapes were essential in spreading the spores of the yeasts that make possible the fermentation of wine. If this is so mankind owes an ancient debt to flies for giving us 'wine that maketh glad the heart of Man'.

Once again I have to dismiss in a few words the activities of Tachinidae, which exert a tremendous regulating influence on the numbers of the many insects they parasitize.

Table 1

A Century of Medical Discoveries Implicating Flies

1869   Raimbert showed by experiment that the bacillus of *Anthrax* could be disseminated by flies.

1878   Manson, working in China, observed the development of the filarial worm *Wuchereria bancrofti* in the body of the mosquito *Culex fatigans*. Later, with others, he proved that the mosquito is an intermediate host of this parasite.

1880   Laveran discovered the malaria parasite.

1895   Bruce showed that the trypanosome that causes nagana in African cattle is present in the tsetse-fly *Glossina morsitans*.

1897   Ronald Ross showed that anopheline mosquitoes could become infested with malaria parasites by biting a malaria patient, and that development of the parasite took place within the mosquito.

1900   Reed and others showed that yellow fever in Cuba was carried by the mosquito *Aedes aegypti*, then called *Stegomyia fasciata*.

1901   Forde observed parasites in the blood of persons suffering from Gambian sleeping sickness, and Dutton gave these the name *Trypanosoma gambiense*.

1902   Graham showed that dengue, or 'break-bone fever' is a mosquito-borne disease.

1903   Bruce and Nabarro showed that Gambian sleeping sickness was caused by the tsetse-fly *Glossina palpalis*.

1906   Bancroft in Brisbane infected two volunteers experimentally with dengue fever.

1909   Doerr, Franz and Tausing showed that sand-fly fever, or papatasi fever is caused by the bite of the psychodid fly *Phlebotomus papatasi*.

1909   Clarke showed that parasitic thydoiditis, or Brazilian trypanosomiasis, caused by *Trypanosoma cruzi* was carried, not by flies like other trypanosomes, but by the Reduviid bug *Triatoma megista*.

1910   Stephens and Fantham showed that Rhodesian sleeping sickness is caused by *Trypanosoma rhodesiense*.

1912   Kinghorn and Yorke showed that the vector of Rhodesian sleeping sickness is *Glossina morsitans*, the same tsetse that carries nagana of cattle.

1913 Mitzmain, in the Philippines, showed that *Trypanosoma evansi*, which causes the disease surra in carabao (*Bubalus bubalis*), as well as in horses and camels in other regions, may be spread mechanically by *Tabanus striatus* and other abundant biting flies.

1913 Leiper, in the West African forest, suggested that the filarial worm *Loa loa* was carried by the 'red flies', *Chrysops silacea* and *C. dimidiata*.

1916 Cleland, Bradley, and Macdonald demonstrated by rigorous experiment that *Aedes aegypti* could be fed upon patients suffering from dengue fever, and used to infect volunteers in another town.

1921 Francis and Mayne showed that the bacillus of tularaemia could be carried mechanically by *Chrysops discalis*.

1922 The Connals confirmed that *Loa loa* is passed from man to man by *Chrysops silacea* and *C. dimidiata*.

1926 Blacklock showed that the filarial worm *Onchocerca volvulus* is carried in Africa by the black-fly *Simulium damnosum*.

1926 Falleroni discovered the races of the mosquito *Anopheles maculipennis* with their different patterns of egg-raft, and thus explained the conflicting accounts of the significance of this fly in relation to malaria. Some races are important vectors, some are not.

1927 Stokes, Bayer, and Hudson showed that monkeys could be experimentally infected with yellow fever. Since then more than a dozen species of mosquito have been shown to be able to transmit this disease from one monkey to another.

1932 Jungle yellow fever first recognized in Espirito Santo, Brazil. This is a disease of monkeys, not tied to human habitations, and spread by a different chain of mosquito vectors.

1933 Kelser passed the virus of equine encephalitis from guinea pig to horse by the bites of *Aedes aegypti*.

1941 Hughes, Jacobs, and Burke showed that many people in Bwamba County, western Uganda, were immune to yellow fever, and so must at some time have been in contact with the virus.

1942 Mahaffy and others isolated the virus from a human case, and also from a wild-caught specimen of the mosquito *A. (Stegomyia) simpsoni*.

1947 Haddow and others showed that a yellow fever cycle exists in forest monkeys, carried by the arboreal mosquito *A. (Stegomyia) africanus*, and that man becomes infected through the domestic mosquito *A. simpsoni* in plantations close to the forest. Links are provided by men going into the forest, and also by monkeys raiding the plantations.

Chapter 20

Swarms of Flies

What . . . spirit can it be that prompts
    The gilded summer-flies to mix and weave
Their sports together in the solar beam
    Or in the gloom of twilight hum their joy?
                                    WORDSWORTH

As when a swarm of gnats at eventide
    Out of the fens of Alla do arise
Their murmuring small trumpets sounded wide
    Whiles in the air their clustery armies flyes:
That as a cloud doth seem to dim the skyes;
    No man nor beast may rest or take repast
For their sharp wounds and noyous injuries
    Till the fierce northern wind, with blust'ry blast
Doth blow them quite away, and in the ocean cast
                SPENSER, 'The Faerie Queene', Book II, Canto 16

Then in a wailful choir the small gnats mourn
    Among the river sallows, borne aloft
Or sinking as the light wind lives or dies
                            KEATS, 'To Autumn'

EVERYBODY TALKS about 'swarms' of flies, but it is only recently that serious attempts have been made to find out just what we mean when we use this expression. A swarm of bees is a corporate body, grouped round its queen, and maintaining its integrity as it moves or rests. Can we say the same about a swarm of midges or mosquitoes, about the flies that buzz round a grazing animal, or 'swarm' into the loft of a house? And what about the aggregations of larvae of flies, the Army worms, the leather-jackets and blow-fly maggots that come out of the soil on wet nights, or the clusters of Bibionid larvae in the soil?

This chapter should really have been headed 'aggregations' of flies but that is an ugly and arid word. Here German is more helpful than English, offering us *der Schwarm* for a true swarm, *der Hauf* for a mere massing of individuals, and *das Gewimmel* for a crowd in the sense of a milling throng. The last of the three seems best to fit most assemblies of flies, which we shall continue to refer to as swarms for simplicity.

The subject of aerial swarming of flies has been discussed by a number of recent authors, notably by Downes [66], by Haddow and Corbet [104], and by Colyer [46]. Haddow and Corbet give an admirable summary of the whole topic, and end their paper with a quotation from Nielsen [215], warning us against fruitlessly speculating about the reasons why flies do certain things when we ought to be finding out much more about what it is that they actually do. I do not intend to try to guess why flies band together as adults or as larvae, but only to compare the manner in which the various assemblies come into being.

The common factor in all these assemblies is that they consist of a number of individuals doing the same thing at the same time. There is no sense of a community, in which various individuals do different things that total up to a community life. Aerial swarms take up a more or less fixed position in relation to a stationary object, and Downes [66] says: 'It is clear . . . that no gregarious factor is necessarily involved in swarm formation, but only the individual's reactions to a common marker.' Other authors have emphasized that a 'swarm' may be as small as one individual. All through this book we have seen how nearly all families of flies have members that sometimes hover. The habit is more common in some families than in others, and generally more typical of males than of females. Hovering necessarily means staying in one place relative to the ground, and an aerial swarm comes into being when a number of flies hover in company.

Downes discusses at some length the use of a marker, a conspicuous fixed object, usually below the hovering flies, but sometimes above. The lesser house-fly, *Fannia canicularis*, seems to use an overhead marker when it circles below the electric light fitting indoors, returning again and again to the same spot in the air. Out of doors it behaves in the same way, and you will often see other small flies hovering under trees. McAlpine describes the flight of a species of *Dasiops*, an acalyptrate fly of the family Lonchaeidae, which 'dance and hover in loose swarms under mesquite trees, a few to up to a hundred, almost entirely males . . . the swarms were usually two to five feet above the ground, and within them the individuals moved slowly back and forth, and up and down, through shafts of sunlight that penetrated the leafy canopy. In this way their glistening bodies easily caught one's attention.' This exactly matches my own experience with the African Stratiomyid *Platyna hastata*, which I described in Chapter 9.

Downes was discussing only the more lightly built and fragile midges and gnats, most of which guided themselves by markers below rather than above. The general practice was to fly against the wind, varying the flying speed so that they alternately flew over the marker, and then allowed themselves to drift downwind of it again, much as seagulls do. According to Downes' observations, the flies vary the length of their 'run' at different heights in such a way that the apparent movement of the marker is always the same. This is only possible if the wind-speed lies between $\frac{1}{2}$ mph and 9 mph. Above the upper limit the fly has too little margin of speed to make sudden accelerations, and so in a gust it is liable to lose touch with the marker; and below the lower limit the method fails because there is nothing to push against, as it were.

We can all observe this behaviour if we watch the summer swarms of non-biting midges. In the open, with a breeze of the right strength, the swarms form upon any convenient marker, even a person walking along, and will formate on him just as faithfully as on a fixed marker. If two people walk together and then separate, the swarm will often divide with them.

If the wind is too strong for them in the open, the swarms will form in the lee of objects, where the eddies provide them with just enough movement to balance themselves against. Most houses have such an eddy under the eaves, and on summer evenings, if you look out of the bedroom window you will probably see such a swarm. I shall mention these eddies again presently in connection with the small Chloropid flies that come into houses in autumn.

Swarms round high buildings such as church spires have often given rise to reports of fire. The 400-ft steeple of Salisbury Cathedral is particularly prone to this sort of false alarm.

Haddow and Corbet raised their high steel tower to 120 ft in the Mpanga Forest of Uganda, to see if they could observe the swarms they expected to find round the tops of the giant trees that emerge here and there from the general level of the forest canopy. They were rewarded for their efforts when the flies chose the tower itself as a marker, and provided the observers with a private display every night throughout the year except in stormy weather. They were concerned with swarming in the open over a marker, and contrast this with what they call 'free ground swarms at low level and not associated with a marker'. The authors do not discuss how the ground swarms keep their position, but these, like the solitary hoverers, are apparently riding the eddies, kept in station as much by the turbulence of the air as by their own efforts. They make clever use of the eddies round trees and bushes, as well as of the convection currents caused by uneven heating of the ground. This is one reason why small insects are often seen dancing in patches of sunlight, apart from the fact that the sunlight makes them conspicuous to us.

One dancing insect is conspicuous to us, and presumably to others of its kind. Each as it approaches becomes subject to the same influences as the first, yet in the main each swarm is made up of insects of one species, which must therefore be able to recognize each other. As we have seen, the sight of most flies is better suited to observing niceties of movement than to forming a clear image of a fixed object. Even a human collector soon learns that many small flies have a characteristic movement, and he can pick them out of the net without needing to use his lens to study them. How much more certainly must the flies themselves be able to recognize each other.

Haddow and Corbet quote a comment by Shannon that the most highly coloured, day-flying mosquitoes do not form swarms. They either have a characteristic courtship behaviour as they walk on the vegetation – as we have seen in Dolichopodids and many Trypetids – or a single male hovers in a likely place. Swarms consist either of drab insects, difficult to distinguish from each other, or of coloured insects that are active only in a poor light.

As everyone knows, mosquitoes make use of sound in mutual recognition, especially the males, which can pick up high-pitched vibrations with the long hairs of their bushy antennae (plate 11). Downes found that in *Aedes hexodontus* the male could respond to a female from a distance of one foot. The antenna is sensitive only when the long hairs are erected. Males of some species keep them permanently erect, and so are ready to mate at any time, even in captivity; others erect the hairs only at certain times of day, and fold them away when they are captured, so that they will not mate in captivity.

The mechanism is not as perfect as it is often thought to be. The male antenna responds to a wide range of frequencies, and according to Downes it serves only to bring the sexes together. It does not guarantee a successful mating. In the fruit-fly, *Dacus*, where the males call and the females respond, Myers [212] found that a male of one species may attract the female of another, but as soon as she touches him the response ceases because the correct specific stimuli are not received. Moreover, sound responses work better if there are no other individuals to cause confusion '. . . attraction of a male to a female [by sound] can only occur in relative isolation and not amongst a crowd or swarm' Haskell [113].

It will be noted that we are assuming here, as we have assumed often through this book, that swarms exist primarily for mating purposes. The common picture is that the males meet in a swarm which is conspicuous to the females, and they visit the swarm to find a mate. While this certainly does occur, and while it is true that most swarms are predominantly male, many modern observers doubt whether any significant proportion of the females of a species find their mates in this way. It is conceded that swarms may have evolved as mating devices, but Haddow and others incline to the view that aerial swarming is possibly a habit that has persisted after its practical use has declined.

Some swarms arise from mass emergences of adults from pupae, and the great preponderance of males after a time is explained because the females go off to mature and lay their eggs, while the males continue their carefree dance. My colleague Mr P. F. Mattingly spoke of such an emergence flight of *Aedes taeniorhynchus* estimated at about 50,000,000 individuals. It has been suggested that activity in the air may raise the body-temperature of the male flies, and render them more potent, and may thus take the place of the courtship behaviour of many flies, which has as much effect in exciting the males as it has in making the females receptive.

The resemblance between aerial swarms and clouds of smoke is not just a trick of the human imagination, since some flies appear to be deceived in the same way. The best known are the 'smoke-flies' of the odd little family Platypezidae, which are seldom seen except when they appear mysteriously in the smoke of a bonfire. Evidently they are attracted either by the swirling movement or by the smell, though Kessel has recently shown that scent attracts them. Similarly the African horse-fly *Chrysops silacea*, which lives in the canopy of the rain-forest, is brought down into the compounds of houses by the smoke of wood fires. In this case the objective is not mating, but feeding, and from Duke's work it seems likely that this attraction by smoke may have led to the association between *Chrysops silacea* and man which permits the fly to transmit the disease of human loiasis. Again the Calliphorid *Stomorina lunata*, which lives at the expense of the egg-pods of locusts, appears suddenly when a locust swarm arrives. Greathead (MS thesis) says of it: '*S. lunata* adults have not been found, despite extensive searching, except during a locust invasion, and then only when a swarm settles and begins copulating. There is a dramatic appearance of large numbers of female flies when this occurs, and they are confined to the immediate area of the swarm. Examination of vegetation in the surrounding country has failed to reveal the presence of any flies.'

The eyes of many male flies are much closer together than those of the females, and swarming seems to be more characteristic of these families than of, say, Asilidae, or most acalyptrates, in which the males have the eyes separated. Some of the latter hover as individuals, as we have noted in *Drosophila* and in *Dasiops*. In *Simulium* and many Tabanidae the upper facets of the eyes may be much bigger than those of the lower half, and in some other Nematocera this difference is greatly exaggerated. This can hardly serve to see the marker better, since this is usually below the swarm. There is some evidence from Haddow's and Wellington's work that the 'zenith light', the illumination directly overhead, is important in determining the time of day at which swarming takes place, and this may be quite critical. Perhaps the enlarged facets are concerned with this factor.

Swarms over a marker necessarily stay in one place. Reports of migration or other 'purposive' movements are few. Hover-flies are among the most migratory of flies, and have been reported as moving steadily through the passes of the Pyrenees, as well as down the eastern coast of England. Occasionally they are taken on ships far out to sea. An isolated example of truly communal behaviour is that of the *Simulium* reported by Muirhead-Thompson, and mentioned in Chapter 6, but the only flies known regularly to move *en masse* are, curiously enough, those that live on the beach, breeding in the wrack, or stranded seaweed.

Mass movements of seaweed flies a foot or two above the beach have been reported by many observers, and studied in detail by

Egglishaw [74]. There are two types of activity; a restricted mass flight just over or near the wrack, provoked by disturbance either by the rising tide or by someone walking on the weed; and a mass migration which takes place at low tide, when large numbers move in a body to new breeding-grounds. The stimuli to mass migration seem to be warmth and perhaps overcrowding, with a raised level of activity such as we see in locusts. This behaviour is characteristic of *Coelopa frigida* and *C. pilipes*, but Egglishaw also noted a mass migration of the small Borborid *Thoracochaeta zosterae*, when one sweep of his net caught 202 of them.

The direction of flight of seaweed flies seems to be generally along the coast, and generally downwind, but they are not very accurate navigators, and if they lose contact with the beach they may suddenly appear inland. Miss Dorothy Jackson records that large numbers of *Coelopa frigida* flew against the windows of her house on the Scottish coast near St Andrews, on a mild night in January. They had risen 60 ft from sea level, and a similar swarm appeared in October round the lantern of a lighthouse in Co. Donegal, Eire.

What now of other aggregations of adult flies? Sometimes foliage is covered with what Séguy calls '*réunions sédentaires*', assemblages of adult flies, either just after emergence, or for the purpose of mating without a mating flight. Roper records such behaviour in *Scatopse picea*, and it has been often seen in the Bibionid *Dilophus febrilis*. Occasionally such swarms find their way indoors, as I have noted of the Sepsid *Themira putris*, which breeds in great numbers in sewage farms, and the Phoridae discussed by Colyer [46], but the notorious 'hibernating flies' that assemble in the lofts and roof-spaces of houses in autumn in temperate countries do not arrive in a mass; they assemble in the building as in a fly-trap.

Pollenia rudis, *Musca autumnalis* and *Dasyphora cyanella* are the usual loft-flies, though sometimes *Musca domestica*, *Calliphora* and even the drone-fly, *Eristalis tenax*, crowd into winter shelter. *Phormia terrae-novae* and *Limnophora humilis* also occasionally do so. *Pollenia rudis* is so confirmed in this habit that it is known as the cluster-fly. The first three species are all to be found throughout the summer in the fields, and the movement towards shelter seems to be initiated by a sudden drop in temperature on the ground at night. In England this occurs in the second half of August in an average year, and from then onwards the behaviour of the flies is characteristic. In the afternoon they settle on the upper parts of walls and on roofs, facing

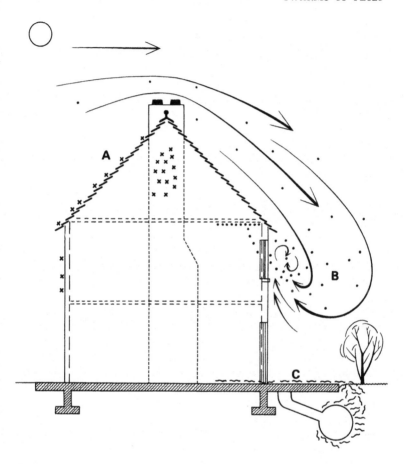

Figure 38. Three ways in which flies 'swarm' into houses: A, cluster-flies and one or two other species settle on the roof on the warm side of the house, crawl inside, and cluster on the rafters and brickwork; B, *Thaumatomyia notata* carried by the wind, and trapped in eddies under the eaves on the lee side, collect on the ceilings of upper rooms; C, larvae of *Lucilia* come out of the soil near a drain when a heavy storm causes temporary flooding.

south or south-west, sunning themselves in the last of the warmth. As the sun sets they crawl into any crevice, and some roofs have so many that the flies simply move into the roof-space, or into crevices in the roof-lining. For a few days they come out in the daytime, and go back at night, but presently this ceases and they spend the winter inside.

Hibernation is very incomplete and easily broken by a mild spell in midwinter. In churches, where the heating often operates only at weekends, the flies may become active in a sluggish sort of way, and fall squirming on to the hair and hymnbooks of the worshippers below.

A house or other building is therefore no more than a large fly-trap. It is found that the same building is infested year after year, while the house next door may be immune. The 'choice' of a house depends on unalterable physical factors of situation and aspect, and probably on the shape of the roof, and at present there is no known remedy for these visitations except to move. There is some slight evidence that *Dasyphora cyanella* prefers more open shelter than the other two, and this species is often found in such places as park band-stands and Dutch barns. This metallic Muscid looks like a blue-bottle, and is often mistakenly reported as such. In Scotland it is sometimes called the 'durn-fly', perhaps from the Old English *derne* or *dyrne*, meaning secret, dark, or private.

These flies normally survive the winter and come out again as soon as the days begin to get warmer. Their behaviour then is the reverse of that of the autumn: they emerge sluggishly and sit about on the walls and nearby vegetation, the males becoming active first. The nuisance is greater than that of the autumn, because they come out *en masse*. Barnes [12] studied *Pollenia rudis* after they had over-wintered, and found the abdomen of the females very brittle and desiccated. After feeding for a few days in captivity they became sleek and fat, and now contained fully matured eggs. He did not see any mating, although many males survived the winter.

It has been thought possible that such a mass of flies in an attic may leave behind a smell which attracts other flies next year. If this were so it might be possible to neutralize this, or even to turn it into a positive deterrent. Dethier found evidence that flies left to accumu-late in a sugar-baited fly-trap generated a substance attractive to other flies, but hibernating flies do not feed. A recent paper by Wiesmann [307] suggests that the accumulation of fly-dung on the ceilings of cowsheds attracts other flies, so that it is possible that there may be some cumulative effect of this nature. Brown and Car-michael report work on lysine as a mosquito attractant, and suggest that it might be possible to isolate the factor in human or animal smell that attracts mosquitoes: perhaps this might lead to some more general result. *Sarcophaga* may sometimes breed in the masses of dead flies.

The other fly that invades houses in the autumn is the Chloropid *Thaumatomyia notata*, which I have already mentioned in Chapter 15. Here the mechanism of entry is different. When the chill of autumn touches the grass the tiny flies take to the air in great numbers, but being much weaker fliers than the muscoid flies, they are carried away by the wind. Eventually they become trapped in the eddies under the eaves of a building, like the midges mentioned earlier in this chapter. They can be seen thus, even outside the Museum building in the heart of London. Rising and falling passively in the eddies, each fly at last reaches the window, which – in England at least – is always open at the top. Crawling inside, the flies walk about on the ceiling, and seldom fly. During the day they disperse across the width of the room, and as daylight fades they gather just over the window.

In spite of observations that these flies arrive fat and lose weight during the winter it does not appear that they really hibernate effectively. They are delicate, easily killed by cold, and in an English unheated bedroom they very soon die and disappear from the ceiling. In some years spring 'swarms' of desiccated individuals appear, but these are not of general occurrence. As we saw in Chapter 15 it seems likely that these autumn adults have arisen by premature emergence from pupae that normally overwinter, and so if they perish during the winter they are a direct loss to next year's population.

Thaumatomyia notata in particular illustrates the astronomical numbers of flies that must exist, dispersed through the countryside. When people bring me jars and bottles packed solid with these flies I find it difficult to convince them that this is not a spectacular outburst from some local focus of breeding, but just a sample of what is flying in the air at the time. Some of the estimates of natural populations of flies are on a scale that supports this view. Hessian-flies of the family Cecidomyiidae have been estimated at 200,000 per acre 'after decimation by parasites', and puparia of *Ephydra gracilis* washed up on the shore of the Great Salt Lake were said to number 370 million per mile of beach. Chironomid larvae in lakes have been thought even more numerous: one lake was said to have 3×10^9, or 3,000 million larvae as its total population, and D. J. Lewis records in one of his papers that: '. . . it is estimated that as many as several hundred million, perhaps a thousand million, Chironomids can emerge from the Blue Nile at Khartoum in one night'. The piece of 'kungu'-cake that I photographed (plate 30) is physical evidence of this, if anyone likes to sample it.

Aggregations and mass movements of larvae are little understood. We have mentioned the Army worm, *Sciara militaris*, which 'marches' across the forest floor, with clear determination but obscure

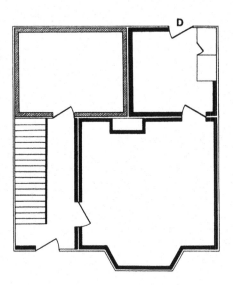

Figure 39. The invasion of an empty house in London by larvae of the greenbottle Lucilia. The heavy black lines mark the rooms where many larvae and pupae were found behind the skirtings; in the other room the infestation was lighter. The larvae apparently came from the soil outside the back door (D).

objective. In temperate countries leather-jackets, the larvae of crane-flies, come out of the garden soil in great numbers and sometimes move across paved and concreted areas, even entering houses. This happens in mild spells in midwinter, and no one really knows what starts them off, or whether they are moving in any particular direction.

I have mentioned the larvae of the blow-flies *Lucilia* and *Calliphora* and the way that they come into houses after heavy rain-storms. They certainly come out of the soil, but we still do not know why they are present there in such numbers. Salt [250] says that *Lucilia sericata* is 'one species that completely saturates its environment', and that it is only kept in restraint by 'competition and self-elimination'. In captivity one female might lay nearly 2,000 eggs, and 156 g of meat exposed in bright sunshine had 5,645 eggs on it after five hours. Holdaway found 48,562 larvae twenty-four hours after exposure, but only 231 flies finally emerged. It seems, therefore, that if the soil, near a drain, or for any other reason is attractive to ovipositing *Lucilia* females it may soon produce these great numbers of larvae. Figure 39 illustrates a particular case that came to my notice, and shows the scale of the problem.

Chapter 21

The Past, Present and Future of Flies

THE PAST OF FLIES is almost as conjectural as their future. A group so clearly distinguished from all others must have been evolving for a long time by human standards, but besides such veterans as cockroaches and crocodiles the flies are comparative youngsters. Perhaps it would be fair to call them 'middle-aged', failing a little in some directions, but still promising in others.

Four-winged insects with a venation like that of crane-flies are found in the Upper Permian of Australia, and have been greeted as 'Protodiptera'. In the Australian Upper Lias there are Tanyderidae, and in the European Lias various families of midges: Mycetophilidae, Bibionidae, Anisopodidae, and also Psychodidae. Ancestral Brachycera are found in the Mesozoic, and Muscoidea appear in the European Cretaceous. Bequaert [18] says: 'In the early Tertiary, particularly during the protracted Eocene, most modern families of Diptera appear to have existed. . . . Few Eocene fossils are at present referable with certainty to the higher Cyclorrhapha, or muscoids . . . but as the muscoids are represented in Oligocene and Miocene deposits by some very specialized types such as tsetse-flies, evidently they appeared before the Eocene, and probably towards the close of the Mesozoic Era.'

By this reckoning, flies are much older than Lepidoptera, a little older than Hymenoptera, and juvenile compared with beetles, which have changed little in appearance or relative abundance since Mesozoic times.

Early in this book we visualized the flies as arising from the Panorpoid Complex of insects, and as originating in swampy conditions, neither fully terrestrial nor fully aquatic. Such conditions existed for a very long time during the Carboniferous period, and there probably are the real ancestors of the Diptera, lost for ever in steamy swamps where no remains were preserved. The Permian was a period of increasing dryness, where suitable breeding-places would become fewer and more isolated. This may have set a premium on more active and efficient flight than was needed in Carboniferous forests, and so encouraged the early flies to discard the hind-wings, perfecting their flight with the fore-wings only.

The basic division of flies into what I have called the 'earthy' and the 'watery' lines of evolution must have arisen quite early. Downes [67] has pointed out that bloodsucking Nematocera all employ the same equipment for piercing the skin of their victims, and that it is unreasonable to imagine that this arose more than once in evolution. Hence he considers that all families of Nematocera that have blood-sucking members belong to one evolutionary stock. 'The non-biting species of groups [of these families] are irregularly scattered within normal, biting [groups], and may reasonably be regarded as secondary.' The discussion in Chapter I shows these to be the 'watery' groups of families. The 'earthy' groups must have had mandibles and maxillae originally, but had all lost them quite early, since the adults of all these families are incapable of biting.

Downes emphasizes that the 'watery' Brachycera (the Tabanoidea of Chapters 8 and 9) have the same sort of piercing mouthparts as the bloodsucking Nematocera, and he rightly assumes that they have inherited them from a common ancestor. I have pointed out the sharp division that occurs between the Tabanoidea and the Asiloidea, and that these other Brachycera and the whole of the Cyclorrhapha are an 'earthy' group comparable to the land midges. Those members of this group that now have aquatic larvae of bloodsucking adults have developed these habits, and the structures that go with them, quite independently, as a new feature.

We are therefore faced with a dilemma. If the Tabanoidea are to be linked with the gnats in one line of evolution, as Downes implies, the Brachycera must be split up, putting the 'watery' Tabanoidea with the gnats and the 'earthy' Asiloidea and Cyclorrhapha with the land midges. Unfortunately there is a sharp division between the larvae of all Nematocera and all Brachycera, in the direction in which the mandibles move, and the division between Nematocera and Brachycera is generally accepted. It would seem, therefore, that Brachycera must have arisen from the common stock before the ancestral land midges lost their mandibles and maxillae, and the subsequent loss of these mouthparts in the Brachycera must have come about independently.

White [305], from the evidence of chromosomes and cell division, found nothing to contradict the customary division into Nematocera and Brachycera + Cyclorrhapha. He concluded that 'all Brachycera' had characters in common with the land midges, Mycetophilidae, etc but his evidence is based on a study of only five families: Asilidae, Calliphoridae, Drosophilidae and Hippoboscidae which we put in our 'earthy' group. It would be interesting and perhaps revealing to know the condition in Tabanoidea.

Carpenter [39] estimated that on the basis of known fossils, Diptera constituted 3% of the insect fauna of the Permian, 5% of the Mesozoic, 27% of Tertiary, and only 10% of recent insects, and that the passage of time showed little or no change in the last three figures, although new fossiliferous beds were constantly being discovered. This must mean that flies as a group have passed their peak as an Order of insects, and are now in decline. How far is this true? When we say that Diptera are 10% of the recent fauna we refer to the number of different species that have been described, but this is not an ideal comparison; we must also take into account the distribution of these species into genera. Thus, as I commented earlier, the Stratiomyidae of the present day have about half as many species as the Tabanidae, but an enormous number of genera by comparison. Stratiomyidae are showing us a burnt-out brilliance, and are beginning to fade away: perhaps their place has been taken by the flourishing Syrphidae. Among Tabanidae, the great majority of the species belong to four genera, which have recently enjoyed a burst of evolution. So a mere counting of heads is as misleading in the affairs of flies as it is in human affairs.

When we come to consider the flies of the present day we ought really to say up to 1850. In the last century man has suddenly begun to alter the face of the earth, to destroy forests, cultivate the prairies and steppes, kill off the game animals at a rate incomparably greater than his ancestors were able to do. Not to mention the deadly insecticides that he now sprays about, so that no family of flies is safe from him. We will come back to man's effect on flies in a moment, but up to 1850 he was only one of the ecological factors that determined the slow evolution of the Diptera.

A list of recently successful flies must include the following families: Culicidae, Ceratopogonidae, perhaps Chironomidae: Mycetophilidae and particularly Sciaridae; Cecidomyiidae; Tabanidae; Asilidae, Bombyliidae, and Dolichopodidae; Phoridae, Syrphidae;

many acalyptrate families; most Muscidae and nearly all Sarcophagidae, Calliphoridae, and Tachinidae.

Tipulidae are a slowly declining group, for all their versatility,
and Alexander has suggested that possibly they reached their peak
as long ago as the Miocene period. Psychodidae have had only a
limited success, though they have had some help from man's addic-
tion to sewage works. Their offspring the Phlebotominae are possibly
a response to the drying up of certain sub-tropical areas since the last
Ice Age, just as Mackerras recently suggested that 'environmental
changes in the Pleistocene provided the conditions for a great deal of
speciation in the Tabanidae'.

Simuliidae have a well-chosen niche of their own, and seem firmly
established at present, but they are vulnerable to human attack, and
do not seem aggressive enough (in the evolutionary sense!) to survive very long, except in the far north where they are inaccessible.
Certain families are obviously on the way to extinction: Tanyderidae,
Thaumaleidae, Xylophagidae, Pantophthalmidae, Mydaidae, Lonchopteridae. Not in our lifetime, of course: all are too widely distributed for that. But there are certain single species of flies that are
known from only one place in the world, and could be extinguished
by an earthquake, a flood or even a new housing estate.

One such is the Asilid *Andrenosoma albopilosum*, a very distinctive
little fly that has been known only from one beach in Corsica. I saw
it breeding there in 1938, without realizing at the time that all the
known specimens came from this one beach. For over twenty years
I have been hoping that someone would bring me one from another
locality, but so far in vain. Recently, however, it has been recorded
from several inland localities in Italy. *Mormotomyia hirsuta*, the grotesquely hairy, wingless fly mentioned earlier, is known only from a
crack in a huge rock in Kenya, where it lives in the dung of bats. It
was discovered only in 1933, and of course other cracks in other
rocks may yield more specimens either of this fly or of even stranger
species. Amani, in Tanganyika, and the nearby Usambara Mountains, shelter a number of such survivors, among them one or two
Tabanidae. Of these the most striking is *Thriambeutes mesembrinoides*, which will probably be saved from extinction by its nocturnal
habits.

All the successful families of flies are worldwide in distribution.
It is fair to say that flies give little or no sign that they prefer one
part of the world to another, provided that the climate is broadly

suitable to their individual taste. On the whole they sensibly prefer warmer countries, and most families reach a greater variety of genera and species in the tropics than in higher latitudes. Some families such as Mycetophilidae, Chironomidae and Empididae seem to do better in cooler countries, but there is some evidence that they are more conspicuous there because of reduced competition from other families less tolerant of the cold. When they are specially looked for and

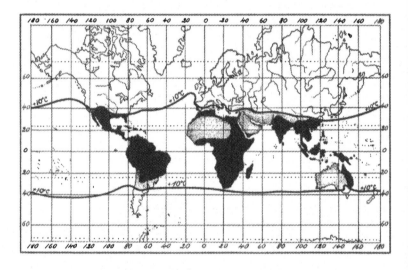

Figure 40. A map of the known distribution of Streblidae (black) and the potential distribution (shaded). The limit of distribution of these bat-parasites seems to be set by the winter isotherm of +10°C, at which temperature the bats hibernate, their body-temperature then falling with that of the atmosphere. From Jobling.

studied, they are often found to be numerous even in the tropics; for example a recent paper by K. G. V. Smith raises the number of Empididae known from Brazil from nine to ninety-nine.

Those flies that can tolerate cooler conditions profit from the reduced competition to produce great numbers of relatively few genera and species. Travellers who are misguided enough to go into the muskeg and tundra of northern latitudes are almost smothered in biting flies, and may have to wear special protective clothing, fastening closely to the neck and wrists. A veil is often needed against

Simulium and *Culicoides*, as well as against mosquitoes. It was at one time a problem to know how these bloodsucking flies could exist in such numbers in regions where warm-blooded animals are few and far between, but recent work has provided one answer. Many biting flies can lay one batch of eggs from the food accumulated by the larva, and so the species can continue to breed. When they can get a blood-meal the extra protein allows them to mature a second batch of eggs, and even a third.

A check through the accounts in this book of the rare families of flies will show that even these are very seldom confined to one area of the world. Panthophthalmidae are an exception, but apart from them and acalyptrates – where family limits are very fluid – rare flies seem to be those that are disappearing, and the family has, as it were, worn thin all over, and leaves scattered remnants in different parts of the world. Such are Tanyderidae, Blepharoceridae, Apioceridae; even Lonchopteridae, which though few in number can be found equally well in northern Europe and in Australia.

The detailed geographical distribution of flies seems to indicate that the continental fauna were derived not only by a distribution from the north, but also by way of some southern route, whether it be land-bridge, or Continental drift. Gressitt [98] gives a most lucid account of what is known of this subject, and underlines the significance of air-currents, which are particularly influential in carrying small flies to islands. The Introduction to Elmo Hardy's first volume on Hawaiian Diptera indicates how the flies may have colonized these islands, with or without human help. The two families Dolichopodidae and Drosophilidae have evolved into many species on the islands, with more success than in any other part of the world. Conversely, some of the flies collected by R. C. L. Perkins in 1892–97 can no longer be found: in particular the flightless Dolichopodid *Emperoptera mirabilis*, which may have fallen a prey to *Pheidole* ants.

And what of the future? Zoologists at the present time are keenly aware that we have discovered organic evolution just at the time when our own activities are putting a stop to it in many directions, or at least violently altering its course. Some have said that future evolution lies in the mind.

We shall not get rid of flies as easily as that, but we shall see great changes in the next few centuries. Firstly, by deliberate attack we shall reduce and possibly eliminate most bloodsucking flies, though recent experiences suggest the disquieting possibility that we may eliminate one species only to encourage another. The biting midges, and others that breed in damp debris will be more difficult, and I think that 'no-see-ums' and *Phlebotomus* will plague us much longer than mosquitoes and black-flies. The horse-flies are also difficult to kill at their breeding-sites because many of them breed in mud or damp soil over extensive areas, but as a family horse-flies are cut short in their prime by the great destruction of game animals all over the world, and drainage will hasten their end.

Of the bloodsuckers at the other end of the evolutionary scale, tsetse-flies are a spectacular relict, and although they fight back and win local victories, they will eventually lose, provided that the African continent becomes agriculturally and industrially developed like the rest of the world. If, for political reasons, there should be a temporary reversal of this trend, then tsetse, and other biting flies too, will get a new lease of life.

Two recent pessimistic comments may be quoted:

'. . . the great body of work on tsetse biology has influenced practical control comparatively little.' Lumsden [325]

'Little or no progress has been made in controlling filariasis, a disease that infects some 200 million people in Asia and the Americas . . . The epidemiology of the disease is not well understood, . . . indeed the situation in some respects may be said to have worsened, especially in regard to urban filariasis.' WORLD HEALTH July-August 1963.

Acalyptrates and calyptrates, I believe, have a prosperous future. They have learned to use decaying, fermenting, or putrefying organic materials, universal media that will always exist. No doubt we shall continue to campaign against the house-fly, but we shall not defeat it by chemicals, because it evolves resistant strains too quickly for us. Hygiene will keep it at bay in superior districts, but there will always be plenty of breeding material left about for it. As quickly as urban areas are denied to the house-fly the tourist and his motor-car make more rural areas attractive to it.

Drosophila, too, will continue to do well, and even now is spreading about the world and evolving new species. Its great radiation in the Hawaiian Islands illustrates this. There is a certain irony in the fact that, while the laboratory geneticist is showing his powers over nature by moulding his fruit-flies as he pleases, the flies themselves are staging a quiet demonstration of their own.

One unpredictable factor in future evolution is the effect of radiation. We cannot tell how far the radiation bogey is a false alarm so far as man himself is concerned, but we can be sure that whatever levels are reached, flies will not die out. Some will always get a sublethal dose and mutate into still more interesting forms for future dipterists to study.

Interesting experiments have already been made in sterilizing male flies by irradiation, and releasing them in an attempt to upset the natural breeding of certain species. Macleod and Donnelly [192] explain the method thus: 'Suppose that to a closed and stable population with 150,000 million females we add sterile but potent males in the proportion of, say, five to every one native male. Since a native female has only a one-in-six chance of fertile impregnation, the next generation should have 25,000 million females. The sterile population is maintained at its original level by replacements. The F_1 female population, reduced to one-sixth, has only a $\frac{1}{30}$ chance of fertilization so the F_2 generation will be only $\frac{1}{186}$ of the original density, 806 million, the F_3 generation 806,000, and the F_4 one female only.'

This method has been dramatically used to eliminate screw-worm, *Callitroga americana*, from the Island of Curaçao, but Macleod and Donnelly report an inexplicable failure when they tried to destroy *Lucilia sericata* on Holy Island off the eastern coast of England.

In any future conditions, I fancy that the flies of the time will be equally evasive.

Much work has been going on lately in the development of chemical sterilants, complex cyclical compounds known shortly under such names as apholate and aphoxide. They can be applied like insecticides, either to be eaten or touched by the insect. Experimental results on *Musca domestica*, *Stomoxys calcitrans*, and several Trypetidae, including the melon fly and the Mediterranean Fruit Fly, have indicated that an irreversible sterility can be produced [324].

It is claimed as an advantage that chemical sterilants can be used directly on the wild population of insects, whereas sterilization by radiation has so far been carried out on specially bred males which were then released. This obviously restricts the radiation method to insects that can be artificially bred.

This distinction seems to rest on the dubious assumption that chemical sterilants are more selective than radiation, and will confine themselves to the insects they are meant to control. Recent misgivings about the cumulative effects of general insecticides will surely apply even more strongly to chemical sterilants. Who can be sure that a particular compound will affect a fly, but not a bird, or a man.

Further Reading about Flies

One of the reasons for deciding to attempt such a book as this was that very little has been published about flies in general.

Although a tremendous number of publications, long and short, deal with flies, nearly all of them discuss identification, and describe detailed differences of shape and structure. Information about biology and habits is usually first published in short notes in one of the hundreds of periodicals that appear throughout the world. The library of the British Museum catalogues over one thousand periodicals devoted entirely or mainly to insects. When we have added the periodicals covering the wider fields of zoology and of general natural history, together with the more specialized journals of medicine, veterinary science, and agriculture it can be seen that biological information about flies is as widely dispersed as the flies themselves.

Authors of monographs dealing with the classification of larger or smaller groups of flies usually give a brief summary of what is known about the living flies, but this is often very little. The only comprehensive works are the *Handbuch*, or introductory volume to Erwin Lindner's *Die Fliegen der Palaearktischen Region* (Stuttgart, 1925–49), and two works by Eugène Séguy: *La Biologie des Diptères* (Paris, 1950) and *Introduction à l'étude biologique et morphologique des insectes diptères* (Rio de Janeiro, 1955). Willi Hennig's *Die Larvenformen der Dipteren* (Berlin, 1948–52), in three volumes, minutely analyses what is on record about immature stages (larvae and pupae) of flies, with references to such detailed descriptions as exist.

The other three works provide an exhaustive digest of the biological information that was available at the time that they were published, and Lindner's *Handbuch* and Séguy's *Biologie* have many fine pictures of flies. All three consist mainly of statements of fact, with little discussion or generalization.

Luther S. West's *The Housefly: its Natural History, Medical Importance and Control* (New York, 1951) is a model study of one species, which goes far beyond its title and is full of ideas that may be applied to other species. Patrick A. Buxton's *The Natural History of Tsetse Flies* is a comprehensive study of a small group of rather peculiar flies, so that little of its content is of direct application elsewhere; but it is written with exceptional clarity and in a critical style which shows how practical problems involving flies should be tackled.

Outside these few works the would-be student of flies is on his own. Any textbook of medical entomology has something to say about flies, but as in all textbooks the information tends to be incomplete and soon out of date. To find out what is known about any particular flies one should approach from two opposite directions. Firstly read what Lindner and Séguy have to say, and follow up the references they give, which in turn will provide other references. Secondly search the classified sections of the *Zoological Record*, and the indexes to the *Review of Applied Entomology* and *Biological Abstracts* for any likely papers of more recent date. The main problem is access to a sufficiently comprehensive library – and of course enough time.

Bibliography

1. ACZÉL, M. (1955) Key to the superfamilies of Schizophora. *Treubia* **23**: 3
2. ADLER, S. & THEODOR, O. (1957) Transmission of disease agents by Phlebotomine sand-flies. *Ann. Rev. Ent.* **2**: 203–26
3. ALLISON, A. C. (1960) Simple methods for identifying hosts of blood-feeding insects. *Trans. R. Soc. trop. Med. Hyg.* **54**: 295
4. ARDÖ, P. (1953) Likflugen, *Conicera tibialis* Schmitz i Sverige (Dipt: Phoridae) i. *Opusc. ent., Lund* **18**: 33–6
5. —— (1957) Studies in the marine shore dune ecosystem, with special reference to Dipterous fauna. *Opusc. ent., Lund*, Suppl. **14**: 255 pp.
6. AUBIN, P. A. (1914) The buzzing of Diptera. *J. R. microsc. Soc.* **1914**: 329–34
7. AUTRUM, H. & STUMPF, H. (1953) Elektrophysiologische Untersuchungen über das Farbsehen von *Calliphora*. *Z. vergl. Physiol., Berlin* **35**: 71–104
8. BAER, H. W. (1960) Winter fänge von *Anopheles*-Mücken. *Angew. Parasit.* **1**: 8–11
9. ——, J. G. (1952) Le mâle d'un Phoride commensal des Achatines de la forêt tropicale africaine. *Rev. suisse Zool., Geneva* **59**: 239
10. BALACHOWSKY, A. & MESNIL, L. (1935–6) *Les insectes nuisibles aux plantes cultivées*. 2 vols. (Paris)
11. BANKS, C. J. (1951) Syrphidae as pests of cucumbers. *Ent. mon. Mag.* **86**: 239–40
12. BARNES, H. F. (1924) Some facts about *Pollenis rudis* Fabr. *Vasculum* **10**: 34–58
13. BASDEN, E. B. (1954) The distribution and biology of Drosophilidae in Scotland, including a new species of *Drosophila*. *Trans. R. Soc. Edinb.* **62**: 603–64
14. BASTOCK, M. & MANNING, A. (1955) The courtship of *Drosophila melanogaster*. *Behaviour*. **8**: 85–111 (Leiden)
15. BATES, Marston (1949) *The natural history of mosquitoes*. 379 pp. (Macmillan: New York)
16. BAUMHOVER, A. H., GRAHAM, A. J., BITTER, B. A., HOPKINS, D. E., NEW, W. D., DUDLEY, F. H. & BUSHLAND, R. C. (1955) Screw-worm control through release of sterilized flies. *J. econ. Ent.* **48**: 462–6
17. BENNETT, G. F. (1960) On some ornithophilic blood-sucking Diptera in Algonquin Park, Ontario, Canada. *Canad. J. Zool.* **38**: 377–89
18. BEQUAERT, J. C. (1954) Evolution and fossil record of Hippoboscidae. *Ent. Amer.* **34**: 38

19. BERG, C. O. (1953) Sciomyzid larvae (Diptera) that feed on snails. *J. Parasit.* **39**: 630–6

20. ——, FOOTE, B. A. & NEFF, S. E. (1960) Evolution of the predator-prey relationships in snail-killing sciomyzid larvae (Diptera). *Amer. Malach. Union Bull.* **25**: 10–11

21. BERTRAND, H. (1955) Les Insectes aquatiques d'Europe. II. *Encycl. ent., Paris* (A) **31**: 161–500

22. BLICKLE, R. L. (1959) Observations on the hovering and mating of *Tabanus bischoppi* Stone (Diptera: Tabanidae). *Ann. ent. Soc. Amer.* **52**: 183–90

23. BODENHEIMER, F. S. (1951) *Citrus entomology in the Middle East.* 663 pp. (Junk: The Hague)

24. BOHART, G. E. & GRESSITT, J. L. (1951) Filth-inhabiting flies of Guam. *Bernice P. Bishop Museum, Honolulu, Bull.* **104**: 152 pp.

25. BRAUNS, A. (1954) *Terricole Dipterenlarven: eine Einführung in die Kenntnis und Ökologie der häufigsten bodenlebenden Zweiflüglerlarven ... der Waldbiozönose auf Systematische Grundlage.* 179 pp. *Puppen terricoler Dipterenlarven.* 156 pp. (Göttingen)

26. BRINDLE, A. (1959) Notes on the larvae of British Stratiomyidae. *Ent. Rec.* **71**: 130–2

27. —— (1960) The larvae and pupae of the British Tipulinae. *Trans. Soc. brit. Ent.* **14**: 63–114

28. ——, and BRYCE, D. (1960) Larvae of the British Hexatomini (Tipulidae). *Ent. Gazette* **11**: 207–17

29. BROCK, E. M. (1960) Mutualism between the midge *Cricotopus* and the alga *Nostoc. Ecology* **41**: 474–83

30. BROWN, A. W. A. (1954) Studies on the responses of the female *Aëdes* mosquito, Part VI: the attractiveness of coloured cloths to Canadian species. *Bull. ent. Res.* **45**: 67–78

31. —— (1958) Insecticide resistance in arthropods. 240 pp. (WHO: Geneva)

32. BRUES, C. T. (1928) Studies on the fauna of hot springs in the Western United States, and the biology of thermophilous animals. *Proc. Amer. Acad. Arts Sci.* **63**: 139–228

33. —— (1930) The food of insects, viewed from the biological and human standpoint. *Psyche* **37**: 1–14

34. —— (1946) *Insect Dietary:* an account of the food habits of insects. Harvard University Press. 466 pp.

35. BUCKHARDT, D. & SCHNEIDER, G. (1957) Die Antennen von *Calliphora* als Anzeiger der Fluggesschwindigkeit. *Z. Naturf. Tübingen* **12B**: 139–43

36. BUSVINE, J. (1951) *Insects and Hygiene.* 482 pp. (Methuen: London)

37. BUXTON, P. A. (1953) *The natural history of Tsetse-flies.* London School of Hygiene and Tropical Medicine, Memoir no. **10**: 816 pp.

38. BYERS, G. W. (1960) Biology and classification of *Chionea* (Diptera: Tipulidae). *XI Int. Congr. Ent. Verha.* **I**: 188–91

39. CARPENTER, G. H. (1930) A review of our present knowledge of the geological history of insects. *Psyche* **37**: 15–34

40. CHAPMAN, J. A. (1954) Observations on snow insects in western Montana. *Canad. Ent.* **86**: 357–63

40A. CHILLCOTT, J. G. (1960) A revision of the Nearctic species of *Fanniinae*. *Canad. Ent.* **92**: Supp. 14, 295 pp.

41. CHOCK, Q. C., DAVIS, C. J. & CHONG, M. (1961) *Sepedon macropus* (Diptera: Sciomyzidae) introduced into Hawaii as a control for the liver fluke snail *Lymnaea ollula*. *J. econ. Ent.* **54**: 1–4

42. CHRISTENSON, L. D. & FOOTE, R. H. (1960) Biology of fruit-flies. *Ann. Rev. Ent.* **5**: 171–92

43. CHRISTOPHERS, Sir S. R. (1960) Aedes aegypti (L.), *the Yellow Fever Mosquito: its Life History, Bionomics and Structure*. 739 pp. (Cambridge University Press)

44. COE, R. L. (1938) Rediscovery of *Callicera yerburyi* Verrall (Diptera: Syrphidae): its breeding habits, with a description of the larva. *Entomologist*. **71**: 97–102

45. COLLESS, D. H. (1956) Environmental factors affecting hairiness in mosquito larvae. *Nature*, London **177**: 229–30

46. COLYER, C. N. (1954) 'Swarming' of Phoridae (Diptera). *J. Soc. Brit. Ent.* **5**: 22–7

47. —— & HAMMOND, C. O. (1951) *Flies of the British Isles*. London, Warne, *Wayside & Woodland* series. 383 pp.

48. COOK, E. F. (1949) The evolution of the head in the larvae of Diptera. *Microentomology* **14**: 1–57

49. CORBET, P. S. (1962) The biological significance of the attachment of immature stages of *Simulium* to mayflies and crabs. *Bull. ent. Res.* **52**: 695–9

50. CRAGG, J. B. (1956) The action of climate on the larvae, prepupae and pupae of certain blow-flies. *Proc. XIV Int. Congr. Zool.* Copenhagen, **1953**: 387–8

51. —— (1956) The olfactory behaviour of *Lucilia* sp. (Diptera) under natural conditions. *Ann. appl. Biol.* **44**: 467–77

52. —— & COLE, P. (1952) Diapause in *Lucilia sericata* (Mg.) *J. exp. Biol.* **29**: 600–4

53. CRISP, G. & LLOYD, LL. (1954) The community of insects in a patch of woodland mud. *Trans. R. ent. Soc. Lond.* **105**: 269–314

54. CUMBER, R. A. (1949) Humble-bee parasites and commensals found within a thirty-mile radius of London. *Proc. R. ent. Soc. Lond.* (A) **24**: 119–27

55. CURRAN, C. H. (1934) *The families and genera of North American Diptera*, New York. 512 pp.

56. —— (1958) Insect acrobats: feats of the house-fly. *Nat. History Magazine* **67**: 83–5 (New York)

57. CUSHING, E. C. (1957) History of Entomology in World War II. *Smithsonian Inst.*, *Publ.* **4294**: 117 pp.

58. DAHL, T. (1959) Studies on Scandinavian Ephydridae. *Opusc. Ent.*, *Suppl.* **15**: 224 pp.

59. DALMAT, H. T. (1958) Biology and control of Simuliid vectors of onchocerciasis in Central America. *Proc. 10th. Int. Congr. Ent.* **3**: 517–33

60. DASGUPTA, B. (1962) On the myiasis of the Indian toad, *Bufo melanostictus*. *Parasitology* **52**: 63–6

61. DAVIES, D. M. (1959) Seasonal variation of Tabanids (Diptera) in Algonquin Park, Ontario. *Canad. Ent.* **91**: 548–53

287

62. DEMERECQ, M. [Ed.] (1950) *Biology of Drosophila*. New York and London, 632 pp.

63. DESCAMPS, M. (1957) Recherches morphologiques et biologiques sur les Diopsidae du Nord-Cameroun. *Min. France d'outre Mer, Bull. Sci.* 7: 154 pp.

64. DOWNES, J. A. (1953) Notes on the life-cycle of *Xylophagus ater* Mg. *Ent. Mon. Mag.* **89**: 136–7

65. —— (1955) The food habits and description of *Atrichopogon pollinivorous*, sp. n. *Trans. R. ent. Soc. Lond.* **106**: 439–53

66. —— (1958) Assembly and mating in the biting Nematocera. *Proc. 10th. Int. Congr. Ent.* **2**: 425–34

67. —— (1958) The feeding habits of biting flies and their significance in classification. *Ann. Rev. Ent.* **3**: 249–66

68. DUNN, L. H. (1930) Rearing the larvae of *Dermatobia hominis* in Man. *Psyche* **37**: 327–42

69. DYTE, C. E. (1959) Some interesting habitats of larval Dolichopodidae. *Ent. mon. Mag.* **95**: 139–43

70. EDWARDS, F. W. (1925) British fungus gnats, with a revised generic classification of the family. *Trans. ent. Soc. Lond.* **73**: 505–670

71. —— (1926) On marine Chironomidae, with descriptions of a new genus and four new species from Samoa. *Proc. zool. Soc. Lond.* **1926**: 779–806

72. —— (1929) British non-biting midges. *Trans. ent. Soc. Lond.* **77**: 279–430

73. —— (1934) The New Zealand Glow-worm. *Proc. Linn. Soc. Lond.* **146**: 3–10

74. EGGLISHAW, H. J. (1961) Mass migrational flights of *Coelopa frigida* (Fabricius) and *C. pilipes* Haliday (Diptera, Coelopidae), and *Thoracochaeta zosterae* Haliday (Diptera, Sphaeroceridae). *Entomologist.* **94**: 11–18

75. EL-ZIADY, S. (1957) A probable effect of moonlight on the vertical distribution of Diptera. *Bull. Soc. ent. Egypte* **41**: 655–62

76. EMDEN, F. I. van (1950) Dipterous parasites of Coleoptera. *Ent. mon. Mag.* **86**: 182–206

77. —— (1957) The taxonomic significance of the characters of immature insects. *Ann. Rev. Ent.* **2**: 91–106

78. —— Evolution of Tachinidae and their parasitism. *XVth Int. Congr. Ent. Proc.* 664–6

79. ENGLISH, K. M. J. (1947) Notes on the morphology and biology of *Apiocera maritima* Hardy. *Proc. Linn. Soc. N.S. Wales* **71**: 296–302

80. ——
[and a number of other papers by Miss English in this journal]

81. EVENS, F. M. J. C. (1953) Dispersion géographique des Glossines au Congo Belge. *Mém. Inst. R. Sci. nat. Belg.* (2) **48**: 70 pp.

82. FALLIS, A. M. & BENNETT, G. F. (1961) Ceratopogonidae as intermediate hosts for *Haemoproteus* and other parasites. *Mosquito News* **21**: 21–8

83. FARR, T. H. (1954) Heleidae [= Ceratopogonidae] attacking blister beetles in Massachusetts and Arizona. *Bull. Brooklyn ent. Soc.* **49**: 88

84. FERON, M. (1960) L'appel sonore du mâle dans le comportement sexuel de *Dacus oleae* Gmelin. *Bull. Soc. ent. Fr.* **65**: 139

85. FINGERMAN, M. & BROWN, F. A. (1953) Color discrimination and physiological duplicity of *Drosophila* vision. *Physiol. Zool.* **26**: 59–67

86. FOOTE, B. A. (1959) Biology and life-history of the snail-killing flies belonging to the genus *Sciomyza*. *Ann. ent. Soc. Amer.* **52**: 31–43

87. FRICK, K. (1959) Synopsis of Agromyzid leaf-miners from N. America. *Proc. U.S. nat. Mus.* **108**: 347–465

88. FROHNE, W. C. (1954) Mosquito distribution in Alaska, with especial reference to a new type of life-cycle. *Mosquito News* **14**: 10–13

89. GARNHAM, P. C. C., HEISCH, R. B. & MINTER, D. M. (1961) The vector of *Hepatocystis* (= *Plasmodium*) *kochi*: the successful conclusion of observations in many parts of tropical Africa. *Trans. R. Soc. trop. Med. Hyg.* **55**: 497–502

90. GILLETT, J. D. (1951) The habits of the mosquito *Aedes* (*Stegomyia*) *simpsoni* in relation to the epidemiology of yellow fever in Uganda. *Ann. trop. Med. Paras.* **45**: 110–21

91. GOODMAN, L. J. (1960) The landing responses of insects. I: the landing response of the fly *Lucilia sericata* and other Calliphorinae. *J. exp. Biol.* **37**: 854–78

92. GORDON, R. M. & LAVOIPIERRE, M. M. J. (1961) *Entomology for Students of Medicine* (Liverpool)

93. GREATHEAD, D. J. (1958) Notes on the life-history of *Symmictus flavopilosus* Bigot (Diptera: Nemestrinidae) as a parasite of *Schistocerca gregaria* (Forskål) (Orthoptera: Acridiidae). *Proc. R. ent. Soc. Lond.* **33**: 107–19

94. GREEN, E. E. (1951) Blow-flies in slaughterhouses. *J. R. san. Inst.* **71**: 138–45

95. GREENE, C. T. (1956) Dipterous larvae parasitic on animals and man. *Trans. Amer. ent. Soc.* **82**: 17–34

96. GRENIER, P. & FERAUD, L. (1960) Étude biométrique et morphologique de la croissance larvaire chez *Simulium damnosum* Theobald. *Bull. Soc. Path. exot.* **53**: 563–81

97. —— & RAGEAU, J. (1960) Simulies de Tahiti: remarques sur la classification das Simuliidae. *Bull. Soc. Path. exot.* **53**: 727–42

98. GRESSITT, J. L. (1958) Zoogeography of insects. *Ann. Rev. Ent.* **3**: 207–30

99. —— (1960) Principles and methods of zoogeography: the field approach. *Canad. Ent.* **92**: 636–9

100. —— (1961) Pacific Insects Monograph 2: *Problems in the zoogeography of Pacific and Antarctic insects* (Honolulu)

101. GRYZBOWSKA, B. (1957) The innervation and sense organs in the wing of *Ornithomyia biloba* Dufour (Diptera: Hippoboscidae). *Bull. ent. Pologne* **26**: 175–85

102. GUIMARÃES, & D'ANDRETTA (1956) Nycteribiidae do Novo Mondo. *Arquivos de Zoologia* **10**: 184 pp.

103. HACKETT, L. W. (1937) *Malaria in Europe: an ecological study.* 336 pp. (London)

104. HADDOW, A. J., CORBET, P. S., GILLETT, J. D. & others (1961) Entomological studies from a high tower in Mpanga Forest, Uganda. *Trans. R. ent. Soc. Lond.* **113**: 249–368

105. HAFEZ, M. & EL-MOURSY, A. A. (1956) Studies on desert insects in Egypt, I, II (Diptera: Rhagionidae). *Bull. Soc. ent. Egypte* **40**: 279–99; 333–48

106. HAINES, T. W. (1953) Breeding media of some common flies. *Amer. J. trop. Med. Hyg.* **2**: 933–40

107. HARANT, H. & BRÈS, A. (1947) Gîtes des larves des Ceratopogonides. *L'Entomologiste* **3** (4)

108. HARDY, G. H. (1951) Theories of the world distribution of Diptera. *Ent. mon. Mag.* **87**: 99–102

109. —— (1956) The superfamily unit in Diptera. *Ent. mon. Mag.* **92**: 213–15

110. HARRISON, R. A. (1956) The Diptera of Auckland and Campbell Islands. Part I. *Rec. Dominion Mus.* **2**: 205–31

111. —— (1959) Acalyptrate Diptera of New Zealand. *N.Z. Dept. Sci. Ind. Res. Bull.* **128**: 382 pp.

112. HASKELL, G. (1961) *Practical heredity with Drosophila.* London.

113. HASKELL, P. T. (1961) *Insect Sounds. Aspects of Zoology* series. 189 pp. (Witherby: London)

114. HAUFE, W. O. (1954) The effect of atmospheric pressure on the flight responses of *Aedes aegypti. Bull. ent. Res.* **45**: 507–26

115. HEISS, ELIZABETH M. (1938) A classification of the larvae and puparia of the Syrphidae of Illinois, exclusive of the aquatic forms. *Illinois Biol. Monogr.* **XVI**, no. 4: 142 pp.

116. HELLMUTH, H. (1906) Untersuchungen zur Bakteriensymbiose der Trypetiden. *Z. Morph. Oekol. Tiere* **44**: 483–517

117. HEMMINGSEN, A. M. (1956) Deep-boring ovipositing habits of some crane-fly species (Tipulidae) of the subgenera *Vestiplex* Bezzi and *Oreomyia* Pok., and some associated phenomena. *Vidensk. Medd. dansk. Naturh. Foren. Kbh.* **118**: 243–315

118. HENNIG, W. (1941) Die Verwandschaftsbezeichnungen der Pupiparen und die Morphologie der Sternalregion des Thorax der Dipteren. *Arb. morph. tax. Ent. Berlin* **8**: 231–49

119. —— (1948–52) *Die Larvenformen der Dipteren* **I–III** (Berlin)

120. —— (1950) *Grundzüge einer Theorie der phylogenetischen Systematik.* 370 pp. (Berlin)

121. —— (1953) Kritische Bemerkungen zum phylogenetische System der Insekten. *Beitr. Ent.* **3** Sonderheft: 1–85

122. —— (1954) Flügelgeäder und System der Dipteren unter Berücksichtung der aus dem Mesozoikum beschreibenen Fossilien. *Beitr Ent.* **4**: 245–388

123. —— (1958) Die Familien der Diptera Schizophora und ihre phylogenetischen Verwandschaftsbeziehungen. *Beitr. Ent.* **8**: 505–688

124. —— (1951) Die Dipteren-Fauna von Neuseeland als systematisches und tiergeographisches Problem. *Beitr. Ent.* **10**: 221–329

125. HERING, E. N. (1951) *Biology of the leaf-miners.* 420 pp. (The Hague)

126. HERTING, B. (1960) Biologie der Westpaläarktischen Raupenfliegen (Diptera: Tachinidae). *Monogr. z. angew. Entomologie* **16**: 188 pp.

127. HINTON, H. E. (1946) A new classification of insect pupae. *Proc. zool. Soc. Lond.* **116**: 282–328

128. —— (1951) A new Chironomid from Africa, the larvae of which can be de-hydrated without injury. *Proc. zool. Soc. Lond.* **121**: 371–80

129. —— (1953) Some adaptations of insects to environments that are alternately dry and flooded, with some notes on the habits of the Stratiomyidae. *Trans. Soc. Brit. Ent.* **11**: 209–27

130. —— (1957) Some little-known respiratory adaptations. *Science Progress* **180**: 692–700

131. —— (1958) The phylogeny of the Panorpoid Orders. *Ann. Rev. Ent.* **3**: 181–206

132. —— (1958) Concealed phases in the metamorphosis of insects. *Science Progress* **182**: 260–75

133. —— (1958) The spiracular gills of insects. *Proc. 10th. Int. Congr. Ent.* **1**: 543–8

134. —— (1958) The pupa of the fly *Simulium* feeds, and spins its own cocoon. *Ent. mon. Mag.* **94**: 14–16

135. —— (1960) Cryptobiosis in the larva of *Polypedilum vanderplanki* Hinton. (Chironomidae). *J. Ins. Physiol.* **5**: 286–300

136. —— (1961) How some insects, especially the egg stages, avoid drowning when it rains. *Proc. S. London ent. nat. Hist. Soc.* **1960**: 138–54

137. HIRVENOJA, M. (1960) Ökologische Studien über die Wasserinsekten in Riihimaki (Süddfinnland). [A series of papers in *Ann. Ent. Fenn.* **26**: 31]

138. HOBBY, B. M. (1931) The British species of Asilidae and their prey. *Trans. ent. Soc. S. England.* **1930**: 1–42

139. HOBSON, R. P. (1932) Studies on the nutrition of blow-fly larvae. III: The liquefaction of muscle. *J. exp. Biol.* **9**: 359–65

140. HOCKING, B. (1953) The intrinsic range and speed of flight of insects. *Trans. R. ent. Soc. Lond.* **104**: 223–345

141. HOLLANDE, A., CACHON, J. & VAILLANT, F. (1952) Recherches sur quelques larves d'insectes termitophiles (Muscidae, Calliphoridae, Oestridae . . .) *Ann. Sci. nat. Paris, Zool.* (**11**) **13** (1951): 365–97

142. HORI, K. (1952) On some flies of medical importance obtained from Korea and adjacent districts. *Oyo-Dobuts- Zasshi* **17** (1952): 77–82 (Tokyo)

143. HORSFALL, W. R. (1955) *Mosquitoes: their bionomics and relation to disease.* 723 pp. (New York)

144. HOTT, C. P. (1954) The evolution of the mouthparts of adult Diptera. *Stanford Univ. Bull.* **27** (1953): 137–8

145. HULL, F. M. (1949) The morphology and inter-relationship of the genera of Syrphid flies, recent and fossil. *Trans. Zool. Soc. Lond.* **26**: 257–408

146. ILSE, D. & NULHERKAR, L. (1954) Mating reactions in the common Indian House Fly, *Musca domestica nebulo. Current Science* **23**: 227–8 (Bangalore)

147. JACOBS, J.-Ch. (1882) De la présence des larves d'oestrides et des muscides dans le corps de l'homme. *C.R. Soc. ent. Belge.* série iii, no. 25: CL–CLX. [English translation by Prof. Cobbold in *The Veterinarian* for 1883.]

148. JAMES, M. T. (1947) The flies that cause myiasis in Man. *U.S. Dept. Agr. Misc. Publ.* 631. 175 pp.

149. JOBLING, B. (1949) Host-parasite relationship between the American Streblidae and the bats. . . . *Parasitology* **39**: 315–29

150. —— (1951) A record of the Streblidae from the Philippines and other Pacific Islands including . . . host-parasite relationship and geographical distribution. *Trans. R. ent. Soc. Lond.* **103**: 211–46

151. —— (1954) Streblidae from the Belgian Congo, with a description of a new genus and three new species. *Rev. Zool. Bot. afr.* **50**: 89–115

19*

152. JOHANNSEN, O. A. (1934–38) Aquatic Diptera. Parts I–V. *Mem. Cornell Univ. Agr. Exp. Stn.*, nos. 164, 177, 205, 210

153. JOHNSTON, W. & VILLENEUVE, G. (1897) On the medico-legal application of entomology. *Montreal Medical J.* [August]

154. JONES, C. M. (1956) Notes on the biology and feeding habits of *Esenbeckia incisuralis* (Say). *J. Kansas ent. Soc.* **29**: 43–6

155. KAHN, M. C. & OFFENHAUSER, W. (1949) The identification of certain West African mosquitoes by sound. *Amer. J. trop. Med.* **29**: 827–36

156. KATO, K. (1953) On the luminous fungus-gnats in Japan. *Sci. Reps. Saitama Univ. Urawa.* (B) **1**: 59–63

157. KEILIN, D. (1915) Recherches sur les larves des Diptères cyclorrhaphes. *Bull. Sci. France et Belgique.* **49**: 14–198

158. —— (1917) Recherches sur les Anthomyiides a larves carnivores. *Parasitology* **9**: 325–450

159. —— (1944) Respiratory systems and respiratory adaptations in larvae and pupae of Diptera. *Parasitology* **36**: 1–66

160. KELLOGG, U. L. (1903) The net-winged midges (Blepharoceridae) of North America. *Proc. Calif. Acad. Sci.* ser. 3, *Zoology* **III**: 187–232

161. KESSEL, E. L. (1952) *Hormopeza brevicornis* (Empididae) attracted to smoke. *Pan-Pacific Ent.* **28**: 56–8

162. —— (1955) The mating activities of balloon flies. *Syst. Zool.* **4**: 97–104

163. —— (1959) Introducing *Hilara wheeleri* Melander as a balloon maker, and notes on other North American balloon flies. *Wasmann J. Biol.* **17** (2): 221–30

164. KETTLE, D. S. & LAWSON, J. W. H. (1952) The early stages of British biting midges, *Culicoides* Latr. and allied forms. *Bull. ent. Res.* **43**: 421–67

165. KHALAF, K. (1954) The speciation of the genus *Culicoides*. *Ann. ent. Soc. Amer.* **47**: 34–51

166. KIM, CHANG-WHAN (1960) On the use of the terms 'larva' and 'nymph' in entomology. *Proc. R. ent. Soc. Lond.* (A) **35**: 61–4

167. KING, E. W. & HIND, A. T. (1960) Activity and abundance in insect light trap sampling. *Ann. ent. Soc. Amer.* **53**: 524–9

168. KNAB, F. (1910) The feeding habits of *Geranomyia* (Diptera: Tipulidae), *Proc. ent. Soc. Washington* **12**: 61–6

169. KNIPLING, E. P. (1955) Possibilities of insect control or eradication through the use of sexually sterile males. *J. econ. Ent.* **48**: 459–62

170. LAURENCE, B. R. (1953) Some Diptera bred from cow dung. *Ent. mon. Mag.* **89**: 281–3

171. —— (1953) On the feeding habits of *Clinocera* (*Wiedemannia*) *bistigma* Curtis (Diptera: Empididae). *Proc. R. ent. Soc. Lond.* (A) **28**: 139–44

172. LAVEN, H. (1956) Induzierte Parthenogenose bei *Culex pipiens*. *Naturwissenschaften.* **43**: 116–17 (Berlin)

173. LEWIS, D. J. (1953) The Tabanids of the Anglo-Egyptian Sudan. *Bull. ent. Res.* **44**: 175–216

174. —— (1957) Observations on Chironomidae at Khartoum. *Bull. ent. Res.* **48**: 155–84

175. —— (1958) Some Diptera of medical interest in the Sudan Republic. *Trans. R. ent. Soc. Lond.* **110**: 81–98

176. —— (1960) Observations on the *Simulium neavei* complex at Amani, in Tanganyika. *Bull. ent. Res.* **51**: 95–113 [and many other papers, too numerous to list, chiefly in *Bull. ent. Res.* and *Ann. trop. Med. Paras.*]

177. LINDNER, E. (1925–49) *Die Fliegen der Palaearktischen Region.* Band I Handbuch [and many later parts on the various families] Stuttgart

178. —— (1957) Vogelblutfliegen (*Protocalliphora*; Calliphorinae, Diptera), ein Beitrag zur Kenntnis ihrer Systematik und Biologie. *Vogelwarte.* Stuttgart **19**: 84–90

179. LINDQUIST, A. W. & KNIPLING, E. F. (1957) Recent advances in veterinary entomology. *Ann. Rev. Ent.* **2**: 181–202

180. LINDSAY, D. R. & SCUDDER, H. I. (1956) Non-biting flies and disease. *Ann. Rev. Ent.* **1**: 323–46

181. LINSLEY, E. G. (1960) Ethology of some bee- and wasp-killing robber-flies of southeastern Arizona and western New Mexico. *Univ. Calif. Publ. Ent.* 16, no. 7: 357–92

182. LLOYD, F. E. (1942) *The carnivorous plants.* 352 pp.

183. LLOYD, LL., GRAHAM, J. F. & REYNOLDSON, T. B. (1940) Materials for a study in animal competition. The fauna of the sewage bacteria beds. *Ann. appl. Biol.* **27**: 122–50

183A. LUDWIG, C. E. (1949) Embryology and morphology of the larval head of *Calliphora crythrocephala* (Meigen). *Microentomology* **14**: 75–111

184. MACDONALD, W. W. (1956) Observations on the biology of chaoborids and chironomids in Lake Victoria, and on the feeding habits of the elephant-snout fish (*Mormyrus kannume* Forsk.). *J. anim. Ecol.* **25**: 36–53

185. MACKERRAS, I. M. (1950) The zoogeography of the Diptera. *Austr. J. Sci.* **12**: 157–61

186. —— (1952) *Zoology and Medicine.* Presidential Address to Section D, Zoology, of the Australian and N.Z. Association for the Advancement of Science.

187. —— (1954) Animal reservoirs of infection in Australia. *Proc. R. Soc. Queensland* **65**: 23

188. —— (1954–55) The classification and distribution of Tabanidae (Diptera), I–III. *Austr. J. Zool.* **2**: 431–54; **3**: 439–511, 583–633

189. —— & MACKERRAS, M. J. (1953) Problems of parasitology and entomology in Australia. Report for the Pan-Indian Ocean Science Congress, 1954. *Austr. J. Sci.* **15**: 185–9

190. MACKERRAS, M. J. (1933) Observations on the life-histories, nutritional requirements and fecundity in blow-flies. *Bull. ent. Res.* **24**: 353–62

191. —— & FRENEY, M. R. (1933) Observations on the nutrition of maggots of Australian blow-flies. *J. exp. Biol.* **10**: 237–46

192. MACLEOD, J. & DONNELLY, J. (1961) Failure to reduce an isolated blowfly population by the sterile males method. *Ent. exp. et appl.* **4**: 101–18

193. MALLOCH, J. R. (1917) A preliminary classification of Diptera, exclusive of Pupipara, based upon larvae and pupal characters, with keys to imagines in certain families. Part I. *Bull. Illinois State Lab. Nat. Hist.* **12**: 161–410

194. MANI, M. S. (1952) Considerations on phylogeny and evolution in some Diptera Nematocera, with special reference to Itonididae [= Cecidomyiidae] *Proc. nat. Acad. Sci. India* **20** (**1950**): 1–147

195. MARCHAND, W. (1917) Notes on the habits of the snow fly (*Chionea*). *Psyche* **24**: 142–53

196. MATTINGLY, P. F. & ADAM, J. P. (1954) A new species of cave-dwelling anopheline [mosquito] from the French Cameroons. *Ann. trop. Med. Paras.* **48**: 55–7

197. McCONNELL, W. R. (1921) Rate of multiplication of the Hessian fly. *Bull. U.S. Dep. Agr. No.* 1008. 8 pp.

198. McMAHON, J. P. (1951) The discovery of the early stages of *Simulium neavei* in phoretic association with crabs, and a description of the pupa of the male. *Bull. ent. Res.* **42**: 419–26

199. MÉDIONI, J. (1956) Réactions phototropiques à des radiations spectrales visibles d'égale énergie chez *D. melanogaster* (race sauvage). *C.R. Soc. Biol. Paris* **150**: 1263–6

200. MÉGNIN, P. (1894) *La faune des cadavres.* 214 pp. (Paris)

201. MEIJERE, J. H. C. DE (1905–06) Im Innern von Farnkräutern parasitierende Insektenformen. *Tijdschr. v. Ent.* **48**: lvi–lviii

202. MELANDER, A. L. (1949) A report on some Miocene Diptera from Florissant, Colorado. *Amer. Mus. Mov.* **1407**: 63

203. MELIN, D. (1923) *Contributions to the knowledge of the biology, metamorphosis and distribution of the Swedish Asilids, in relation to the whole family of Asilids.* 317 pp. (Uppsala)

204. MÖHN, E. (1960) Studien an paedogenetischen Gallmückenarten (Diptera: Itonididae). I *Stuttgarter Beitr. zur Naturkunde.* **31**: 11 pp.

205. MORGE, G. (1959) Monographie der Palaearktischen Lonchaeidae. *Beitr. Ent.* **9**: 1–92, 323–71; **10**: 62–134, 909–45, 135–71

206. MORRIS, H. M. (1921–22) The larval and pupal stages of the Bibionidae. *Bull. ent. Res.* **12**: 221–32; **13**: 189–95

207. MOTTER, M. G. (1898) A contribution to the study of the fauna of the grave. A study of one hundred and fifty disinterments, with some experimental observations. *J. N. York ent. Soc.* **6**: 201–31

208. MUIRHEAD-THOMSON, R. C. (1937) Observations on the biology and larvae of the Anthomyidae. *Parasitology* **29**: 273–358

209. —— (1951) *Mosquito behaviour in relation to Malaria transmission and control in the tropics.* 219 pp. (London)

210. —— (1956) Communal oviposition in *Simulium damnosum. Nature* **178**: 1297–1299

211. MUNRO, H. K. (1947) African Trypetidae (Diptera). *Mem. ent. Soc. S. Africa.* 284 pp.

212. MYERS, K. (1952) Oviposition and mating behaviour of the Queensland fruit-fly [*Dacus (Strumeta) tryoni*] and the Solanum fruit-fly [*Dacus (Strumeta) cacuminatus*]. *Austr. J. sci. Res. Melbourne* (B) **5**: 264–81

213. NAGATOMI, A. (1961) Studies in the aquatic snipe-flies of Japan (Diptera: Rhagionidae). *Mushi* **32**: 47–67; **33**: 1–3; **35**: 11–27, 29–38

214. NEFF, S. E. & BERG, C. O. (1961) Observations on the immature stages of *Protodictya hondurana* (Diptera: Sciomyzidae). *Bull. Brooklyn ent. Soc.* **56**: 46–55

215. NIELSEN, E. T. & HAEGER, J. S. (1960) Swarming and mating in mosquitoes. *Misc. publ. ent. Soc. Amer.* **1** (3): 71–95

216. NIJVELDT, W. (1954) The range of prey of *Phaenobremia urticariae* Kffr. and *Ph. aphidivora* Rübs. (Diptera: Itonididae [= Cecidomyiidae]) *Ent. Ber.* **15**: 207–11

217. NJORTEVA, M. (1959) [*Medetera* spp. (Dolichopodidae) which spend their larval stage in the galleries of bark beetles]. *Ann. ent. fenn.* **25**: 192–210

218. OLDROYD, H. (1949) *Handbooks for the Identification of British Insects* IX (1). *Diptera: Introduction and Key to Families.* 49 pp.

219. —— (1954–57) *Horseflies of the Ethiopian Region I–III.* (Brit. Mus. (N.H.): London)

220. —— (1958) *Collecting, preserving and studying insects.* 327 pp. (London & New York)

221. —— (1960) *Insects and their world.* 139 pp. (London)

222. ORI, S., SHIMOGAMA, M. & TAKATSUKI, Y. (1960) Studies of the methods 4. On the effect of colored cagetrap. *End. Dis. Bull. Nagasaki Univ.* **2** (3): 229–35

223. ORMEROD, W. E. (1961) The epidemic spread of Rhodesian sleeping sickness, 1908–1960. *Trans. R. Soc. trop. Med. Hyg.* **55**: 525–38

224. O'ROURKE, F. J. (1956) Observations on pool and capillary feeding in *Aedes aegypti* L. *Nature* **177**: 1087–8

225. PALMÉN, E. (1956) Periodic emergence in some Chironomids – an adaptation to nocturnalism. In Wingstrand, *Zoological papers in honour of Bertil Hanström* 348–56·

226. PATTERSON, J. T. (1943) Studies in the genetics of *Drosophila*. III: the Drosophilidae of the Southwest [of the USA]. *Univ. Texas Publ.* no. 4313, 327 pp.

227. PEARSON, R. G. & PEARSON, J. L. (1961) The flight activity of Diptera Nematocera at Brucebyen, West Spitzbergen. *Ent. mon. Mag.* **96**: 181–2

228. PEFFLY, R. L. (1953) Crosses and sexual isolation of Egyptian forms of *Musca domestica*. *Evolution* **7**: 65–75

229. PENN, G. H. (1954) Introduced pitcher plant mosquitoes in Louisiana. *Proc. Louisiana Acad. Sci.* **17**: 89–90 (Baton Rouge)

230. PETERSON, B. V. & DAVIES, D. M. (1960) Observations on some insect predators of Black Flies (Diptera: Simuliidae) of Algonquin Park, Ontario. *Canad. J. Zool.* **38**: 9–18

231. PETERSON, D. G. & WOLFE, L. S. (1958) The biology and control of blackflies (Diptera: Simuliidae) in Canada. *Proc. 10th. Int. Congr. Ent., Montreal* **3**: 551–64

232. PHILIP, C. B. (1931) The Tabanidae (horse-flies) of Minnesota, with special reference to their biologies . . . *Tech. Bull. Univ. Minnesota Agric. Exp. Stn.* **80**: 1–128

233. PHILLIPS, V. T. (1946) The biology and identification of Trypetid larvae. *Mem. Amer. ent. Soc.* **12**: 161

234. PITCHER, R. S. (1957) The abrasion of the sternal spatula of the larva of *Dasyneura tetensi* (Rubs.) [Diptera: Cecidomyiidae] during the post-feeding stage. *Proc. R. ent. Soc. Lond.* (A) **32**: 83–9

235. PITTERDRIGH, C. S. (1954) On temperature independence in the clock system

controlling emergence time in *Drosophila*. *Proc. nat. Acad. Sci. Chicago* **40**: 1018–29

236. POMEISL, E. (1953) Studien an Dipterenlarven des Mauerbaches. In *Pleskot, G. Beiträge zur Limnologie der Wienerwaldbäche. Wetter und Leben* **II**: 165–176

237. PROVOST, M. W. (1953) Motives behind mosquito flights. *Mosquito News* **13**: 106–9

238. QUARTERMAN, K. D. & SCHOOF, N. F. (1958) The status of insecticide resistance in arthropods of public health importance in 1956. *Amer. J. trop. Med. Hyg.* **7**: 74–83

239. RAPP, W. (1954) *Culiseta inornata* (Culicidae) attacking man. *Ent. News.* **65**: 96–7

240. REFAL, F. Y., MILLER, B. S., JONES, T. & WOLFE, J. E. (1956) The feeding mechanism of Hessian-fly larvae. *J. econ. Ent.* **49**: 182–4

241. REID, J. A. (1953) Notes on house-flies and blow-flies in Malaya. *Inst. Med. Res. Malaya, Bull.* **7**: 26

242. ROBACK, S. S. (1951) A classification of the muscoid calyptrate Diptera. *Ann. ent. Soc. Amer.* **44**: 327–61

243. ROTHSCHILD, M. & CLAY, T. (1952) *Fleas, flukes and cuckoos: a study of bird parasites.* Collins' *New Naturalist* series. 394 pp. (London)

244. ROUSSET, A. (1957) Diptères brachycères cavernicoles du Côte d'Or. (Contributions à la faune de Bourgogne no. 26). *Trav. Lab. Zool. Dijon.* **22**: 4–9

245. RÜBSAAMEN, E. H. & HEDICKE, H. (1926–38) Die Cecidomyiden (Gallmücken) und ihre Cecidien. *Zoologia* **29**: 1–350

246. RUSS. K. (1954) Beiträge zur Atmungsphysiologie und Biologie von *Calliophrys* und *Atherix*, etc. *Oest. zool. Z., Wien* **4**: 531–55

247. RUSSELL, P. F. (1955) *Man's mastery of malaria.* 308 pp. (Oxford University Press)

248. RYCKMAN, R. E. & ARAKAWA, K. Y. (1952) Additional collections of mosquitoes from wood rats' nests. *Pan-Pacific Ent.* **28**: 105–6

249. SABROSKY, C. W. (1952) House-flies in Egypt. *Amer. J. trop. Med. Hyg.* **1** (**2**)-333–6

250. SALT, G. (1932) The natural control of the sheep blow-fly, *Lucilia sericata*. *Bull. ent. Res.* **23**: 235–45

251. SATÔ, S. (1957) On the dimensional characters of the compound eye of *Culex pipiens* var. *pallens* Coquillet. *Sci. Rep. Tôhoken Univ., Sendai* (**4**) **23**: 83–90

252. ——, KATÔ, M. & TORIUMI, M. (1957) Structural changes of the compound eye of *Culex pipiens* var. *pallens* Coquillet in the process of dark adaptation. *Sci. Rep. Tôhoken Univ., Sendai* (**4**) **23**: 91–100

253. SAUNDERS, L. G. (1930) The early stages of *Geranomyia unicolor* Hal., a marine Tipulid. *Ent. mon. Mag.* **60**: 185–7

254. SAUNDERS, D. S. (1960) The ovulation cycle in *Glossina morsitans* Westwood, and a possible method of age determination for female tsetse flies by examining their ovaries. *Trans. R. ent. Soc. Lond.* **112**: 221–38

255. SCHMALL, J. (1954) [Oviposition of *Atherix ibis*]. *Mitt. naturw. Arb. Gemeinsch.* **3–4** Zool.: 22–3

256. SCHOLL, H. (1956) Die Chromosomen parthenogenetischer Mücken. *Naturwissenschaften* **43**: 91–2 (Berlin)

257. SCHULTZ, W. A. (1904) Dipteren als Ektoparasiten an südamerikanischen Tagfaltern. *Zool. Anz.* **28**: 42–3

258. SCOTT, H. (1953) Discrimination of colours by *Bombylius*. *Ent. mon. Mag.* **89**: 259–60

259. SÉGUY, E. (1941) Étude biologique et systematique des Sarcophagines myiasifères du genre *Wohlfartia*. *Ann. Paras. hum. comp.* **18**: 220–32

260. —— (1950) *La biologie des Diptères*. 609 pp. (Paris)

261. SCHUURMANS-STEKHOVEN, J. H. (1954) Biologische Beobachtungen an *Stenepteryx hirundinis* L. (Diptera: Pupipara). *Z. Parasitenk.* **16**: 313–21

262. SENIOR-WHITE, R. (1954) Adult anopheline behaviour patterns: a suggested classification. *Nature* **173**: 730

263. SHTAKEL'BERG, A. A. (1956) Synanthropic Diptera of the fauna of the U.S.S.R. [in Russian]. *Opred. Faune SSSR.* **60**: 1–164 (Moscow)

264. SLEIGH, M. A. (1954) Survival of a dehydrated Chironomid larva in pure nitrogen. *Entomologist.* **86**: 298–300

265. SMITH, G. F. & ISAAK, L. W. (1956) Investigations of a recurrent flight pattern of flood water *Aedes* mosquitoes in Kern County, California. *Mosquito News* **16**: 251–6

266. SMITH, M. E. (1952) Immature stages of the marine fly *Hypocharasus pruinosus*, with a review of the biology of immature Dolichopodidae. *Amer. midl. Nat.* **48**: 421–32

267. SNODGRASS, R. E. (1943) The feeding apparatus of biting and disease-carrying flies: a wartime contribution to medical entomology. *Smiths. misc. Coll.* **104**: 51

268. —— (1961) Insect metamorphosis and retrometamorphosis. *Trans. Amer. ent. Soc.* **87**: 273–80

269. SØMME, L. (1961) On the overwintering of house-flies (*Musca domestica*) and stable flies (*Stomoxys calcitrans*) in Norway. *Norsk. ent. Tidsskrift* **11**: 191–223

270. STALKER, H. D. (1956) On the evolution of parthenogenesis in *Lonchoptera* *Evolution*. **10**: 345–59

271. STATZ, G. (1940) Neue Dipteren (Brachycera + Cyclorrhapha) aus dem Oberoligozän von Hott. *Palaeontographica* **91**: 119–74

272. —— (1941) Plagegeister aus dem rhienischen Braunkohlenwalde. *Rheinischer Naturfreund* **5** (1): 16 pp.

273. STEVENSON, R. & JAMES, H. St. C. (1953) Temperature as a factor in size and activity of wild populations of *Drosophila*. *J. Tenn. Acad. Sci.* **28**: 43–8

274. STEYSKAL, G. C. (1953) A suggested classification of the lower Brachycerous Diptera. *Ann. ent. Soc. Amer.* **46**: 237–42

275. STOKES, B. M. (1956) A chemotactic response in wheat-bulb fly larvae. *Nature.* **178**: 801

276. STRICKLAND, A. H. (1960) [losses in agriculture from pests and diseases] *New Scientist* **8**: 660–2

277. STUCKENBERG, B. R. (1960) Diptera (Brachycera): Rhagionidae in *South African Animal Life*: results of the Lund University expedition in 1950–51. **VII**: 216–308

278. SUKHOVA, M. N. (1952) Synanthropic flies. [in Russian] *Pom. med. Rabot* no. 3. **1952**: 60 pp. (Moscow)

279. TALICE, R. V. (1952) Tropismos de *Musca domestica* y de otros moscas domésticos. *Arch. Soc. Biol. Montevideo* **19**: 15–17

280. TAYLOR, L. R. (1960) The distribution of insects at low levels in the air. *J. anim. Ecol.* **29**: 45–63

281. TESCHNER, D. (1961) Zur Dipterenfauna an Kindeskot. *Deutsche ent. Z.* **8**: 63–72

282. THEODOR, O. (1936) On the relation of *Phlebotomus pappatasii* to the temperature and humidity of the environment. *Bull. ent. Res.* **27**: 653

283. —— (1952) On the zoogeography of some groups of Diptera in the Middle East. *Rev. Fac. Sci. Univ. Istanbul* (B) **17**: 107–19

284. THEOWALD, B. (1957) Die Entwicklungsstadien der Tipuliden inbesondere der westpalaearktischen Arten. *Tidschr. Ent.* Leiden **100**: 195–308

285. THIENEMANN, A. (1954) *Chironomus* – Leben, Verbreitung und wirtschaftliche Bedeutung der Chironomiden. *Die Binnengewässer*. Stuttgart Bd. **20**: 834 pp.

286. THOMSEN, M. (1935) A comparative study of the development of the Stomoxydinae (especially *Haematobia stimulans* Meigen) with remarks on other coprophagous muscoids. *Proc. Zool. Soc. Lond.* **1935**: 531–50

287. THORPE, W. H. (1930) The biology of the petroleum fly (*Psilopa petrolei* Coq.) *Trans. ent. Soc. Lond.* **78**: 331–43

288. —— (1931) Miscellaneous records of insects inhabiting the saline waters of the Californian desert regions. *Pan-Pacific Ent.* **7**: 145–53

288A. —— (1941) The biology of *Cryptochaetum . . . Parasitology* **33**: 149–68

289. —— (1934) Observations on the structure, biology and systematic position of *Pantophthalmus tabaninus* Thunb. (Diptera: Pantophthalmidae). *Trans. R. ent. Soc. Lond.* **82**: 5–22

290. —— (1942) Observations on *Stomoxys ochrosoma* (Diptera: Muscidae) as an associate of army ants (Dorylinae) in East Africa. *Proc. R. ent. Soc. Lond.* (A) **17**: 38–41

291. TILLYARD, R. J. (1935) The evolution of the scorpion-flies and their derivatives (Order Mecoptera). *Ann. ent. Soc. Amer.* **28**: 1

292. TWOHY, D. W. & ROZEBOOM, L. E. (1957) A comparison of food reserves in autogenous and anautogenous *Culex pipiens* populations. *Amer. J. Hyg.* **65**: 316–24

293. VANSKAYA, R. A. (1942) Hibernation of *Musca domestica* L. *Med. Paras.* **11**: 87–90 (in Russian; English summary in 1944, *Rev. appl. Biol.* (B) **32**: 53)

294. VARLEY, G. C. (1937) Aquatic insect larvae which obtain oxygen from the roots of plants. *Proc. R. ent. Soc. Lond.* (A) **12**: 55–60

295. —— (1947) The natural control of population balance in the knapweed gall-fly (*Urophora jaceana*). *J. anim. Ecol.* **16**: 139–87

296. VOGEL, G. (1954) Das optische Weibchenschema bei *Musca domestica*. *Naturwissenschaften* **41**: 482–3 (Berlin)

297. VOROBJEV, B. A. (1960) (in Russian; title translated as 'The inhabitation of larvae in waterbodies in the axils of the leaves of *Dipsacus*'). *Rev. Ent. U.R.S.S.* **39**: 799–801

298. WALSHE, B. M. (1950) The function of haemoglobin in *Chironomus plumosus* under natural conditions. *J. exp. Biol.* **27**: 73–95

299. WARDLE, R. A. (1926) The respiratory system of contrasting types of crane-fly larvae. *Proc. zool. Soc. Lond.* **1926**: 25–48

300. WATERHOUSE, D. E. & PARAMONOV, S. J. (1950) The status of the two species of *Lucilia* (Diptera: Calliphoridae) attacking sheep in Australia. *Austr. J. Sci.* **3** (3): 310–36

301. WEBER, N. A. (1934) Arctic Alaskan Diptera. *Proc. ent. Soc. Wash.* **56**: 86–91

302. WEEREKOON, A. C. J. (1953) On the behaviour of certain Ceratopogonidae. *Proc. R. ent. Soc. Lond.* (A) **28**: 85–92

303. WELLINGTON, W. G. (1954) Atmospheric circulation processes and insect ecology. *Canad. Ent.* **86**: 312–33

304. WENTGES, H. (1952) Zur Biologie von *Tabanus sudeticus sudeticus. Nachr. Bl. Bayer. Ent, München.* **1** (1952): 78–9

305. WHITE, M. J. D. (1949) Cytological evidence on the phylogeny and classification of the Diptera. *Evolution* **3** (3): 252–61

306. WHITTEN, J. M. (1955) A comparative morphological study of the tracheal system of larval Diptera. Part I. *Quart. J. Micro. Sci.* **96**: 257–78

307. WIESMANN, R. (1961) Untersuchungen über Fliegenkotkonzentrationen und Fliegenansammlungen in Viehställen. *Mitt. Schweiz ent. Ges.* **34**: 187–209

308. WIGGLESWORTH, V. B. (1938) The regulation of the osmotic pressure and chloride concentration in the haemolymph of mosquito larvae. *J. exp. Biol.* **15**: 235–47

309. WILLIAMS, C. B. & OSMAN, M. F. H. (1960) A new approach to the problem of the optimum temperature for insect activity. *J. anim. Ecol.* **29**: 187–99

310. WIRTH, W. W. & STONE, A. (1956) *Aquatic Diptera of California*, in Usinger, R. L. *Aquatic insects of California*: 372–82

311. WISHART, G. & RIORDAN, D. F. (1959) Flight responses to various sounds by adult males of *Aedes aegypti. Canad. Ent.* **91**: 181–91

312. WYATT, J. J. (1961) Pupal paedogenesis in the Cecidomyiidae (Diptera). *Proc. R. ent. Soc. Lond.* (A) **36**: 133–43

313. WYNIGER, R. (1953) Beiträge zur Oekologie, Biologie und Zucht einiger europaischer Tabaniden. *Acta trop.* **10**: 310–37 (Basle)

314. ZHUKOVSKII, A. V. (1957) The diapause in *Myetiola destructor* Say (Diptera: Itonididae [= Cecidomyiidae]). *Ent. Obozr.* (in Russian, English summary) **36**: 28–43 (Moscow)

315. ZUMPT, F. (1951) Myiasis in Man and animals in South Africa. *S. Afr. J. Clinical Science* **2** (1): 38–68

316. —— (1957) Some remarks on the classification of the Oestridae. *J. ent. Soc. S. Africa.* **20**: 154–61

Additional References while this book was in the Press

317. ADAMOVIC, Z. R. (1963) (in Russian, with English précis) Ecology of some Asilid species and their relation to the honey bee. *Mus. Hist. nat. Beograd, Hors série* **30**: 104 pp.
318. APTED, F. I. C. & others (1962) Sleeping sickness in Tanganyika, past, present and future. *Trans. R. Soc. trop. Med. Hyg.* **56**: 24–9
319. BRINDLE, ALLAN (1963) Terrestrial Diptera larvae. *Ent. Record* **75**: 47–62
320. CHOYCE, D. P. (1962) [Onchocerciasis & blindness] letter in *New Scientist* **16**: 106
321. COLLESS, D. (1962) (new family Perissommatidae) *Austr. J. Zool.* **10**: 519–35
322. GARCIA, R. & RADOVSKY, F. J. (1962) Haematophagy by two non-biting Muscid flies and its relationship to Tabanid feeding. *Canadian Entomologist* **94**: 110–16
323. HARTLEY, J. C. (1961) A taxonomic account of the larvae of some British Syrphidae. *Proc. zool. Soc. Lond.* **136**: 505–73
324. KNIPLING, E. F. (1962) Potentialities and progress in the development of chemosterilants for insect control. *J. econ. Ent.* **55**: 782–6
325. LUMSDEN, W. H. R. (1962) Development of Trypanosomiasis Research in East Africa. *Nature* **195**: 1139–41
326. NOWAKOWSKI, J. T. (1962) Introduction to a systematic revision of the family Agromyzidae, with some remarks on host selection by these flies. *Annales Zool. Warsaw.* **XX**, no. 8: 67–183
327. OKADA, T. (1962) Bleeding sap preferences of the Drosophilid flies. *Japanese J. appl. Ent. & Zool.* **6**: 216–29
328. ROHDENDORF, B. B. (1961) Neue Angaben über das System der Dipteren. *Verhandlungen des Internationalen Kongresses für Entomologie in Wien*, 1960 **I**: 153–8
329. SABROSKY, C. W. (1961) The present status of the systematics of Diptera. *ibid.*: 159–66
330. SERGENT, E. & ROUGEBIEF, H. (1926) Des rapports entre les moucherons du genre *Drosophilia* et les microbes du raisin. *Ann. Inst. Paris* **40**: 901–21
331. WHITFIELD, F. G. SAREL (1925) The relation between the feeding-habits and the structure of the mouthparts in the Asilidae (Diptera). *Proc. zool. Soc. Lond.* **1925**: 599–638

Index

Abbreviated larval life, 152
abdominal membranes, distension of, 151
Acalyptrata, 11, 170, 280
  evolution of, 171
  terrestrial habits, 171
  larvae of, 16, 20, 22
acanthophorites, 126, 127
accidental myiasis, 251
Achatina, 151
Achias, stalked eyes, 190, 198
Acletoxenus, 200
Acroceridae, 136
adaptation to environment, 74, 105
adult flies, feeding of, 201
  functions of, 14
Aedes, 94
  eggs of, 85
  egg-laying of, 90
  in Bromeliads, 91
  at light, 80
  detritus, 88
  aegypti, 244, 245
  hexodontus, 265
  nubilis, biting cycle of, 81
aestivation, 87
Africa, development obstructed by tsetse-
  flies, 210, 249
Afrocneros mundus, 198
aggregations of flies, 111, 113, 262
Agromyzidae, 195, 258
  larvae of, 176
air currents, dispersal by, 4
  raid shelters, *Culex molestus* in, 80
  stores in tracheal system, 50, 93, fig. 13
  transport, and plant pests, 258
  — and yellow fever, 245
Alamira termitoxenizans, 152
alcohols attractive to fruit-flies, 197
algae, larvae feeding on, 52, 185
  Dolichopodidae feeding on, 145
  Forcipomyia feeding on, 57
Allolobophora, 226
Amani, Tanganyika, rare flies at, 277

Amitermes atlanticus and *Termitometopia*, 227
ammonia in excreta of blow-flies, 215
Ammophila parasitized by Conopidae, 168
amphibia attacked by *Lucilia*, 17, 225
  — by *Sycorax*, 53
amphipneustic larva, 38
anal gills, 88, 92
Ancala, larva of, 103
ancestral flies, 274
  larval feeding of, 18, 33
Andrenosoma albopilosum, 277
animal protein, 119
Anisopodidae, 39
Anomalopteryx maritima, 95
Anopheles claviger, 87, 92
  labranchiae atroparvus, 88
  maculipennis, 242
anopheline mosquitoes, 78, 244
  eggs of, 15, 85
  larvae of, at surface, 91, 93
  — in Bromeliads, 91
  nocturnal habits of, 80
antelopes and horse-flies, 98
  nose bots in, 220
antennae, fig. 5
  of male midges, 58, 59, fig. 14
  of Phoridae, 148
  of *Simulium*, 66
  of stalk-eyed flies, 190
Anthomyia, larval habits of, 205, 206
anti-malarial forces, 243
Antocha, larva and pupa of, 35, 36
ant-lions, Tabanidae living with, 106
ants, larvae living in nests of, 43
  and Calliphoridae, 226
  and *Forcipomyia*, 57
  and *Microdon*, 160
  and mosquitoes, 84
  and Mydaidae, 129
  and Phoridae, 149, 152
  and Scatopsidae, 46
  and *Stomoxys ochrosoma*, 208
  and Stratiomyidae, 116

aphid-feeding, origin of, 165
  by Chamaemyiidae, 182
  by Drosophilidae, 200
  by Syrphidae, 163, 164
Aphidoletes, 47
apholates and aphoxides, 281
Aphrosylus, larval habitat of, 145
Apiocera maritima, larva of, 129
Apioceridae, 279
  distribution of, 129
Apion attacked by *Chlorops*, 193
apneustic larvae, 44, 56
Apocephalus, attacking ants, 152
apricots, damage to, 258
Apterina, 141
Apterodromia evansi, 95
'aquarium' of Bromeliads, 90
aquatic behaviour of *Simulium*, 68
  flies, 15
  larvae, 16, 17, 169
  larva of *Atherix*, 112
  — of biting midges, 57
  — of Blepharocerids, 73
  — of Chironomidae, 61
  — of *Limnophora*, 205
  — of mosquitoes, 77
  — of Psychodinae, 51
  — of Thaumaleidae, 72
  — of Stratiomyidae, 115
  — Syrphidae, 116
  larvae of acalyptrates, 172
  pupae of Tabanidae, 105
  snipe-flies, 113
  status of mosquito larvae, 84
Arachnida, 17
Arachnocampa luminosa, 43
archedictyon, 135
Archischiza, 168
Ardö, P., 175
Arctic Circle, 98
  aquatic larvae in, 64
Argyra, 144, 145
arista of antenna, 13, fig. 5
Armigeres, 90, 94
army worm (*Sciara militaris*), 43, 272
  (*Laphygma exempta*), 133
arum-spaths, flies in, 93
Arundo, galls in, 193
Ascodipteron, 61, 234, 236, pl. 27
Asilidae, 8, 127, 136
  compared with Empididae, 138
  oviposition of, 126, 196
  pupae of, 128
Asilus crabroniformis, 128
  barbarus, 128
asphyxiation of Phoridae in pupa, 150
assemblies of flies, 263
asthma caused by flies, 53

Asyndetus, larva of, 145, 146
Atherix, 104
  aggregation of adults in, 113
  egg-laying of, 111
  bloodsucking by, 109
  larva of, 117
Atrichomelina pubera, 179
Atrichopogon, 56
Atrichops, biting frogs, 109
Atta, and Mydaid larvae, 129
attics, flies in, 255
Aubin, P. A., 157
Auchmeromyia luteola, 217
Aulacigaster, 200
Austen, E. E., 184, 218, 222
Australia, primitive flies in, 106, 115
Austroleptis, bloodsucking by, 109
autumn-fly (*Musca autumnalis*), 255
auxiliary lung (plastron), 15
avoidance of horse-flies by pastoral people, 97

Bacon, *Piophila casei* in, 176, 257
bacteria carried by flies, 255
Baer, Prof., 152
balloon, in Empid mating, 140
balloon-flies, 140
bamboos breeding mosquitoes, 86
Bancroftia larvae in Bromeliads, 91
banded house mosquito, 87
Banks, C. J., 156
Baridius attacked by *Chlorops*, 193
bark, larvae under, 42, 48, 146
barley, 259
Barnes, H. F., 24, 46, 258, 270
Basden, E. B., 199
Bates, Marston, 82, 85, 88, 91
bats, flies parasitizing, 10, 11, 229, 235
  Mormotomyia in dung of, 183
  and Streblidae, 235
beaches, horse-flies on, 106
bed-pans, *Drosophila repleta* in, 199
bee-flies, 8, 131, 134
bee-louse (*Braula coeca*), 182
bees, flies resembling, 104
  and Asilidae, 125, 126
  and Bombyliidae, 132
  and *Cacoxenus indagator*, 200
  and Conopid larvae, 169
  and Miltogramminae, 227
  and Phoridae, 152
  and Stratiomyidae, 114
  and Syrphidae, 158, 159
beetle-fly, 172, 173
beetle-larvae attacked by larval Mydaidae, 129
beetles, aquatic, 15
beet fly, 207

behaviour of bluebottles, 214
of crane-flies, 32
beneficial activities of flies, 259
Bengalia and ants, 226
Bequaert, J. C., 274
Berg, C. O., 179, 188
Beris morrissii, 114
Bertrand, H., 145
Bezzi, Mario, 168
Bibio marci, 45
Bibionidae, 45
fertilizing flowers, 259, pl. 9
pupating in larval skin, 49
big flies, decline of, 5
binocular vision in robber-flies, 121
Biological Abstracts, 284
biological clocks, 82
birds attacked by flies, 11, 17, 229
— by mosquitoes, 80
— by *Neottiophilum*, 182
— by *Protocalliphora*, 217
— by *Simulium*, 66, 67, 246
nests, larvae in, 43, 129, 205
biting cycles, 79, 81
flies, 242
of Man by horse-fly larvae, 102
by *Simulium*, 208
midges, 55, 247
Bittacomorpha, larva of, 38
black-flies, 18, 19, 66, 69
Blepharoceridae, 72, 279
blindness, *Onchocerca* and, 246
blister-beetle and *Forcipomyia*, 56
blood attractive to *Sarcophaga*, 227
taken by Congo floor maggot, 218
larvae in, 252
loss of, to biting flies, 97
mopping up of, 202, 248
blood-gills, 50, fig. 13
blood-meal, a significant expression, 83
essential or not, 80, 279
bloodsucking by Brachycera, 247
— by calyptrates, 248
— by horse-flies, 101
— by Pyschodidae, 53
— by pupipara, 230
— by tsetse-flies, 209
bloodsucking flies, 11, 14, 201, 202, 280
and Man, 242
and poliomyelitis, 248
primitive, 275
bloodsucking larvae, 21
of *Neottiophilum praeustum*, 182
of *Passeromyia*, 205
bloodworms, 62
blotch mines, 195
blow-flies, 201, 213
'blown' meat, 213

bluebells, *Rhingia* feeding on, 158
bluebottle, 171, 213
as advanced fly, 5
evolution of, 11
feeding of adult, 214
noisy flight of, 214
'bobbing' of crane-flies, 32
Bodega gnat, 57
bodily secretions, flies attracted to, 250
boggy places, larvae in, 162
Bohart, G. E. and Gressitt, J. L., 185
boils caused by fly-larvae, 218
Bombomima, 126
Bombus species imitated by *Volucella bombylans*, 159
orientalis, mimicked by *Lycastris*, 159
Bombyliidae, 8, 119, 136, 145, 157
feeding at flowers, 132
habits of, 131
parasitic larvae of, 133
Bombylius major, 131
bones, *Piophila* larvae on, 177
bonfires, Platypezidae attracted to, 266
Borboridae, 141, 177
bots, 16, 213, 221
Brachycera, 7, 8
larvae of, 16, 19
Brachydeutera, 186
Brachyopa, larva of, 161
Brachypteromyia, 231
brackish water, larvae in, 19
mosquitoes in, 88
Braula coeca, 182, 183
breathing in water, problems of, 50
tube, 16, 38
Bremse, 96
breeze-flies, 96
bristles, function of
in robber-flies, 121
in Phoridae, 148
Brock, E. M., 64
Bromeliads, 19, 86, 90, 91, 178
brooding of eggs by *Atherix*, 111
— by *Goniops*, 104
Bruce, James, 249
Bruchomyia, 54
Brues, C. T., 117
Brugia malayi, 245
buffalo-gnats, 66
buffer-stage of life-history, 85, 86
Bufolucilia attacking toads, 225
bugs, aquatic, 15
buildings, flies invading, 255
bulbs, larva attacking, 161, 258, pl. 32
bumblebees and Conopidae, 168
burrows of mammals, flies in, 172
Burton, Sir Richard, 249
bushy antenna of male midge, 59, fig. 14

buttress-roots, *Phlebotomus* among, 54
Buxton, P. A., 32, 60, 116, 184, 211, 284
Byers, G. W., 179

Cabbage root fly, 207, 259
Cacoxenus indagator, 200
caddis-flies, 36
Calabar swellings, 247
calcium carbonate in Stratiomyid larvae, 116
Calobatidae, 177
Calotermes irregularis, 197
Callan, E. Mc. C., 104
Callicera, larva of, 162
Calliopum, 172
Calliphora, 254
    in houses, 268
    larval feeding habits, 214
    erythrocephala (=*vicina*), 213
    vomitoria, 213
Calliphoridae, 251, 213
    evolution of feeding habits, 225
Calliphoroidea, 201
Callitroga, 217
    larva of, 21
    americana, eliminated from Curaçao, 281
calypters, 11, 170
calyptrates, 11, 16, 20, 22, 170, 201
    larval food of, 22
    origin of, 201
cannibalism, of Asilidae, 125
    of horse-flies, 94
    of *Megarhinus* larvae, 93
    of *Scatophaga* larvae, 212
camels, head-bot of, 220
    Hippobosca on, 232
Camilla, 200
Camponotus, Phorids parasitic in, 152
Campsicnemus, 141, 144, 146
canopy of rain forest, 80, 101, 166, 244
capture dates of Asilidae, 121
carbohydrates in diet, 14
Carboniferous Period, 274
Carnidae, 180
carnivorous flies, 8, 14, 119
carnivorous habits, evolution of, 94
    of Blepharoceridae, 73
    of Empididae, 139
    of *Ochthera mantis*, 189
    of *Scatophaga*, 211
carnivorous larvae, 19, 22, 50, 51, 52, 144,
        162, 171, 175, 179
    of *Apiocera*, 129
    of Asilidae, 125
    of Bombyliidae, 132
    of Cecidomyiidae, 47
    of Chamaemyiidae, 182
    of *Chaoborus*, 75
    of Chloropidae, 191

    of Coenomyidae, 108
    of Drosophilidae, 200
    of mosquitoes, 93, 94
    of Mydaidae, 129
    of Rhagionidae, 109
    of Stratiomyidae, 116
    of Tabanidae, 102
    of Tanypodinae, 64
Carnus hemapterus, 180, 181, 184, 233
Carpenter, G. H., 276
carpenter-bees and Asilidae, 125
    and Nemestrinidae, 136
carpets, larva of *Scenopinus* in, 129
carrion as a larval food, 175
carrion-feeding a stage towards parasitism in
        blowflies, 217
    larvae, 13, 144, 252
    of acalyptrates, 171
    of Empididae, 141
    of Phorids, 150
    of *Piophila*, 177
    of *Sarcophaga*, 227
    of Thyreophoridae, 181
carrot fly, 258
cart-ruts, breeding mosquitoes, 89
Casuarina trees, 107
caterpillars attacked by *Forcipomyia*, 56
    parasitized by Bombyliidae, 133
cattle, attacked by *Chrysomyia bezziana*, 217
— by *Dermatobia hominis*, 219
— by *Hippobosca*, 232
— by Tabanidae, 98
— by tsetse-flies, 209
— harassed by *Musca autumnalis*
caves, flies living in, 172
Cecidomyiidae, 46, 47, 195, 258
celery fly, 258
    leaf-mines in, 197
Celyphidae, 172, 173
Centaurea, 196
Central Africa, effect of tsetse on, 249
Cephalopina titillator in camels, 220
Cephenomyia, 220, 251
    pratti, reputed high speed of, 220
Cephenomyiinae, 218
Cerambycid beetles, 135
Ceratitini, 196
Ceratitis capitata, 196, 253
Ceratopogonidae, 55
    attacking other insects, 55
    larvae of, 19
    method of biting, 55
    sizes of, 55
ceilings, *Thaumatomyia notata* on, 255, 271
Cercopidae attacked by Pipunculidae, 165
cereals attacked by larvae, 191, 259
Cerioidinae, 159
Ceroplatinae, 43

Ceroplatus nipponensis, luminous larva of, 44
chalk downs, horse-flies on, 103
Chalybosoma casuarinae, 107
Chamaemyiidae, 182
 larva of, 164
Chaoborinae, 74, 75
Chaoborus, larva of, 18, 75, fig. 13
 edulis, 75
 plumicornis, 59
chamois, *Melophagus rupicaprinus* on, 233
cheese, larva of blowflies in, 215
cheese skipper, 176, 257
Cheesman, Miss Evelyn, 32
Cheilosiinae, larvae of, 161
chemical sterilants, 281
Chenopodium, mines in, 188
Chersodromia cursitans, 141
chicken-houses, *Fannia* in, 207
Chionea, 29
Chiromyia oppidana, 173
Chironomidae, 58, 172, 195
 larvae of, 19, 62
 — in sewage filters, 52
 — terrestrial, 61
 length of adult life, 60
 as prey of *Arachnocampa*, 44
Chironomus grimmi, egg-laying by pharate
 adult, 61
 plumosus, feeding of larva of, 62
Chiromyzini, 115
Chlorichaeta tuberculosa, 185
Chloropidae, 11, 190, 201
 larvae of, 22
 number of generations per year, 192
Chlorops pumilionis (= *taeniopus*), 191
 speciosa, larva of, 193
 strigula, larva of, 193
Chortoicetes, 135
Chortomyia, 49
Choyce, D. P., 246
chromosomes as clue to evolution, 276
chromosomal races of *Lonchoptera*, 154
Chrysidimyia, 159
Chrysogaster piercing water-plants, 162, 163
Chrysomyia, 217
 bezziana, 217
 megacephala, 206
Chrysopa attacked by *Forcipomyia*, 56
Chrysopilus, 109
Chrysops, 80, 97, 99
 biting cycle of, 82
 recent evolution of, 98
 dimidiata, 247
 discalis and tularaemia, 247
 silacea, 103, 247
 — attracted by smoke, 266
churches, flies hibernating in, 270
cider, 254

citrus fruits attacked by larvae of Trypetidae,
 258
claws of larvae, 16
clearings, activities of flies in, 80
clegs, 96
 in Europe and N. America, 96
Clinocera, 140
 larva of, 145
'close season' in tropics, 105
clover, mines in, 172
Clunio adriaticus, in deep water, 60
Clunioninae, 60
Clusiidae, 176
Cluster fly, 226, 268
Clythiidae, see Platypezidae
Cnephia dacotensis, non-bloodsucking, 68
Cobboldia, ill-effects of, 224, 225
coccids eaten by Drosophilidae, 200
cocoon, 23, 24, 43
 of Dolichopdoidae, 146
 of *Simulium*, 71
Coe, R. L., 147, 162
Coelopa, 174, 205
 frigida, 6, 141, 174, 268
Coelopidae, 11
Coenomyidae, 108
coffee, 258
coffin-flies, 9, 147
colder-regions, flies of, 278
Cole, P., 137
Colless, D., 49
Collin, J. E., 166
colour of aphid-feeding larvae, 164
 of Stratiomyidae, 113
 of Tabanidae, 103
 of Hippoboscidae, 164
Colyer, C. N., 226, 263
 and Hammond, C. O., 46, 139
commensals, 45, 63, 152
commercial damage by fruit-flies, 258
communal behaviour in *Simulium*, 69, 267
Compositae, larvae in flowers of, 196
compost-like substances as a breeding-
 medium for flies, 42
 larvae in, 19, 20, 22, 41, 116, 170
 and acalyptrates, 171
 and Chloropidae, 192
 and Diopsidae, 190
 and *Forcipomyia*, 57
 and Syrphidae, 160
Congo floor maggot, 217, 241
Conicera tibialis, 147
Conopidae, 168
 variation in larval structure, 169
concealment of pattern by wings when at
 rest, 114
contagious diseases spread by flies, 255
container habitats, 18, 19, 161, 162

continental faunas, derivation of, 279
continuous breeding of house-flies, 203
control measures, uncertain future of, 280
contusions, blowflies attracted to, 217
convection effects, 166
convergence of Conopidae with Syrphidae, 168
Corbet, P. S., 71, 263, 264
Cordylobia anthropophaga, 218, 251
Cordyluridae, 212
Corethra larvae, 75, fig. 13
corpses, flies breeding in, 147, 185
  Piophila in, 176, 177, pl. 25
  sequence of larvae in, 215
cot-blankets, *Fannia* breeding in, 207
cottony-cushion scale controlled by *Cryptochaetum*, 200
couch-grass, 192
courtship in Empididae, 139
  in Dolichopodidae, 143
crabs and *Simulium neavei*, 70
crane-flies, 7, 29
  adult habits of, 29
  carnivorous larvae of, 37
  dispersal of, 40
  evolution of, 40
  larvae of, 19
  larval spiracles of, 33
  numbers of, 40
  primitive, 5
  size of, 34
Crataerina pallida, 231
creeping welts, 38
Cremastogaster ants, flies soliciting from, 189
  and *Milichia*, 180
  and mosquitoes, 84
Crewe, W., 103
Cricetomys attacked by *Stasisia*, 218
Cricotopus nostoicola, in alga, 64
Crisp, G. and Lloyd, Ll., 38, 51, 178, 185
crochets, 145
crops attacked by larvae of flies, 45, 258
Crustacea, flies and, 17, 70
cryptobiosis of Chironomid larva, 65
Cryptochaetum, 200
Ctenacroscelis, biggest crane-flies, 34
Ctenophora larvae in rotting wood, 35
cuckoo-like behaviour of Phoridae, 152
cucumbers, pollination of, 156
Cucurbitaceae, Trypetids in, 197
  damage to, 258
Culex, eggs of, 85
  fatigans, 245
  pipiens, 80, 87
  — breeding in *Dipsacus*, 91
  larvae in Bromeliads, 91
  molestus, 80
  nocturnal habits of, 80

Culicini, 78, 244
  larvae of, 93
  eggs of, 16
Culicidae, three sub-families of, 74
Culicoides, 55, 57, 247
  effects of bites on Man, 56
  numerous in Arctic, 279
  anophelis, attacking mosquitoes, 56
  austeni, 247
  grahami, 247
cultures of *Drosophila melanogaster*, 199
Cumber, R. A., 169
cursorial habits of Empididae, 140
Curtonotum feeding on eggs of locusts, 226
Cushing, E. C., 243, 245
Cuterebrinae, 219
Cyclorrhapha, 7, 9, 11
  larvae of, 16
  origin of name, 23
Cylindrotominae, larvae of, 33
cylinders of mud made by Tabanid larvae, 105
Cyclopodia greefi, feeding of, 238
Cymatopus, larval habitat of, 145
Cymbopogon, attacked by Diopsid larvae, 190
Cypselidae, see Borboridae
Cynonycteris straminea, 238
Cyrtandria paludosa, leaf-mines in, 35
Cyrtidae, 133, 136

Dacinae, 197
Dacus, sound production by, 266
daddy-long-legs, 5, 29, 34
dance-flies, 138
dancing of Bibionidae, 45
  of *Hilara*, over water, 139
  of *Platyna hastata*, 113
  of Platypezidae, 166
Danube, Golubatz fly on, 246
darkness, flies active during, 79
Dasiops, 263
Dasyhelea confinis in *Nepenthes*, 57
Dasyllis, a bee-like Asilid, 126
Dasyomma, bloodsucking by, 109
Dasybasis, a littoral horse-fly, 106
Dasyphora cyanella, 268, 270, 276
DDT against resting mosquitoes, 83
dead animals, larvae in, 22
  insects eaten by larvae of Stratiomyidae, 116
  leaves, mines in, 195
debris, larvae in, 43
decapitation of ants by Phoridae, 152
decline of flies since Tertiary times, 276
declining families of flies, 279
decaying vegetable matter, 172
  on seashore, 145
  Ceratopogonid larvae in, 57

decaying wood, 18, 34, 39
deer-flies (*Chrysops*), 97, 99
deer, Hippoboscidae on, 233
 ked (*Lipoptena cervi*), 181
deep litter, *Fannia* in, 206
dehydration of Stratiomyid larvae, 92
Delia antiqua, 207
dens of animals, breeding-places for flies,
 54
dengue, virus of, 245
Dermatobia hominis, 218, 251
Descamps, m., 189
Desmometopa, 180
destruction of human tissue, 252
detritus, larvae in, 18, 19
Deuterophlebiidae, 72
Diachlorini, 107
diagonal vein of Nemestrinidae, 135
Diamesinae, 64
diapause, 86, 87
Diasemopsis, larval feeding of, 190
Diastata, 199
Dicranomyia, leaf-mining larva of, 35
Dictenidia, larva in rotting wood, 34
Dictya, 179
diet of adult flies, 14
 of larvae, 14
digestion of muscles by maggots, 214.
Dilophus febrilis, 45
 massing of, 268
dimethylphthalate (DMP), 67
Dipetalonema perstans, 247
Diptera, definition of, 3
Dipsacus silvestris, 57
Dioctria, 122
 prey of, 124
Diopsidae, 177, 189
 feeding of adults, 190
 number of generations per year, 190
 replacing Dolichopodidae in tropics, 143
discharges from sores attractive to flies, 252
Discomyza, 185, 186
dishcloths, larvae in, 39
distribution of Empididae compared with
 Asilidae, 139
disease carried by horse-flies, 97
— by mosquitoes, 77
— mechanically by flies, 97
display of male Dolichopodidae, 143, 144
distances covered by flies, 67
distribution, vertical, 79
diurnal mosquitoes, 83
Dixinae, 74, 75
dog days, 207
dog fly, 232
dogs attacked by *Oestrus ovis*, 220
 flies breeding in dead, 147
Dolichopeza, larva and pupa of, 34

Dolichopodidae, 8, 9, fig. 3, 120, 138, 141,
 154, 166, 168, 185, 189, 248
 in Hawaii, 279
 larva of, 19, 110
 larval habits of, 144
 larval habitats of, 145
 legs compared with Phoridae, 150
 method of feeding of, 142
 sizes of, 144
domestic animals killed by *Simulium*, 66
donkeys, *Gasterophilus* in, 224
Donisthorpe, H. St. J. K., 160
Dorilaidae, see Pipunculidae
Downes, J. A., 55, 96, 102, 110, 263, 269, 275
downlooker fly, 109
dragonflies and Chironomidae, 63
drains and *Lucilia* larvae, 217
drinking by male Tabanidae, 100
drone-flies, 159
Drosophila, 11, 22, 209, 254
 evolutionary vitality of, 5
 future of, 280
 in milk bottles, 150, 199
 natural habitats of, 199
 and the yeasts of grapes, 259
 busckii, 199
 funebris, 199
 hydei, 254
 melanogaster, 198, 199
 repleta, 199, 254
Drosophilidae, 11, 155, 195
 in Hawaii, 279
drought as hazard to aquatic insects, 87
drone-fly, 161
Dryomyzidae, 22, 172, 185
ducks, attacked by *Simulium*, 66
duckweed, mines in, 188
Duke, B. O. L., 267
dung, flies attracted to, 84
 as larval food, 21, 175
 larvae in, 17, 22, 45, 177
 acalyptrate larvae in, 171
 Dolichopodidae in, 146
 Ephydridae in, 189
 Psychodidae in, 53
 Sepsidae in, 175
 Scatophaga and, 211
dung-flies, 96, 170
dustbins, *Lucilia* and, 215
Dysmachus, egg-laying of, 127
Dyte, C. E., 22

Eagle attacking osprey, 232
'earthy' Brachycera, 8
— groups of larvae, 7, 19
— of adult flies, 175
Echestypus, 233
ecology of Streblidae, 236

eddies, use of by flies, 264, 265
Edwards, F. W., 40, 61
Egglishaw, H. J., 268
egg-laying by bluebottles, 213
— by Bombyliidae, 132
— by Chaoborinae, 76
— by *Gasterophilus*, 223, 224
— by mosquitoes, 85
— by *Simulium*, 69
— by Tabanidae, 104
   on ground, 16
   in holes, 90
   in plants, 16
   in soft media, 16
eggs of *Calliphora*, 214
   of Chironomidae, 61
   of Cyrtidae, 137
   of *Gasterophilus intestinalis*, 223
   large numbers of, 50
   locusts, flies feeding on, 226
   numbers laid by flies, 15
   ornamental shell of, 15, fig. 6
   production of, 15
   protection against rain, 15
   large and few, 151
   respiration of, 15, fig. 6
   robber-flies, 127
   Scatophaga, 211
Egypt, plagues of, 255
Elachiptera, larva of, 192
elephantiasis, 245
elephants, 99
   Cobboldia in stomach of, 224
   fly larvae attacking, 218
   Neocuterebra in, 224
   Ruttenia in, 224
   throat-bots of, 220
Elephantoloemus indicus, 218
Elephantomyia, 30
elongation of proboscis in plant-feeding flies, 30
emergence flights, numbers involved in, 266
Emperoptera mirabilis, 279
Empididae, 8, 120, 138, 189
   distribution of, 139
   eyes of, 138
   habits of, 139
   larvae of, 19, 110, 141, 144
   South American, 278
   wingless, 95, fig. 17
Empis aerobatica, 140
   tessellata, 139
English, Miss Kathleen, 106, 129
environment, rapidly changed by Man, 242
Eocene, flies of, 274
Ephydra hians, 186
   riparia, 188

Ephydridae, 11, 38, 185, 195, 201, 258
   carrion-feeding, 141
   as family in active evolution, 189
   larvae of, 16, 21, fig. 11
   reduced wings in, 95
Eristalis, 161, 162
   proboscis of, 156, fig. 27
   tenax, hibernating as adult, 268
Erioischia brassicae, 207, 259
esters attractive to fruit-flies, 197
Eucampsipoda hyrtlei, 238
Eulalia, 115
Eumerus, 161, 162, 258, pl. 32
Eurynogaster, larval habitat of, 145
Eusimulium, 68
evolution of flies, 5, 6
   of larval feeding-habits, 201
   of Diopsidae, 190
   in Ephydridae, 189
Exoprosopa, 134
extensible breathing-tube, 16
'eye-flies', 188, 191
'eye-gnats', 255
eye, human, *Loa loa* in, 247
eyes of flies, 10
   ability to detect movement, 121
   of Empididae, 138, 141
   enlarged in male flies, 59, fig. 14, 166
   extreme reduction of, 151, 237
   meeting both above and below antennae, 141
   meeting in both sexes, 71
   nearly spherical in Pipunculidae, 165
   relation to swarming habits, 267
   in robber-flies, 120
   separated in male Conopidae and acalyptrates, 169
   separated in male *Microdon*, 160
   stalked, 189
   of Tabanidae, 97
eyestalks of Diopsidae, unrolling of, 189

'Face-fly' (*Musca autumnalis*), 205, 255
facets of male eyes, 45, 97
fallen logs, larvae in, 34
false legs, 16, 38, 62, 145, 161, 186
Fannia, larva of, 149, 206, fig. 33
   canicularis, 206, 254, 263
   scalaris, 206
farms, flies around, 51
Fasciola gigantea, 180
fat-flies, 137
fecundity of house-flies, 203, 204
feeding by pupa of *Simulium*, 23
female phase of adult Termitoxeniinae, 153
Ferdinandea, 161
fermentation, larvae feeding on products of, 171

fermentation—*cont.*
  and *Drosophila*, 199
  and Trypetidae, 197
fermenting sap, 39
fever, break-bone, 245
  malarial, 243
  sand-fly, 245, 246
  yellow, 244
'fighting' of males of *Calobata*, 177
filariasis, 245
  and *Culicoides*, 247
  and mosquitoes, 245
  control of, 280
filter-feeding by *Chironomus* larvae, 62
  by mosquito larvae, 93
fire, flies attracted by, 166
first-stage larva of Bombyliidae, 133
'fishing' for mosquito larvae by Lispinae, 207
'flat-flies,' 230
fleas, comparison with Psychodidae, 54
flesh-feeding by larvae of *Sarcophaga*, 227
flesh-flies, 213
flies transmitting disease, Table 1, 260
flies entering houses, 268
flight of *Coelopa frigida*, 174
  of Cyrtidae, 137
  of *Fannia*, 206
  of Mydaidae, 128
  powers of, 14
'flies proper', 7
flotation device, 206
flower-feeding by bee-flies, 132, 134
— by crane-flies, 30
— by hover-flies, 101
— by mosquitoes, 78
— by male horse-flies, 100
— by Mydaidae, 128
— by *Neottiophilum praeustum*, 182
— by Nemestrinidae, 134
— by Stratiomyidae, 113
— by Vermileoninae, 110
— by larvae, 196
  general in flies, 8, 155, 180
flowers, deep, feeding from, 157
  fertilisation of, 259
floats of eggs of mosquitoes, 85
fly-belts of tsetse, 210
'fly-blown', 213
flying foxes, Nycteribiidae on, 238
fodder-plants, 258
Fonseca, E. A. d'Assis, 166, 226
food, flies as, 75
  flies destroying human, 257
  requirements of adult flies, 14
footprints as container habitat, 89
Forcipomyia, 55, 57
formalin, larvae resistant to, 151
Formica fuliginosa, 116

fossil, flies, 274
freshwater larvae of Dolichopodidae, 145
frit-fly, 11, 191, 259
froghoppers parasitized by Pipunculidae, 165
frogs attacked by Ceratopogonidae, 55
— by *Lucilia*, 225
— by Rhagionidae, 109
frost, effect of on mosquito larvae, 87
fruit-flies, 199
  egg-laying of, 16
  large, 11, 196
  small, 11, 199
  small, evolutionary vitality of, 5
fruits, larvae attacking, 196
  Trypetidae in, 197
Fucellia, a seashore fly, 205
Fulgoridae parasitized by Pipunculidae, 165
Fuller, Miss Mary, 135
Fungivoridae, see Mycetophilidae
fungus-feeding larvae, 19, 43, 44, 47, 166
fungus-gnats, 7
further reading about flies, 283
future of flies, 279

'Gadding' of cattle, 222
gad-flies, 222
gall-making larvae, 17, 19, 46
  arisen many times, 36
  specialization of, 193
  of acalyptrates, 171
  of Chamaemyidae, 182
gall-midges, 7, 46, 258
  emergence of adult, 24
galls, biological significance of, 193
game destruction and tsetse control, 209
Garcia, R. and Radovsky, F. J., 202
Garnham, P. C. C., 247
Gasterophilidae, 219, 224
Gasterophilus, 251, 253
  intestinalis, 223
  nasalis, 224
gastro-enteritis spread by flies, 255
geological age of Diptera, 274
Geranomyia, 30, 36
Geomyza, 193
gift to female during mating, 140
gills, respiratory, 162
  anal, of mosquito larvae, 88
  tracheal, of Chironomid larvae, 62
Gitona, carnivorous Drosophilid, 200
glacier, Chironomid larvae in, 65
glands, labial, 43
Glossina, 15, 248, 249
  habits and distribution of, 209
  hybrids of, 211
  fusca-group, 209
  morsitans-group, 209

Glossina morsitans, 211
  palpalis-group, 209
  palpalis, 250
  swynnertoni, 211
Glyceria aquatica, 162
glycogen in diet of mosquitoes, 81
gnat, house (*Culex pipiens*), 80, 87
gnats, 6
goats attacked by *Oestrus ovis*, 220
  and *Melophagus ovinus*, 233
Gressitt, J. L., 279
Golubatz fly, 66, 246
Goniops chrysocoma broods eggs, 104
Gorgonopsis, stalked eyes of, 190
gourds, larvae in, 197, 258
gout fly, 11, 191, 192, 259
grasses, Chloropid larvae in, 191
  mines in, 188
grassland, larvae in, 18
grapes, yeasts of, 259
Graptomyza, 159
Greathead, D. J., 136, 226, 267, 276
greenbottle, 213, 215
green pigment in Hippoboscidae, 164
  in Syrphid larvae, 164
  in Tabanidae, 103
greenhouse pests, 61
gregarious larvae of *Teichomyza fusca*, 186
ground pools, mosquitoes in, 89
ground temperature and hibernating flies,
  268
growing plants, damaged by Bibionidae, 45
— by Cecidomyiidae, 47
Guam Island, 185
guano and *Fannia*, 254
gynandromorphs, 84
Gyrostigma, 224, 251

Habitations of animals and flies, 154, 242
habits of Empididae and Asilidae, 140
Haddow, A. J., 80, 83, 100, 244, 263, 264,
  267
Haematobia, 208, 222, 248
Haematopota, 96
  recent evolution of, 98
  world distribution of, 99
  larva of, 103
haemoglobin in *Chironomus*, 62
  in *Gasterophilus*, 223
Haemoproteus and *Culicoides*, 247
Haines, T. W., 254
hairs of antenna, erection, 265
Halidayella, 172
Halophila ovalis, 61
halteres, 3, 157, 170
  loss of, 95, 151, fig. 17
— by *Melophagus*, 233
hams, *Piophila* in, 176, 257

Hardy, D. Elmo, vii, 36, 54, 279
Hardy, G. H., 126
Harpagomyia (= *Malaya*), 84
Hartley, J. C., 161
Haskell, P. T., 266
hatching stimulus of eggs, 87
  of *Gasterophilus*, 223
Hawaiian Islands, 36
  flies of, 279
  radiation in, 280
heads of flies, 6, 9, 10
Hecamede persimilis, 185
Helcomyza ustulata, 175
Heleidae, see Ceratopogonidae
Helicella virgata, attacked by *Melinda
  cognata*, 225
Helix, larvae parasitizing, 186
Helomyzidae, 172
Hemipyrellia, 217
Hemerodromia, 140, 145
hemipupa, 49
Hennig, W., 16, 168, 283
Henria psalliotae, 49
Hepatocystis and *Culicoides*, 247
Hercostomus, larval habitat of, 145
Hering, M., 194
Hermetia illucens, 116
Herms, W. B., 242
Hershkovitzia, 238
Herting, B., 228
Hesse, A. J., 133
Hessian fly, 47
Hexatomini, 37
hibernation in egg stage, 87
  as a larva, 87
  as adult flies, 87, 268, 270
Hidalgo County, Texas, polio in, 257
Highland Britain, 231
'higher flies', 11
  moults of, 18
Hilara, aerobatics of, 139
Hill, D. S., 230
hind legs of robber-flies, 122
  wings, loss of, 275
Hinton, H. E., 15, 23, 65, 211, 214
Hippelates ('eye-fly'), 191, 255
Hippobosca, 232
Hippoboscidae, 11, 153, 181, 229, 249
  compared with Phoridae, 149
  green pigment in, 164
  originally on birds?, 232, 233
hippopotamus, nosebots in, 220
  and Tabanidae, 99
Hirmoneura, eggs of, 136
  larvae of, 135
hive-bees preyed on by robber-flies, 126
hives, *Braula coeca* in, 182
Hobby, B. M., 124

Hobson, R. P., 214
Holy Island, *Lucilia* in, 281
Homeophthalmae, 132
Homoptera, parasitized by Pipunculidae, 165
honey-dew, and flies, 157, 191
  Bombyliidae, 132
  Chamaemyiidae, 182
  Diopsidae, 190
  Drosophila, 199
  mosquitoes, 84
  Tabanidae, 101
hoof-prints as a 'ground pool', 89
hoofed mammals, evolution of Tabanidae along with, 98
horn-fly, 208, 248
hornets mimicked by Syrphidae, 159
'horns' of *Phytalmia*, 198
  of pupae, 23
horse-flies, 8, 19, 96
  and game animals, 280
  habits of male, 100
  larvae of, 102
  mating habits of, 100
  not really aquatic, 102
  numbers of, 102
  size of, 97
  without mouthparts, 102
horses, *Hippobosca* on, 232
  nose-bots in, 220
  stomach-bots in, 223
host-selection by Hippoboscidae, 231
— by Nycteribiidae, 238
— by Streblidae, 235
hot springs, larvae in, 117
house-fly (*Musca domestica*), 171, 201, 202
  adaptability of, 203
  advanced type of fly, 5
  alleged decline of, 254
  breeding requirements of, 254
  distribution of, 202
  eaten by Syrphid larvae, 164
  evolution of, 11
  fecundity of, 204
  future of, 280
  larval food of, 203
  relation to epidemics, 255
  overwintering of, 203
  range of flight of, 204
  resistant strains of, 256, 280
house-fly, biting (*Stomoxys calcitrans*), 207
house-fly, lesser (*Fannia canicularis*), 254, 286
house-gnat (*Culex pipiens*), 80, 87
houses entered by *Crataerina*, 231
— by blowfly larvae, 216, 273, fig. 39
— by Helomyzidae, 172
— by hibernating flies, 268, 270, fig. 38
— smoke attracting flies to, 267

hover-flies, 9, 155
  fertilizing flowers, 259, pl. 9
hovering of flies, 155, 157
  of *Nemestrinus*, 134
  of Pipunculidae, 165
human bot fly, 218
humidity, effect on biting flies, 82
  on eggs of *Calliphora*, 214
  on tsetse-flies, 210
humming of Syrphidae, 156
humus, larvae in, 146
hunting methods of Asilidae, 122, 123
Hybos, 140
hybridization of tsetse-flies, 211
Hydrellia, 188
  larva of, 21, fig. 11
  nasturtii, 258
Hydrophorus, 141
  larval habitat of, 145
hydrostatic organs of *Chaoborus*, larva of, 50,
Hydrotaea, 205, 255   [fig. 13
hygiene, flies and, 256
Hygroceleuthus, larval habitat, 145
Hyperechia, 125
hypermetamorphosis, definition of, 133
  in Bombyliidae, 133
  in Nemestrinidae, 136
hyperparasites, 133
Hypoderma, 221, 224, 251
  larval migrations of, 221
  bovis, 222
  lineatum, 222
Hypodermatinae, 219, 221
Hypocharasus, 146
hypopharynx, 119

Icerya purchasi, 200
ichneumon-flies and robber-flies, 124
immunity to polio, derived from flies ?, 257
infants, larvae swallowed by, 253
infrabuccal pockets of ants, 160
injury from larvae, 252
inland occurrence of seaweed flies, 174
instars, number of, 18
intestine, damaged by maggots, 253
  myiasis of, 251, 252
  Phorids in, 151
  Piophila in, 176
Imms, A. D., 183
Iron Gates of Danube, 246
irradiation, sterilization by, 281
  of *Callitroga*, 281
  of *Lucilia*, 281
  of tsetse-flies, 211
irreversibility of evolution, 33
isotherm of 10°C, 235
Itoniidae, see Cecidomyiidae
Ixodes ricinus, 233

'Jackals of the insect world', 180
Jackson, Miss Dorothy, 268
Jacobs, J.-Ch., 251
James, M. T., 225, 251
Jardinea congoiensis, 190
Jassidae, parasitized by Pipunculidae, 165
Jobling, B., 142, 234, 235
Johnston, Sir Harry, 249
jungle yellow fever, 80

Kal-azar, 246
kangaroo, Hippoboscidae on, 233
    throat bot of, 221
Kato, K., 44
keds, 181, 230, 233
Keilin, D., 17, 18, 20, 21, 33, 53, 150, 205,
    222, 225, 226
Kellogg, V. L., 73, 74
kelp-flies, 173
Kessel, E. L., 140, 266
kleptoparasite, 224
Kim, C.-W., 12
kitchens, flies breeding in, 39, 199
Knab, F., 30
knapweed gall-fly, 196
'kungu-cake', 75, 272, pl. 30

Labella used as mandibles, 142, 144
labial glands and silk, 43
labium, 10, 202
    of Dolichopodidae, 142
    of Syrphidae, 101
laboratory rearing of *Coelopa*, 174
    of *Drosophila*, 174
Lake Victoria, sleeping sickness at, 250
lamellicorn beetles, parasitized by *Neme-
    strinus*, 135
Lampromyia, larva of, 111
land-midges, 7, 19, 35, 41
lantern-flies parasitized by Pipunculidae, 165
Laphria resembling bees, 126
Laphriinae, 135
    habits of, 126
    larvae of, 127
    pupae of, 128
largest flies, 4
larvae of flies, 12
    adaptation to environment, 17
    attacking Amphibia, 17, 55, 109, 225
    ancestral habits of, 18
    aquatic, 17, 35, 50, 68, 85, 103, 112, 161,
        186, 208
    aphids eaten by, 163, 164, 182
    attacking birds, 17, 181, 217
    attacking Man, 17, 37, 217
    breeding media of, 17

carnivorous, 19, 22, 50, 144, 162, 171, 175,
    179
in compost-like substances, 20, 42, 145
on Crustacea, 17, 70
distinct life from adult, 13
in dung, 17
evolutionary significance of, 24
tapping aquatic plants, 85, 92, 188
evolution of, 12
fed internally by mother, 17, 209, 228
at holes made by another insect, 190
feeding period of, 18
food-reserves of, 23
in fungus, 19
identification of, 16
in intertidal zone, 37, 145, 205
leaf-feeding, 33
leaf-mining, 34, 195
in leaf-mould, 19
losses among, 51
in Man, 242
parasitic, 17
in plant-galls, 19
as principal feeding-stage, 14
attacking reptiles, 17
as scavengers, 17
search for richer food, 6
in snails, 17, 150
terrestrial, 17
living on vegetation, 18
in worms, 17
of acalyptrates, 22, 171
of *Anthomyia*, 205
of *Apiocera*, 129
of Asilidae, 127, 128
of *Atherix*, 112
of Bibionidae, 45
of Blepharoceridae, 73, 77, pl. 4
of Bombyliidae, 132
of Brachycera, 16, 19
of *Braula coeca*, 182
of *Calliphora*, 214
of calyptrates, 22
of Cecidomyiidae, 47
of Ceratopogonidae, 19
of *Chaoborus*, 75
of Chironomidae, 19
of Coelopidae, 173
of Conopidae, 168
of crane-flies, 19, 33
of Cyclorrhapha, 16
of *Dermatobia*, 219
of Dolichopodidae, 144, 145
of *Drosophila*, 199
of Empididae, 141, 144
of Ephydridae, 16
of *Fannia*, 206
of *Gasterophilus*, 224

Larvae of flies—*cont.*
  of *Geranomyia*, 36
  of *Lonchoptera*, 154
  of mosquitoes, 93
  of *Musca domestica*, 202
  of Muscidae, 205
  of Mycetophilidae and Sciaridae, 43
  of Mydaidae, 129
  of Nematocera, 16
  of Pantophthalmidae, 108
  of Phoridae, 149
  of *Piophila*, 176
  of Rhagionidae, 110
  of *Sarcophaga*, 227
  of *Simulium*, 70, 71
  of Syrphidae, 16, 20, 157, 159
  of Tabanidae, 102, 103
  of Thaumaleidae, 72
  of Therevidae, and Scenopinidae, 129
  of Tipulidae, 18
  of *Termitoxenia*, 152
Larvaevoridae, see Tachinidae
larvo-viviparous habits, 227
Lasiohelea, 55
Lasius flavus and Phoridae, 152
latrine-flies, 206
Laurence, B. R., 140, 141, 144
Lauxaniidae, 172, 195
lawns attacked by larvae, 34, 258
leaf-axils of plants, 90
leaf-feeding larvae of crane-flies, 33
leaf-galls, 195
leaf-hoppers, parasitized by Pipunculidae, 165
leaf-miners, 258
  mines, structure of, 194, fig. 32
  mining, families of flies concerned in, 195
  — evolved several times, 36
  mould, larvae in, 18, 19
leaping (skipping) of larvae, 138
leather-jackets, 34, 258
Leguminosae, 258
leishmaniasis, 246
lemurs, Hippoboscidae on, 234
Lepidoptera, flies attacking, 55
Lepiselaga, larva of, 103
Leptidae (=Rhagionidae), 108
Leptoconops, 55, 57
Leptogaster, 111, 122, 123
Leptohylemyia coarctata, 207
lesser house-fly, 206, 254
lettuce, leaf-miner in, 197
Leucocytozoon carried by *Simulium*, 246
Leucostola, larval habitat of, 145
Leuresthes, eggs eaten by *Fucellia*, 205
Lewis, D. J., 66, 67, 92, 98, 271
Lias deposits, flies in, 274
lilies attacked by *Merodon*, 161

light, behaviour of bluebottle to,
  Coelopa attracted to, 268
  intensity and biting flies, 82
  — and mosquitoes, 83
  — and swarms, 75
  production by larvae, 44
  — by pupae, 44
  — by adults, 44
light-traps, 43
life-history of flies, 12
  four stages of, 15
Limoniinae, 29, 30
  carnivorous larvae of, 37
Limnophora, 205
  humilis hibernating indoors, 268
Liogma, larval habits of, 33
Lipara lucens, gall-making by, 193
lipids, in diet of mosquitoes, 81
Lipoptena cervi, 233
liquid food, sponging up of, 170, fig. 29
Lindner, E., viii, 283
Lindquist, A. W. and Knipling, E. F., 256
Lindsay, D. R. and Scudder, H. I., 256
liver-fluke, 180
liverworts, larvae in, 47
Livingstone, David, 75
Liriopeidae, see Ptychopteridae
Lispinae, 207
Lloyd, Ll., 52
Loa loa carried by *Chrysops*, 247
local species, 277
Locust, Australian Plague, 135
locusts, and Bombyliidae, 133
  and *Stomorina lunata*, 226, 267
Locusta pardalina, 136
loiasis, and horse-flies, 97
Lonchaeidae, 198, 263
Lonchopteridae, 153, 279
  larva of, 154
long-headed flies, 138, 141
loss of mandibles in female horse-flies, 102, 104
louse-flies, 230
low temperatures, 203
Lowland Britain, 231
Lucilia, 215, 254
  larva of, 21
  — food-requirements, 215
  — emerging from soil, 216, 269, fig. 38
  — invading a house, 272, fig. 39
  bufonivora, 225
  cuprina, 215
  porphyrina, 225
  sericata, 213, 215, 273, 281
luminous larvae of *Ceroplatus*, 44
Lund, larva of, 218
Lundbeck, W., 228
lung, auxiliary, 15

Lumsden, W. R., 244
Lutzia, 94
Lycastris austeni, 157, 159
lymphatic glands and filariasis, 245
Lyperosia, 208, 222, 248
lysine attractive to mosquitoes, 270

Macdonald, W. W., 245
Machimus, egg-laying of, 127
Mackerras, I. M., 5, 101, 106, 134, 135
Macleod, J. and Donnelly, J., 281
Machaerium, larva of, 145
McMahon, J. P., 70
Madiza glabra, 180
maggot, 12, 16
    as a type of larvae, 20
    evolution of, 21
    extensible cuticle of, 18
    range of food materials, 252
    'rat-tailed', 20
maggot-like larva of *Symmictus*, 136
maize, and *Mesogramma*, 165
malaria as human problem, 242
    and fighting troops, 243
    epidemiology of, 243
    outlook concerning, 244
    in birds, 244
    — and *Culicoides*, 247
    in monkeys, 244
Malaya (*Harpagomyia*), 84
'male phase' of *Termitoxenia*, 153
males of biting muscids, 248
    of mosquitoes, 84
    of Dolichopodidae, 143
    of Stratiomyidae, 114
Mallophora, a bee-like Asilid, 126
Malpighian tubules, 88
mammals, flies feeding on, 11, 229
    Hippoboscidae on, 232
    larvae attacking, 17
Man attacked by flies, 241, 242, 150
    major agent of change, 242, 276
    and Congo floor maggot, 217, 218
    and *Crataerina pallida*, 231
    and dipterous larvae, 17
    and larvae of blowflies, 251
    and Phoridae, 151
    and sheep nostril fly, 220
mandibles, in horse-flies, 102
    in Stratiomyid larvae, 116
    in woodboring larvae, 128
mangold fly, 207
Mansonia mosquitoes, 85
    eggs of, 85
    larval respiration of, 92
    and *Brugia malayi*, 245
manure, larvae in, 38
Marchand, W., 30

March flies, 45, pl. 9
marine Dolichopodidae, 145
    flies, 36
marker for swarm, 263
    overhead used by *Fannia*, 206
Marleyimyia, 133
marshes, larvae in, 162
    malarious areas, 88
    salt, Ephydrid larvae in, 186
Maruina, 52
mass emergence of horse-flies, 100, 101
massing of maggots in meat, 215
Mastotermes darwiniensis, 197
maternal incubation of larvae, 228, 229
mating behaviour of Chaoborinae, 75
— of Chironomidae, 59
    of crane-flies, 31
    of Dolichopodidae, 143
    of Empididae, 139, 140
    of mosquitoes, 83
    of *Piophila*, 176
    of Sepsidae, 175
    of Syrphidae, 155
    of Tabanidae, 100
    of Trypetidae, 198
    of tsetse-flies, 211
mating-lure offered by Empididae, 139
    by Trypetidae, 198
Mattingly, P. F., 266
Mayetiola, 49
mayflies and Chironomidae, 63
mechanical transmission of disease, 97
    of trypanosomiasis, 210, 250
Medeterus on tree-trunks, 144
medical discoveries involving flies, 260
Mediterranean fruit-fly, 196, 253, pl. 20
    chemical sterilization of, 281
    larval habitat of, 146
Megalybus crassus, 137, fig. 23
Megarhinus, 91
    food of larvae, 93
    larval habitat of, 90
    larvae in Bromeliads, 91
Megaselia, larva of, 149
    halterata in mushrooms, 150
    meconicera, 148
    rufipes, 151
    scalaris, 151
Megistostylus, 9, fig. 3
Melander, A. L., 139
Melanochelia, no larval moults, 18
Melanostoma mellinum, 164
Melin, D., 121
Melinda cognata attacking snails, 225
Meliponinae mimicked by *Graptomyza*, 159
Meloidae attacked by *Atrichopogon*, 56
melon fly, chemical sterilization of, 281
melons, larvae in, 158, 197

Melophagus, 233
Melusinidae, see Simuliidae
Meoneura, 180
Merodon, 101, 258, pl. 32
Mesembrinellinae, systematic position of, 228
Mesogramma, larvae of, 159, 164
mesothorax, 3
Mesozoic Period, flies of, 274
Metopina stealing food from ants, 152
Miastor, paedogentic larva of, 48
Microdon, larva of, 159, 163
Micropezidae, 177
Microphorus steals food from spiders' webs, 141
Microsania, 166
Microtipula, smallest crane-fly, 34
midge, weight of individual, 59
midges, 6
  biting, 55, 247
  land, 7, 19, 41
  non-biting, 14, 58, 264
  water, 7, 15, 19, 50
migrations of flies, 267
  of fully-fed larvae, 215
  of people to avoid horse-flies, 97
Milesiini, 161
Milichiidae, 180
milk bottles, *Drosophila* in, 199
  Phoridae in, 150, pl. 30
millet and Diopsid larvae, 190
Miltogramminae, 227
mimicry, theory of, 125
  in Asilidae, 125
  in Conopidae, 168
  in Syrphidae, 158, 159
Mochlonyx, 75
Möhn, E., 49
molluscs, flies and, 17
monkeys and yellow fever, 244
  mosquitoes feeding on, 80
Mono Lake, California, 186
Mormotomyia, 172, 184, 236
mosquitoes, 7, 77
  abundant in Arctic, 279
  antennae of, 79, 265, fig. 16, pl. 11
  aquatic habits of, 77
  biting habits of, 77
  definition of, 77
  eggs of, 16, pl. 10
  feeding from flowers, 78
  larvae can exist away from surface, 92
  larvae, carnivorous habits of, 94
  larvae preyed on by *Phaonia* larvae, 208
  and Man, 242
  numbers of, 78
  number of larval moults, 18
  pupae of, 94

related to other Nematocera, 74
  resting habits of, 83
  transporting eggs of *Dermatobia*, 218
moss, larvae in, 18, 47, 57
  ancestral flies probably in, 33
moth-flies, 51
moths and Syrphid larvae, 164
moulds, 'larvae feeding on, 43, 47
  Drosophila and, 199
moults, number of, 18
Mt Mlanje, 124
'mountain midges', 72
moustache of robber-flies, 121
mouthbrushes, of mosquito larvae, 93
  of *Simulium*, 69
mouth-hooks, 16, 20, 22
mouthparts of Asilidae, 119
  of Bombyliidae, 132
  chewing, 45
  of Cyrtidae, 137
  as guide to evolution, 275
  loss of in Oestridae, 219
  of Syrphidae, 157
movement attractive to robber-flies, 121
mucus, flies attracted to, 84
mud-flats, larvae in, 145
mud, larvae on surface of, 52
Muirhead-Thompson, R. C., 69, 267
mules, *Hippobosca* on, 232
  Gasterophilus in, 224
mummified carrion, Phoridae in, 148
Munro, H. K., 196
Musca autumnalis, 202, 204, 226
  first appearance in US, 204
  in houses, 204, 255, 268
  larval food of, 205
  temperature tolerance of, 204
Musca domestica, 170, 202
  chemical sterilization of, 281
  hibernating in houses, 268
  see also house-fly
Muscaria, 170
Muscidae, 195, 205
Muscina, 205
Muscoidea, 201
mushrooms, larvae in, 42
  Cecidomyiidae in, 49
  Phoridae in, 150
'mushroom flies', 150
Musidoridae see Lonchopteridae
mutualism between larva and alga, 64
Mutillimyia, 159
Mycetophilidae, 41, 42
Mycetobia, relationships of, 39
Mycetophiloidea, 41
Mydaidae, acanthophorites in, 127
  antiquity of, 129
  distribution of, 129

Mydaidae—*cont.*
  habits of, 128
  larvae of, 129
Myers, K., 266
myiasis, 116, 251
  by Phorid larvae, 151
  by *Teichomyza fusca*, 186
Myiophthiria, 231
Myiospila, 205
Myolepta, larva of, 161
Myriapoda, 17

Nagana, 249
Nagatomi, A., 113 ,
nappies, *Fannia* breeding in, 207
narcissus fly, 161
Neave, S. A., 101, 103, 124
nectar, 157
nectar-feeding by Bombyliidae, 132
— by Tabanidae, 101
Neff, S. E., 179
Nematocera, 6
  antenna of, 59, fig. 14
  evolution of, 275
  larvae of, 16
  moults of, 18
Nemestrinidae, 133, 134, 157
Nemestrinus, wing of, 135, fig. 22
Nemopalpus, 54
Nemotelus, snout of, 113
Neocuterebra, 224
Neodixa, 75
Neoitamus, 127
Neorhynchocephala, 136
Neottiophilum praeustum, 181, 182, 217, 251
Nepenthes, larvae in, 19, 57, 91
nestlings killed by *Protocalliphora*, 217
nests of ants, 116
  Forcipomyia in, 57
  larvae in, 43
  Microdon, in, 160
  Miltogramminae in, 227
  Phoridae in, 149, 152
  of bees, Miltogramminae in, 227
  — Bombyliidae in, 132
  of birds, larvae in, 42, 43, 129, 181
  — *Anthomyia* in, 205
  — *Neottiophilum* in, 182, 217, 251
  — *Protocalliphora* in, 217
  — *Passeromyia* in, 205
  of termites, larvae in, 43
  — Miltogramminae in, 227
  — Phoridae and Termitoxeniinae in, 152
  of wasps, larvae in, 43
  — Syrphidae in, 158
net, Pipunculidae hovering in, 165
  spinning of by Chironomid larvae, 63
Neurigona, larval habitat of, 146

Neuroctena amilis, 172
Neuroptera, Ceratopogonidae and, 55
newts attacked by *Lucilia*, 225
New Zealand glow-worm, 43
'nibbling' by mosquito larvae, 93
Nielsen, E. T., 263
night activity, 67, 79
Nile cabbage (*Pistia stratiotes*), 103, 164
Nile, exploration of, 249
nimitti (*Simulium*), 67
nitrogen in diet, 81, 206
nocturnal mosquitoes, 83
noise made by feeding larvae, 47
non-biting flies and Man, 250
non-biting midges, 58
Northern Rhodesia, sleeping sickness in, 250
nose-bots, 219, 220
Nostoc parmeloides and *Cricotopus*, 64
Notiphila, 187, 188
numbers of flies, 5, 185, 203, 271
Nyasa, Lake, 75
Nycteribiidae, 3, 11, 153, 229, 236, fig. 37
  of Old and New Worlds, 238
  origin of, 236
  rate of reproduction, 238
Nycterophila coxata, like a winged flea, 234

Oats, 259
Ochthera mantis, 189
Ocydromia, 140
Odontomyia, 115
odour emitted by female Sepsidae, 175
  left behind by flies, 270
Odysseus, 98
Oedemagena tarandi, 222
Oestranthrax pagdeni, 133
Oestridae, 213, 219, 221, 251
  adults of, 220
Oestrus ovis, 220, 224, 251
offals, preference of *Lucilia* for, 215
oil, against mosquito larvae, 92
olfactory pits, 13
Olfersia fumipennis on osprey, 232
  spinifera on frigate bird, 231
Oligochaetes, 17
Omphralidae, see Scenopinidae
Onchocerca gibbonsi, 247
  volvulus, 246
onchocerciasis, 70, 246
Oncodidae, 136
onion fly, 207, pl. 31
open sea, flies on, 60
operculum of snails, 180
Opomyza, 193
orange-juice, larvae in, 253
Ormerod, W. E., 250
Ornithomyia biloba, 231
  ecology of British, 230

Oropeza, larva and pupa, 34
Orphnephilidae, 71
Ortalidae, 190, 198
Oscinella frit, 191
Oscinosoma, parasitic larva of, 193
osmotic pressure, 35, 92
osprey, *Olfersia fumipennis* on, 232
ostrich, *Hippobosca struthionis* on, 232
Otitidae, 178, 198, fig. 31
oviposition in treeholes, 90
  of *Stomorina lunata*, 226
Ovipositor, 16
  of Asilidae, 19, 121, fig. 9
  of Cyrtidae, 137
  of plant-feeding flies, 22
  of Trypetidae, 196, 197
  of *Xylodiplosis*, 47
overwintering of *Musca*, 203, 170
  of *Scatophaga stercoraria*, 212
  of *Thaumatomyia notata*, 191, 271
owl-midges, 51
oxygen essential to *Simulium* larvae, 68
  and respiration of *Chironomus*, 63
Ozaenina diversa, 178, fig. 31

Paedogenesis, larval, 48
  pupal, 49
pain of bites by flies, 242, 251
Pallopteridae, 198
palpi of Phoridae, 147, 148, fig. 26
Panama Canal, yellow fever and, 244
Pangoniinae, 101
Panicum repens attacked by Diopsid larvae, 190
Panorpoid Complex, 36, 274
Pantophthalmidae, 5, 108, 279
Paragus, 163
paralytic polio, and house-flies, 257
Paramonov, S. J., 130, 174
Parascaptomyza, 220
parasitic flies, 14
  larvae, 17
parasitic larvae of Bombyliidae, 132
  of Chironomidae, 63
  of Chloropidae, 190
  of Conopidae, 168
  of Hippoboscidae, 230
  of Phoridae, 152
  of Pipunculidae, 165
  of Tachinidae, 228
parasitic life, extreme adaptation to, 233
parasitoids, 169
parasol ant, 129
Paraspiniphora in milk bottles, 150, pl. 30
parenchyma of leaf, 194
parsnips, 258
parsley, 258
parthenogenesis, 152

in Chironomidae, 61
  in *Lonchoptera*, 154
past, present and future of flies, 274
Paterson, H. E., 198
Patton, W. S., 220, 221, 225
peaches, damaged by larvae, 258
peaks of biting activity, 81
pear-midge, 258
Pediciini, 37
Peffly, R. L., 255
Pegomyia hyoscyami, 207
pelagic flies, 60
Pelecorrhynchus, 108
Pemphigus bursarius, 191
Penthetria holosericea, 49
Petauristidae, see Trichoceridae
percentage of flies among fossil insects, 276
Pericoma, 52
Perissommatidae, 49
Perkins, R. C. L., 279
permanent water, mosquitoes in, 85, 86
Permian Period, flies of, 274
person, flies following a, 264
petroleum fly, 186, 187
Phaenicia (= *Lucilia*) *sericata*, 213
Phallus impudicus, 172
'phantom larva', 50, fig. 13, 75
Phaonia keilini, 208, fig. 34
pharate adult, 49
  of Asilidae, 128
  of *Cricotopus*, 64
  of *Simulium*, 71
Pharygobolus africanus, 220
Pheidole ants, 279
Philip, C. B., 97
Philognathus silenus, 128
Philoliche longirostris, feeding of, 101
  magrettii, 249
Philonicus, egg-laying of, 126
Phlebotominae, 53, 55, 245
Phalacrocera, larval habits of, 33
Pholcid spiders and crane-flies, 32
phoresy, 63, 70
Phoridae, 9, 154, 168
  alleged bloodsucking of, 153
  compared with Hippoboscidae, 149
  evolutionary origin of, 149
  head of, 148, fig. 26
  in human intestine, 253
  larvae, 100, 149, 151
  legs compared with Dolichopodidae, 150
  reproducing underground, 147
  size of, 148
  swarming into houses, 268
Phormia terrae-novae, hibernating, 268
Phragmites communis, 182
  larvae of *Thrypticus* in, 145
Phryneidae, see Anisopodidae

Phyllophylla heraceli, 196, 197
Physocephala, 168
Physogastric condition, 151
  in *Ascodipteron*, 236
  in *Carnus hemapterus*, 181
Phytalmia, 198
Phthiria, 134
piercing mouthparts, 11, 119, 201
pigeon fly, 231
pigs, attacked by *Auchmeromyia*, 218
  nose-bots in, 220
Piophila casei, 176, 253
Pipiza, 163
Pipunculidae, 155, 165
Pistia stratiotes, 103, 164
Pitcher, R. S., 48
pitcher-plant, larvae in, 91
pits, olfactory, 13
pits of Vermileonine larvae, 111
plant-axils, mosquitoes in, 86
  cavities, larvae in, 19
plants, decaying, larvae in, 258
plant-bugs, Drosophilidae feeding on, 200
plant-feeding by *Mesogramma* larvae, 165
— by Muscid larvae, 207
— by Scatophagid larvae, 212
— by various acalyptrate larvae, 171
plant-galls, larvae in, 19
plant-reservoirs, 17
  Ceratopogonidae in, 57
Plastophora parasitic in ants, 152
plastron, respiratory, 15, 35, 115, 211
  in eggs of *Calliphora*, 214
  over posterior spiracles of larvae, 37
Platyna hastata, 114, 264
Platychirus, larva of, 164
Platypezidae, 166, 168, 169, 266
Platystomidae, 190
Pocota, 162
poliomyelitis (polio; infantile paralysis), 248
  house-flies and, 256, 257
pollen, licking of by *Desmometopa*, 180
  of maize eaten by *Mesogramma*, 164
Pollenia rudis (cluster fly), 226, 268
pollination by flies, 45, 56, 156
Polypedilum vanderplanki, 65
Polietes, 205
pomace flies, 199
ponds, larvae in, 19
Pontomyia natans, 60
pools, ground, 86, 89
populations of flies, 271
Porphyrops, larval habitat of, 145
posterior spiracles of larva of *Atherix*, 112
  of Coelopidae, 175
  of Ephydridae, 186
  of flesh-feeding flies, 218
  of *Gasterophilus*, 224

  of *Hydrellia*, 188
  of *Hypoderma*, 222
Potamida ephippium, 116
Potamogeton, mines in, 187, 188
Potamon niloticus, and *Simulium neavei*, 70
potatoes, larvae in rotting, 172
pre-adaptation, 22
predatory flies, 8, 14, 119
  legs and proboscis of, 95
  — in Dolichopodidae, 141
  — in Ephydridae, 189
  prestomal teeth, 202, 207, 212
prey, birds of, and Hippoboscidae, 232
prey of robber-flies, 124
  of *Scatophaga*, 212
primitive genera of horse-flies, 106
Proagonistes athletes, 122, fig. 21, 124
proboscis, 10, 11
  of Asilidae, 121, 122, fig. 21
  of Cyrtidae, 137
  of Muscidae, 248
  of Mydaidae, 128
  of Pupipara, 230
  of *Scatophaga*, 212
  of Syrphidae, 156, 157, 158
  of Tabanidae, 101
Procapra gutturosa, 233
  picticaudata, 223, pl. 26
prolegs, 38, 62, 145, 161, 186
Prosimulium, 68
protective clothing against biting flies, 278
protein requirements of larvae, 20, 175, 213
  of adult flies, 119
  of Bombyliidae, 132
  of *Piophila* larvae, 176
prothoracic spiracles of pupae, 23.
  in mosquitoes, 94
  in Tabanidae, 105
Protocalliphora, 217
Protodiptera, 27
Protoplasa fitchii, 38
provision-shops, *Piophila* in, 176
Pseudacteon formicarium, 152
Pseudolynchia canariensis, 232
pseudopods, 16, 38, 62, 145, 161, 186
pseudotracheae, 201
pseudotracheal teeth, 142, 202
Psila rosae attacking carrots, 258
Psilidae, 195
Psilopa leucostoma, mines of, 188
  petrolei, 186
Psorophora, 94
Psychoda, 52, 85
Psychodidae, 51
Pterobosca, 55
Pteropus, Nycteribiids on, 238
ptilinum, 23, 170
Ptychomyza, 38

Ptychoptera, larva of, 38
pupae, activity of, 23
   within last larval skin, 49, 117, 149
   of Asilidae, 128
   of Brachycera, 23
   of Chaoborinae, 76
   of Lepidoptera, attacked by flies, 129
   of *Microdon*, 160
   of mosquitoes, 22, 94
   of Nematocera, 23
   of *Simulium*, 23, 145
   of Stratiomyidae, 24, 117
   of *Symmictus*, 136
   of Tabanidae, 105
   of tsetse-flies, 210
pupal paedogenesis, 49
   period, 15
puparium, 23
   ptilinum used in breaking out of, 170
   of Phoridae, 149
pupation of blowflies in unlikely places, 215
   of bot-fly larvae, 221
   of Hippoboscidae, 231
   inside last larval skin, 24, 49, 117, 149
   of Streblidae, 236
   of Nycteribiidae, 239
Pupipara, 11, 15, 229
pursuit of other insects, by Asilidae, 122
— by Conopidae, 169
— by Empididae, 138
pus, larvae in, 151, 250
   blowflies attracted to, 225

Queensland, onchocerciasis in, 247

Radiation, effect of on flies, 281
rafts of mosquito eggs, 85, pl. 10
rapids, larvae in, 52
rare species, 277
'rat-tailed maggots', 20, 38, 160, 162, fig. 10
recognition of other flies, 266
reeds, galls in, 182, 193
   larvae of *Thrypticus* in, 145
   mines in, 187
Reid, J. A., 203
reindeer warbles, 222, 223
reproduction rate in house-flies, 204
   in tsetse-flies, 211
reptiles, flies and, 17
   and horse-flies, 99
resistance to insecticides, 256
respiration underwater, by plastron, 35
   of Sciomyzidae, 179
   of Stratiomyidae, 115
respiratory gills, 162
   pigment of *Gasterophilus*, 224
   tubes of parasitic Bombyliid larvae, 136
restaurants, flies in, 199, 254

resting habits of mosquitoes, 83
réunions sédentaires, 268
Review of Applied Entomology, 284
Rhagionidae, a primitive family, 109
   alleged bloodsucking by, 109
   larvae of, 110
   piercing mouthparts of, 110
rheophile and rheophobe larvae, 52, 53
Rhinomyzini, 104
Rhizotrogus attacked by Nemestrinidae, 136
Rhynchopsilopa and ants, 189
Rhingia, 158, 161
rhinoceros, *Gyrostigma* in stomach of, 224,
   pl. 26
Rhyphidae, 39
rhythms, inherited, 82
Richards, O. W., 179
Richardiinae, 178, fig. 31
Rioxa termitoxena, 197
'river blindness', 246
robber-flies, 8, 19, 119, pl. 15
   adaptation for hunting, 120
   egg-laying of, 126
   eyes of, 120
   larva of, 119
   methods of attack, 122
   pupae of, 128
Rodhain, E. and Bequaert, J. C., 237
rodents, and Cuterebrinae, 219
Roederoides, 145
roof-spaces, flies in, 255
roots, damage to, 45
   larvae of Cecidomyiidae in, 47
   of underwater plants, as source of air, 92
root-aphis, Chloropids feeding on, 191
rot-holes in trees, 116, 161
rotting vegetation, larvae in, 53
   wood, larvae in, 161, 177
— Asilidae in, 127
— *Atherix* in, 112
— Clusiidae in, 176
— *Forcipomyia* in, 57
Rousettus, Nycteribiidae on, 238
running of Empididae and Dolichopodidae,
   140, 143
Ruttenia in elephant's skin, 224
rye, 192

Sabethes, 94
   and light intensity, 80
   larvae in Bromeliads, 91
   adults projecting eggs, 90
safes, *Sciara* larvae in, 42
St Mark's fly, 45
salamanders attacked by *Lucilia*, 225
salinity and mosquitoes, 85, 88
   and Ephydridae, 186
salivary duct, 119

Salt, G., 273
salt-balance, 19, 35, 50, fig. 13
sand, egg-laying in, 126
sand-flies (*Phlebotomus*), 245
　distribution of, 53
　habits of, 53
　larvae of, 55
　and disease (sand-fly fever), 245
'sand-flies' (Ceratopogonidae), 55
sap of wounded trees, 39, 161, 171, 199
Sapromyzidae, 172, 185
saprophagous larvae of acalyptrates, 172
　of Syrphidae, 160
Sarcophaga, viviparous, 227
　breeding in dead cluster flies, 271
Sarcophagidae, 213, 227, 251
Sarracenia, larvae in, 57, 91
sausages, maggots in, 253
sawflies attacked by Ceratopogonidae, 56
Scaeva pyrastri, 164
scales, of Asilidae, 120
　of Bombyliidae, 132
　of mosquitoes, 78
Scaptia, 106
Scaptomyza, leaf-mining by, 200, 258
Scatella, 185
Scatophaga, prey of, 212
　proboscis of, 211
　stercoraria, 211
Scatophagidae, 195, 212
Scatopse notata indoors, 46
　picea, massing of, 268
Scatopsidae, 45, 49
scavengers, 17, 116, 152, 162
　Bombyliidae as, 132
　Syrphidae as, 160
Scenopinidae, larvae of, 129
　no larval moults, 129
Scepsidinae, 104
Schmitz, Father H., 153
Sciapus, larval habitat of, 146
Sciadocera, 167, 168, fig. 28
Sciomyzidae, and snails, 22, 179
Sciara, larva of, 42
　militaris, 272
Sciaridae, 195
　numbers of, 41
Scott, Hugh, 150
screw-worms, 217
　number of eggs laid, 217
　eliminated from Curaçao, 281
scutellum of Celyphidae, 172, fig. 30
　of Thyreophoridae, 181
seashore, Dolichopodid larvae on, 145
　Empididae on, 141
　flies on, 36, 60, 178, 185, 205
　larvae on, 145
　Tabanidae on, 102, 104, 106

seasonal fluctuations in rain forest, 106
seaweed, Dolichopodidae in, 145
seaweed-flies (Coelopidae), 6, 11, 141, 173, 178, 205
　ability to take off from water, 24
　massed flights of, 267
　secretions, larvae feeding in, 71, 152
secondary sexual characters in Dolichopodidae, 143
　in Sepsidae, 175
seeds, larvae attacking, 196
segmentation of thorax, 4, fig. 1
Séguy, Eugène, 283
selective clearing of vegetation, 211
semi-specific myiasis, 251
Sepedon, 179, 180
Sepsidae, 22, 175
septic tanks, *Teichomyza* in, 186
Sergent, E. and Rougebief, H., 259
Sericomyia, 162
serpentine mines, 195
sewage sludge, larvae in, 52, 72, 175, 185
sexual display in Dolichopodidae, 143
　in Empididae, 139, 140
shagreened larval skin, 116
Shannon, R. P., 83
shedding of wings by *Lipoptena*, 181, 233
sheep blowfly, 21
　range of, 215, 216
sheep ked, evolution of, 233
　nostril fly, 220
　strike by blowflies, 215, 216
　tick, *Ixodes ricinus*, 233
ships at sea, flies on, 267
shore-flies (Ephydridae), 11, 185
shore-insects (*Clinocera*), 140
short-palped crane-flies, 29, 31, 36
'short-tailed' larvae of Syrphidae, 161
Sicily, horse-flies in, 98
Sicodus, 140
sight, acuteness of, 14
silk, spinning of by *Hilara*, 139
— by *Simulium*, 71
silk-glands of larvae, 43
silken cases of larvae, 36
　tubes of *Antocha* larvae, 37
Simuliidae, 66
　comparison with *Antocha*, 36
　emergence of adult under water, 24
　pupae preyed on by Empididae, 145
　most fully aquatic Diptera, 68
　numerous in Arctic, 279
Simulium adersi, 71
　columbaschense, 66
　damnosum, 68, 246, fig. 15
　griseicolle, 67
　neavei, 70, 246
sink, flies breeding around, 39

'singing' of Syrphidae, 156
siphon, respiratory, 38, 52, 88
  of Stratiomyidae, 115
  of Syrphidae, 20, 38, 160, 162
  of Tabanidae, 102
Siphona, 248
Siphunculina, 191, 255
Siphonella, parasitic larvae of, 193
size of flies, 4
skin of larvae, breathing through, 165, 200
  of vertebrates, myiasis of, 218, 251
skipping of larvae, 138, 176
skull, larva of *Dermatobia* scratching on
  man's, 219
sleeping in exposed places, danger of, 252
sleeping sickness, 210, 250
smallest flies, 4
smell, sense of, 14
smoke, attractive to flies, 166, 266
'smoke-flies' (Platypezidae), 166, 266
Smith, K. G. V., 158, 278
snails, association of Chironomid larvae
    with, 63
  Giant (*Achatina*), 151
  larvae attacking, 22, 186
  attacked by *Melinda cognata*, 225
  and Phoridae, 150
  and Sciomyzidae, 179
  selection of one species of, 225
sneezing of bot-larvae by camels, 220
snipe-flies, 108, 109
Snodgrass, R. E., 24
snout of *Nemotelus*, 113
snow-fly (*Chionea*), 29
soil, larvae in, 18
  Lucilia emerging from, 216
  ptilinum used for escaping from, 170
soldier-flies, 113
Solenopsis, phorids parasitic in, 152
Solva, 117
Somme, L., 203
sores, blowflies living in, 151, 217
sorghum, 190
sound, recognition by, 265
Spaniopsis, bloodsucking by, 109
Spaniotoma furcata, in greenhouses, 61
Spathulina, 196
spatula of cecidomyiid larvae, 48
specific myiasis, 251
speed of flight, 220
Sphaeroceridae, 177
Sphecodemyia, 107
spiders, Cyrtidae on, 138
  hunted by *Leptogaster*, 123
  larvae in egg-cases of, 193
  webs, flies living among, 32
  — flies stealing food from, 141
spinach fly, 207

spines, of pupae, 23, pl. 15
  of Stratiomyidae, 113
  used to pierce underwater stems, 188
spiracles, posterior, of larvae, 16, 22
  prothoracic, of pupae, 23
  of Bibionid larvae, 45
  of Syrphid larvae, 159
Sphyracephala brevicornis, 190
sponge-like proboscis of Syrphid, 156, fig. 27
  of Sciomyzid, 170, fig. 29
spray, larvae living in, 44, 52
spread of flies over the world, 279
spring emergence from hibernation, 220
springs, hot, 117
squama (= calypter), 11, 170, 157
squashes, damage to, 197, 258
stable-fly, 11, 248, 254
stagnant water, larvae in, 38
stains caused by *Stomoxys*, 208
stalk-eyed flies (Diopsidae), 177, 189
  other than Diopsidae, 190, 198
Stalker, H. D., 154
stationary prey hunted by *Leptogaster*, 123
stealing of food, 152
stem-borers, 196
  Dolichopodidae, 145, 146
stem-mining by Chironomidae, 64
stenogastric condition, 152
Stenepteryx hirundinis, 231
stereoscopic vision, 121
sterile males, control by, 281
  of tsetse, 211
Stichopogon, habits of, 123
still water, larvae in, 19
stinkhorn fungus, 172
stomach-bots, 220, 223
Stomorina lunata, 226, 267
Stomoxys, 11, 222, 248, 249
  calcitrans, 207, 254, 281
  ochrosoma, 168, 208
stouts, 96
Stratiomyidae, 108, 113
  compared with mosquitoes, 85
  a declining family ?, 115, 276
  distribution of, 115
  evolution of, 115
  larva of, 92
  length of larval life, 116
  number of, 114
  pupation of, 24, 117
  systematic position of, 115
Stratiomys, 115
  larva of, and desiccation, 117
streams, larvae in, 19
Streblidae, 10, 11, 153, 229, 234, 278, fig. 40
Strickland, A. H., 259
Stuckenberg, B. R., 113
surface of water, taking flight from, 174

Stylogaster and ants, 168
sub-Arctic regions, adaptation to, 179
Sub-Orders of Diptera, 6, 7
successful families, list of, 276, 277
suckers, of larvae, 16
sucking mouthparts, 31
sugar in diet, 14
sulphur, water containing, 117
sunlight and dancing flies, 265
supersonic speed, discounted, 220
Suragina, 113
surface film, flies walking on, 188
    mosquitoes living at, 85, 87, 145
    Ochthera mantis alighting on, 189
surface tension, 15, 88
surra, 210
    carried by Tabanidae, 247
swallowing of larvae by Man, 251, 252
swamps, larvae in, 18
swarms of flies, definition of, 6, 262
    significance of, 101, 262, 266
    factors governing, 263
    Bibionidae, 45
    Chaoborinae, 75
    Chloropidae, 191
    Coelopidae, 174
    crane-flies, 31
    midges, 58, 60
    mosquitoes, 83
    Platypezidae, 166
    Simulium, 67, 69, 70
    Stratiomyidae, 114
    Tabanidae, 100
'swarms' of flies indoors, 255
    of *Leptocera caenosa*, 180
    of *Madiza glabra*, 180
    of Phoridae, 150
    of Scatopsidae, 46
    of *Themira putris*, 175
sweat, flies and, 84
sweat-flies, 255
swifts and *Crataerina pallida*, 231
swimming, 50
Sycorax attacking frogs, 53
symbiosis between fly and alga, 64
Symbiocladius, parasitic larva of, 63
Symmictus, 136
Symphoromyia, bloodsucking by, 109
synanthropic flies, 241
Syrphidae, 9, 155, 168, 195
    evolution among, 161, 165
    flower-feeding by, 156, 157
    a flourishing family, 276
    larvae of, 16, 20
    — aphid-feeding, 163
    — damaging bulbs, 258, pl. 32
    — eating larvae of Chamaemyiidae, 182
    — rat-tailed, 38

Syrphus, 163
    luniger, 164
Systoechus, parasitic larva of, 133
Systropus, 134

Tabanidae, 22, 96
    evolution of feeding habits of, 102
    length of larval life, 105
    pupae of, 105
Tabanoidea, 8, 96
    relationships of, 275
Tabanocella, 104
Tabanus, 96
    recent evolution of, 98, 99
    larva of, 103
    glaucopis, 103
    kingi, larva of, 103
Tachinidae, 213, 227
    larval habits of, 227, 228
    importance of, 259
Tachydromia, 140
Tachypeza, 140
Taeniorhynchus, 92
tanneries, *Piophila* in, 177
Tanyderidae, 38, 279
Tanypodinae, 64
Tanyptera, egg-laying of, 35
Tanytarsus boiemicus, egg-laying by pharate
    female, 61
taons, 96
tea-plant, 258
teasel, larva in water in, 57, 91
teeth, prestomal and pseudotracheal, 144
Teichomyza fusca, 186
Tekomyia populi, pupal paedogenesis of, 49
Telmatoscopus, 52
temporary water, eggs of mosquitoes in, 85,
    86, 89
temperature, effect on biting flies, 82
    on tsetse, 210
Tendipedidae, see Chironomidae
Tephrochlamys, 172
'terminal galls', 196
Termitometopia skaifei, 227
Termitoxeniinae, 152
    male and female phases of, 153
termites, larvae in nests of, 43
    Calliphoridae and, 226
    Phoridae and, 149, 152
    Trypetidae and, 197
terrestrial flies, 17, 42, 45, 57, 120
    of acalyptrates, 171
    of crane-flies, 33, 61
    of Dolichopodidae, 145, 146
    of Rhagionidae, 110
    of Sciomyzidae, 179
    of Syrphidae, 163
Tetanocerinae, 179

Thaumaleidae, 71
  evolution of, 72
Thaumatomyia notata, 271
  breeding of, 191
  origin of overwintering adults, 191
  trapped by eddies, 269
Thambemyia pagdeni, 142, fig. 24
theca, 10
Themira putris, 175
  massing of, 268
Theobaldia annulata, 87
  eggs of, 85
Therevidae, larvae of, 129
  relationships of, 130
Thompson, W. R., 228
Thoracochaeta zosterae, 178
  massed flights of, 268
  thorax of flies, 3, 4, fig. 1
Thorpe, W. H., 108, 186, 200, 208
Thriambeutes mesembrinoides, 277
thrips, caught by Dolichopodidae, 144
Thrypticus, 145
Thyreophoridae, 181
Thyridanthrax and tsetse-flies, 133
tidal waters, 88
tidemarks, larvae between, 36, 37, 145
tiges oculaires of Diopsidae, 189
Tipulidae, larvae of, 18, 19, 34, 35, 37
Tipulinae, 34
tissues invaded by *Hypoderma* larvae, 221
toads attacked by *Lucilia*, 225
— by Tabanid larva, 103
tomato ketchup breeding *Drosophila*, 254
Tomophthalmae, 132
tongue of horse, and *Gasterophilus*, 224
torrents, larvae living in, 19, 52, 73
tower, observations from a high, 264
Toxophora, 134
Toxorhynchites, 91, 93
Toxorhynchitini (Megarhinini), 78
tracheal gills of *Cryptochaetum* larva, 200
Tracheomyia macropi, 221
trees, fallen, larvae in, 18
  horse-flies living in, 107
  growing larvae in, 18
  wounded, 161
tree-holes, 89, 116
  horse-flies in, 104, 106
  mosquitoes in, 86
  larvae living in, 161
treetop platforms, 244
tree-trunks, *Medeterus* on, 144
Trentopohlia and spiders, 32
tribes of mosquitoes, 78
Trichacantha, aggregations of, 113
trichlorethylene, attractive to *Coelopa frigida*, 174
Trichoceridae, 39

Trichomyia, larvae in rotting wood, 53
Trichoprosopa, egg-laying of, 86
Trichopsidea, 135, 136
Trichoptera, 36
Triogma, larval habits of, 33
Tripteroides, 91
trophallaxis, 160
tropics, behaviour of house-flies in, 203
Tropidia, 162
trumpets of mosquito larvae, 94
trypanosomes carried by tsetse, 210
trypanosomiasis, 249
Trypetidae, 11, 195, 196
  larvae of, 196, 258
  larval spiracles of, 22, fig. 12
'Tsaltsalya' fly, 249
tsetse-flies, 14, 15, 153, 208, 209, 248
  antennae of, 13, fig. 5
  biology of, 280
  compared with Hippoboscidae, 229
  distribution of, 248
  effect on development of Africa, 210
  fatal to beasts of burden, 249
  future of, 280
  hybridization among, 211
  puparia parasitized by Bombyliidae, 133
tubular flowers, feeding from 157
tularaemia, spread of, 247
Tumbu fly, 218, 251
tunnels of Laphriine larvae, 128
  silken, 43
turkeys attacked by *Simulium*, 66
Twohy, D. W. and Rozeboom, L. E., 81
Tylidae, see Micropezidae

Ubiquity of flies, 5
Ukazzi Hill, Kenya, 183
Ulidiidae, 198
Ulomyia, 52
underground, larvae able to live, 147
underwater habitat of *Atherix* larva, 112
  emergence of *Simulium* adult, 24, 71
  movement of adult *Clunio*, 60
  plants, method of piercing, 92
  plants, pupae attached to, 94
urban yellow fever, 244
urns of *Nepenthes*, 19, 57, 91
urine as a larval food, 21, 186
  and *Fannia*, 206, 207, 254
Urophora jaceana, 196
uterus of tsetse-flies, 17

Valley black gnat, 57
van Emden, F. I., 184, 228
van Someren, V. G. L., 184
Varley, G. C., 163, 196
vegetable-feeding larvae, 18, 23
  of Tabanidae, 103

vegetables, larvae attacking, 196
vegetation and tsetse-flies, 209
Vermileoninae, larvae of, 111
  long proboscis of, 110
verruga peruana, 246
vertical distribution of flies, 79
Vespula parasitized by Conopidae, 168
vinegar flies, 199
viola, galls in, 172
vision, all-round in Pipunculidae, 165
  of Empididae, 138
  of robber-flies, 121
viviparity in *Sarcophaga*, 227
Volucella, 158, 162, 163

Wallabies, Hippoboscidae on, 233
Wallacea, 115
Walshe, Miss B. M., 62
Wandolleckia, 151, 152, 153
warble-flies, 14, 219, 221
  failure to spread to other regions, 221
warbles, 16, 213, 222
  of reindeer, 222
warmer climates, flies of, 278
wart-hogs attacked by *Auchmeromyia*, 218
  nose-bots in, 220
wasps, larvae in nests of, 43
  Cacoxenus in nests of, 200
  Miltogramminae, and, 227
  Phoridae in nests of, 152
  mimicked by Bombyliidae, 134
  — by Conopidae, 168
  — by Syrphidae, 159
water, flies breeding in, 7, 17, 19, 51, 68, 85, 103, 112, 161, 186, 208
  flies walking on, 188
watercress, mines in, 188, 258
waterlilies, larvae mining in, 34, 187
water-midges, 7, 15, 19, 50
water-plants, air from, 85
  mines in, 187
water-beetles and Chloropidae, 63
water-bugs and Chironomidae, 63
waterside vegetation, flies on, 143
water suitable for mosquito-larvae, 88
water-surface, larvae at, 145
waving of wings, 32, 175
wax, digestion of by *Braula coeca*, 183
webs of fly-larvae, 43
Wellington, W. G., 267
West, Luther S., 284
wetting of eggs, 15
whame-flies, 96
wheat-bulb fly, 207
White, M. J. D., 276
white-flies eaten by *Drosophila*, 200
Wiedemannia, 145
Wiesmann, R., 270

Wuchereria bancrofti, 245
wind, effect on dispersal, 204
  speed, limiting swarms, 264
windows, flies on, 51, 173
  Drosophilidae on, 199
  Anisopus on, 39
  Scenopinus on, 130
wine, 254
  made possible by *Drosophila*, 259
wine-flies, 198
wing-beat of *Hypoderma*, 223
wings of flies, 3
  folding of, 159, 235
  pattern of, 198
  reduction or loss of, 29, 60, 61, 95, 115, 141, 149, 151, 153, 178, 180, 181, 184, 230, 233
  of *Culicoides*, 57
  of Cyrtidae, 136
  of Hippoboscidae, 231
  of Lonchopteridae, 154
  of *Nemestrinus*, 135
wing-spots, 176
wing-veins of Phoridae, 148
  Ceratopogonidae sucking from, 55
wing-waving, in Sepsidae, 175
  in Dolichopodidae, 143
  in Trypetidae, 198
winter, flies active during, 61
winter-gnats, 39
winter, house-flies in, 203
woodland mud, insects in, 38
World Wars, First and Second, 243
worm-lions, 111
wounds, maggots used to cleanse, 252
  Calliphoridae attacking, 217, 225, 250
  flies feeding at, 202
wounded trees, sap of, 19, 171
wrack on seashore, 173
Wyatt, J. J., 49
Wyeomyia, 90

Xylocopa and Asilidae, 125
  and Nemestrinidae, 136
Xylodiplosis, egg-laying of, 47
Xylomyia, 117
Xylophagidae, 108
Xylota, larva of, 161

Yeast-feeding larvae, 23
yellow dung-fly, 211
yellow fever, 80, 244

Zebra, *Gasterophilus* in, 224
zenith light, 267
Zeren, 233
Zeugnomyia, 94
Zoological Record, 284
Zumpt, F., 219, 221, 224

Lightning Source UK Ltd.
Milton Keynes UK
UKOW01f2106311017
311966UK00001B/2/P

9 780393 003758